AF556713

The WBF Book Series: Volume 4

ISA-88 and ISA-95 in the Life Science Industries

The WBF Book Series: Volume 4

ISA-88 and ISA-95 in the Life Science Industries

By WBF

Edited by
William Hawkins
Dennis Brandl
Walt Boyes

Momentum Press, LLC, New York

ISA-88 and ISA-95 in the Life Science Industries

First published in 2011 by
Momentum Press®, LLC
222 East 46th Street, New York, N.Y. 10017
www.momentumpress.net

ISBN-13: 978-1-60650-203-7 (hard back, case bound)
ISBN-10: 1-60650-203-4 (hard back, case bound)

ISBN-13: 978-1-60650-205-1 (e-book)
ISBN-10: 1-60650-205-0 (e-book)

DOI forthcoming

Volume 4 in the WBF Book Series by WBF, edited by William Hawkins, Dennis Brandl, and Walt Boyes, published by Momentum Press®, LLC

Cover Design by Jonathan Pennell
Interior Design by Scribe, Inc. (www.scribenet.com)

First Edition: March 2011

10 9 8 7 6 5 4 3 2 1

Printed in Taiwan

Contents

List of Figures vii

List of Tables xv

WBF Foreword xvii

Foreword by Walt Boyes xix

Preface xxi

1 ISA-88 Provides a Framework for a Pharmaceutical Process Module Library 1

2 ISA-88 Design and Implementation Case Study for a Complex Bulk Pharmaceutical Batch Process 13

3 Managing Complex Equipment Status in a GMP Environment 29

4 Impact of Batch Software Upgrades on Validated Batch Applications 39

5 Is It Possible to Build a Pharmaceutical Plant in 18 Months or Less Using ISA-88? 47

6 Batch On-line Analytics: A Solution Beyond Six Sigma 57

7 Implementing ISA-88 across Life Science Development Operations 71

8 Process Definition Management: Using ISA-88 and BatchML as a Basis for Process Definitions and Recipe Normalization 85

9 ISA-88 Design and Implementation Case Study for a Pharmaceutical Batch Process 95

10 Product Life-cycle Stages Linked Using ISA-88 and ISA-95 113

11 Jazz Up Your Batch Projects 129

12 The Challenge of Integrating Multiple Batch Systems to Global Business Systems 143

13 The Road to Full MES Integration: Practical Experience from the Pharmaceutical Industry 155

14 MES Roll-out in a Regulated Environment: Reducing the Costs of Validation Based on Risk Assessment 173

15 Fast and Efficient Configuration and Integration of Automation Solutions from a Global Perspective: A Practical Approach 183

16 Risk-based Engineering Assessment and Qualification: A Case Study 199

17 Lean Computer Validation through a Risk-based Approach: A Case Study 209

18 Multiple Products in a Monoclonal Antibody ISA-88.01 Batch Plant 221

19 Considerations for Managing Global Recipe Development 231

20 General Recipes as Contracts with Manufacturing 241

21 Using General Recipes for Standardized Multiple Plant Manufacturing Science 251

22 Manufacturing Science Model Extensions to Address Lean Manufacturing and Supply Chain Optimization 261

23 Manufacturing Science Model Extensions to Address Product and Process Sustainability 273

24 Manufacturing Science Model Extensions to Address Quality by Design and Risk Assessments 293

25 Batch Release and Material Use Reporting: A Case Study 315

Index 323

Figures

1.1	The process module library workflow.	2
1.2	GR library.	3
1.3	Harvest vessel P&ID.	5
1.4	Mechanical features of the harvest vessel.	6
1.5	Harvest unit EMs.	7
1.6	EM library with expansion of "Clean Air Supply."	8
1.7	Transfer unit example.	9
1.8	Harvest vessel unit process operations.	11
1.9	The process module library, with units for Product Alpha.	12
2.1	Example process model diagram.	15
2.2	Example physical model drawing.	17
2.3	Example requirements outline.	18
2.4	Example procedural model database.	19
2.5	Example of radio channels for coordination control.	23
2.6	Example batch reporting strategy.	26
3.1	Equipment model.	32
3.2	Equipment status architecture.	36
5.1	The construction period.	49
5.2	Parallel project phases in a construction period.	49
5.3	Preassembled facilities.	50

5.4 Skid mount units. 51

5.5 Preassembled units. 52

5.6 PM. 53

5.7 A set of PMs connected as LEGO bricks. 53

5.8 The PM adapted into the ISA-88 structure. 54

5.9 Project activity model and the modules for each phase. 54

5.10 Test phases for parallel activities. 55

6.1 PAT workstation used for a bioreactor application. 60

6.2 Data trends: original (left) and aligned (right). 61

6.3 Illustration of the raw trajectories R and X and the aligning optimal path c(k). 62

6.4 Hybrid data unfolding and PCA model development. 63

6.5 Data transformation from the original space to the normalized principal component space. 64

6.6 PLS, **T**, maximizing variances both from operational data, **X**, and quality data, **Y**. **P** and **Q** are PLS model matrices. 65

6.7 PCA on-line functionality. 66

6.8 PCA on-line statistical trends for a mammalian cell simulated bioreactor. 67

6.9 Process variables contribution plot. 67

6.10 Bioreactor temperature perturbation. 68

6.11 PCA on-line process tag contribution chart for related statistical trends. 69

7.1 General recipes in a typical development function. 74

7.2 Example laboratory procedure. 78

7.3 Drug product ISA-88 operations. 79

8.1 86

8.2 88

8.3 91

9.1	Area layout.	97
9.2	System architecture.	100
9.3	Design model.	101
9.4	MBMA data flow.	103
9.5	Main display.	104
9.6	Work instruction.	109
9.7	Electronic batch record.	110
10.1	High-level product life cycle.	116
10.2	Recipe life-cycle phases.	118
10.3	Recipe hierarchy for multiple sites.	120
10.4	Testing.	124
11.1	It is obvious that these two cars are different, yet they are both "handmade" according to specifications.	131
11.2	The early decisions have the most impact on the investment.	132
11.3	Front-end definition locks down 85% of the design.	132
11.4	Objectives for defining objectives.	133
11.5	Map of product planning, with a feedback loop.	133
11.6	Sample of a PLCD.	135
11.7		136
11.8	Total computer-controlled production process.	137
11.9	Now the FS is in place.	140
11.10	Tank temperature control.	141
12.1	R&D spending and number of approved NMEs.	144
12.2	Forecast GDP for 2020.	145
12.3	The regulatory process today.	145
12.4	The possible regulatory process in 2020.	147

12.5 The transition to modular process construction. 148

12.6 A farm module. 149

12.7 Analysis of downtime. 150

12.8 Degrees of automation. 151

12.9 The ISA-95 functional hierarchy and its common description. 152

13.1 Aerial view of the IBP. 156

13.2 NNE MES architecture. 158

13.3 Parallel PCS architecture. 159

13.4 CIP recipe procedure, CIP_1_CONSUMER. (Icons on the far left refer to the state of the parent compound operation.) 161

13.5 Part of CIP unit procedure, CIP_RINSE_SEQUENCE. (Icons on the far left refer to the state of the parent compound operation.) 162

13.6 Document life cycle. 164

14.1 The ISA-95 manufacturing operations management model as a basis for MES. 175

14.2 The GAMP V-Model. 176

14.3 These metrics can be used to classify the risk according to severity. 178

14.4 Risks should be addressed in order of decreasing severity. 178

14.5 Classifying risk by probability of detection. 179

14.6 A risk assessment form. 179

14.7 Steps in risk assessment. 180

15.1 Fast launching of products is vital in the pharmaceutical industry. 184

15.2 Modular engineering from conceptual design to handover and support. 185

15.3 Decomposing a complex plant into simple modules based on recognized engineering standards. 187

15.4 The main building block is the process module. 188

15.5 Standardization of process modules within the project and across projects. 188

15.6 A-PAM, based on GAMP 4 terms. 189

15.7 Examples of computer-based project management and engineering tools. 191

15.8 Example of global network for supply of configurable modules. 192

15.9 Overall facts and time frame for project implementation. 193

15.10 From a 3-D model to final implementation of the pharmaceutical plant. 194

15.11 Process flow diagrams transferred into process modules. 194

15.12 Example of process module from 3-D design to installation. 195

15.13 Modular engineering approach based on ISA-88. 196

16.1 QSIT. 201

16.2 Architectural block diagram of the BAS. 202

16.3 RW system. 204

17.1 Interfaces to SCADA. 211

17.2 System model. 214

17.3 Logarithmic ALARP scheme. 215

18.1 Process overview. 223

18.2 Impact of multiple products. 225

18.3 Process transfer. 226

19.1 Versioning of recipes. 235

19.2 Reusable components. 236

19.3 Different versions for each component. 237

19.4 All-in-one architecture. 238

19.5 Distributed architecture. 239

20.1 A GR representation. 243

20.2 Process definition within a GR. 244

20.3 Elements of a GR. 244

20.4 Basic processes generally used for process actions. 246

21.1 Quality attributes and process parameters. 254

21.2 A process report definition. 255

21.3 Documenting observed modes. 255

21.4 Elements of multiple-site manufacturing science. 257

21.5 Corporate and site knowledge. 257

21.6 Production history and manufacturing science investigations. 258

21.7 Elements of manufacturing science. 259

22.1 DMAIC elements and ISA-88 elements. 264

22.2 Process reports for Lean optimization. 265

22.3 Alternate process operations in a GR. 268

22.4 Alternate sourcing decisions from GR information. 269

22.5 Alternate materials listed in a GR. 270

23.1 Where sustainability decisions are made. 281

23.2 Common process stages in pharmaceutical manufacturing. 282

23.3 Recovered and waste material in a GR. 285

23.4 Additional process parameter types for sustainability attributes. 287

23.5 Actual sustainable attributes values in process reports. 288

23.6 Sustainability attributes added to GR information. 290

23.7 Sustainability factors in recipe development. 291

24.1 Design space development and risk assessment in the development cycle. 296

24.2 Example of a multidimensional design space from ICH Q8. 296

24.3 Examples of a complex from ICH Q8. 298

24.4 Sample DoE with cooperating factors, from Umetrics. 299

24.5 Example of a risk assessment Ishikawa diagram from ICH Q8. 301

24.6 Example of a fault tree, identification of critical sets, and probability calculation. 302

24.7 Design space examples from ICH Q8. 304

24.8 Figure 8 from ANSI/ISA-88.00.03 2003. 306

24.9 Figure 20 from ISA-88.00.04. 307

24.10 Section 5.16.14 from ISA-88.00.04. 308

24.11 Process operation for granulation. 310

24.12 An example process operation definition with design space information. 311

25.1 Example procedural model with integrated CPP/CA reporting. 318

25.2 Example CPP Monitoring function block. 318

25.3 Example procedural model with integrated material tracking. 320

25.4 Transfer report summary sheet output. 321

25.5 Material transfer sheet output. 322

Tables

2.1 Example list of common and unique phases 20

2.2 Example documentation list 27

3.1 Class-based status types and values 33

3.2 Status value rules 35

10.1 Project teams 126

11.1 FS specifications 138

16.1 RW subsystems 205

16.2 RW system analysis 205

17.1 GAMP 4 software categories 213

17.2 Summary of the risk assessment flow—step by step 216

17.3 Definitions of severities used in the risk assessment 217

19.1 Comparing central and local engineering 233

20.1 Sample process actions 247

20.2 Sample equipment constraints 249

20.3 Example definition states 250

24.1 Required attributes for the GROI value 308

WBF Foreword

The purpose of this series of books from WBF, The Organization for Production Technology, is to publish papers that were given at WBF conferences so that a wider audience may benefit from them.

The chapters in this series are based on projects that have used worldwide standards—especially ISA 88 and 95—to reduce product variability, increase production throughput, reduce operator errors, and simplify automation projects. In this series, you will find the best practices for design, implementation, and operation and the pitfalls to avoid. The chapters cover large and small projects in a wide variety of industries.

The chapters are a collection of many of the best papers presented at the North American and European WBF conferences. They are selected from hundreds of papers that have been presented since 2003. They contain information that is relevant to manufacturing companies that are trying to improve their productivity and remain competitive in the now highly competitive world markets. Companies that have applied these lessons have learned the value of training their technical staff in relevant ISA standards, and this series provides a valuable addition to that training.

The World Batch Forum was created in 1993 as a way to start the public education process for the ISA 88 batch control standard. The first forum was held in Phoenix, Arizona in March of 1994. The next few years saw growth and the ability to support the annual conference sessions with sponsors and fees.

The real benefit of these conference sessions was the opportunity to network and talk about or around problems shared by others. Papers presented at the conferences were reviewed for original technical content and lack of commercialism. Members could not leave without learning something new, possibly from a field thought to be unrelated to their work. This series is the opportunity for anyone unable to attend the conferences to participate in the information-sharing network and learn from the experiences of others.

ISA 88 was finally published in 1995 as *ISA-88.01-1995 Batch Control Part 1: Models and Terminology*. That same year, partially due to discussions at the WBF conference, ISA chartered ISA 95 to counter the idea that business people should be able to give commands to manufacturing equipment. The concern was that business people had no training in the safe operation of the equipment, so boardroom control of a plant's fuel oil valve was really not a good idea. There were enough CEOs smitten with the idea of "lights-out" factories to make a firewall between business and manufacturing necessary. At the time, there was a gap between business computers and the computers that had infiltrated manufacturing control systems. There was no standard for communication, so ISA 95 set out to fill that need.

As ISA 95 began to firm up, interest in ISA 88 began to wane. Batch control vendors made large investments in designing control systems that incorporated the models, terminology, and practices set forth in ISA 88.01 and were ready to move on. ISA 95 had the attention of vendors and users at high levels (project-funding levels), so the World Batch Forum began de-emphasizing batch control and emphasizing manufacturing automation capabilities in general. This was the beginning of the transformation of WBF into "The Organization for Production Technology." Production technology includes batch control.

The WBF logo included the letters "WBF" on a map of the world, and since this well-known image was trademarked, the organization dropped the small words "World Batch Forum" entirely from the logo after the 2004 conference in Europe. WBF is no longer an acronym. Conferences continued annually until the economic crash of 2008. There was no conference in 2009 because many companies, including WBF, were conserving their resources.

WBF remained active and solvent despite the recession, so a successful conference was held in 2010 using facilities at the University of Texas in Austin. Several papers spoke of the need for procedural control for continuous and discrete processes. The formation of a new ISA standards committee (ISA 106) to address this need was announced as well. Batch control is not normally associated with such processes, but ISA 88 has a large section on the design of procedural control. There is a need for a way to apply that knowledge to continuous and discrete processes, and some of those discussions will no doubt be held at WBF conferences, especially if the economy recovers. We would like to invite you to attend our conference and participate in those discussions.

WBF has always been an organization with an interest in production technologies beyond batch processing, even when it was officially "World Batch Forum." Over the years, as user interests changed, so has WBF. We have not lost our focus on batch; we have widened our view to include other related technologies such as procedural automation. We hope you will find these volumes useful and applicable to your needs, whatever type of process you have, and if you would like more information about WBF, we are only a simple click away at http://www.wbf.org.

William D. Wray, Chairman, WBF
Dennis L. Brandl, Program Chair, WBF
August 2010

Foreword by Walt Boyes

Many years ago, some dedicated visionaries realized that procedure-controlled automation would be able to codify and regularize the principles of batch processing. They set out on a journey that eventually arrived at the publication of the batch standard ISA 88 and the development of the manufacturing language standard ISA 95.

Many end users have benefited from the work of these visionaries, who founded not only the ISA88 Standard Committee but also the WBF. WBF has been an unsung hero in the conversion of manufacturing- to standards-based systems.

Today, WBF continues as the voice of procedure-controlled automation in the process and hybrid and batch processing industries. The chapters that make up this book series provide a clear indication of the power and knowledge of the members of WBF.

I have been proud to be associated with this group of visionaries for many years. *Control* magazine and ControlGlobal.com are and will continue to be supporters of WBF and its aims and activities.

I would like to invite you to come and participate in WBF, both online and at the WBF conferences in North America and Europe that are held annually. You will be glad you did. You can get more information at http://www.wbf.org.

Walt Boyes, ISA Fellow
Editor in Chief
Control magazine and ControlGlobal.com

Preface

The twenty-five chapters in this book are concerned with processes and systems that operate in a government-regulated environment. What is quite all right in the petrochemical industries is inadequate in the life science industries.

Modular design has proven to be useful. Chapters 1, 5, 11, and 15 apply modular design to building ISA-88 modules and even process modules that are installed with the aid of a crane.

Implementation stories can be very useful, especially when they are written by users. Chapters 2, 7, 9, and 18 discuss implementations of pharmaceutical processes. Chapter 2 introduces a novel concept for coordination control among units. Chapter 3 describes the implementation of clean and sterile states in a petrochemical control system.

Validation is a tool used by government agencies to assure that a system or process either will operate or is currently operating according to its approved design specifications. Chapter 4 discusses the impact of software upgrades on validated systems. Chapters 16 and 17 describe the use of risk analysis to reduce the amount of time and paper work required for validation. Chapter 25 explores the application of ISA-88 principles to a reporting system that must be validated.

Process Analytical Technology (PAT) has been adopted by regulatory agencies to assure that a process is operating properly. It is also useful for diagnosing process problems. Chapter 6 provides a mathematical analysis of techniques that can be applied to on-line operation, with illustrations of the results. If the math is daunting, then skip it and read about the results.

Chapter 8 discusses process definition management and recipe normalization as a means to reduce the time between identifying the need for a new process and the first of many deliveries of a new product.

A manufacturing process needs to be connected to business systems such as Enterprise Resource Planning (ERP), usually by a Manufacturing Execution System (MES). This is the subject of Chapters 10, 12, 13, and 14. If the MES can affect the quality of an Active Pharmaceutical Ingredient (API), then it must be validated. The ERP may also be involved.

An enterprise may be global in scope, with manufacturing facilities in many regions and countries. All the facilities must make the same product in order to avoid validation nightmares. The ISA-88.01 recipe model has the General Recipe (GR) at the top of a hierarchy that gets progressively more detailed. ISA-88.03 applies to general and site recipes, where a site may be a region or country, as well as a single facility. Chapters 19 and 20 discuss the use of GRs for global facilities. Chapters 21 to 24 introduce and extend the concept of a manufacturing science model that is documented with GRs.

Chapter 21 introduces the use of GRs as common documents for processes in different facilities, making it possible to compare diverse processes. Chapter 22 applies GRs to Lean manufacturing and the supply chain. Chapter 23 is about sustainability. Chapter 24 addresses risk assessments and quality by design.

Once again in this volume, the style requires a minimal, consistent use of capitalized words. This tends to increase the use of acronyms, but very few had to be created, given the regulator's propensity for using them. See the "Style" section in the preface to *ISA-95 Implementation Experiences*, volume 3 in the WBF series.

Basics of Life Science Industries

Those who have been through validation have no need to read this section. It is here for the innocents that have not previously encountered regulated projects or worried about killing all the leftover microbes from the previous process.

The name *life science* refers to both the lives of humans and the lives of genetically engineered cellular molecule factories. Most of the equipment and control systems used by the life science industries may also be found in other process industries. One difference is that, in life science industries, it must be possible to clean and perhaps sterilize equipment in place, so that bad bugs don't reproduce. Processes and products that affect human health are typically regulated by a division of government, if that government exists to protect and nurture a society. For example, Clean In Place (CIP) and Sterilize In Place (SIP) are required to eliminate any possibility of contamination of pharmaceutical, food, beverage, or cosmetic products; the requirement is not just a good idea, it is the law.

In the United States, regulation is governed by the Food and Drug Administration (FDA), an agency of the U.S. Department of Health and Human Services. Elsewhere, documents pertaining to ISO 9001 provide regulations. The U.S. Code of Federal Regulations (CFR) contains the detailed regulations for all executive departments and agencies, subdivided into fifty titles. These regulations are considered too detailed to be passed through the U.S. Congress, a body of elected representatives and senators who do not generally have the specialized knowledge necessary to judge each regulation. New regulations are published for public comment for a period of time and may then be added to the CFR, which is published annually in the *Federal Register*. See http://en.wikipedia.org/wiki/Code_of_Federal_Regulations for further details.

A CFR citation, such as 21 CFR 820, should be read as "title 21, part 820 of the CFR." Additional information may point to a section of the part or a paragraph within the section. This particular citation (21 CFR 820) refers to the FDA quality system regulation for medical devices, including design validation in 820.30(g). The citation 21 CFR 11 refers to the part of the code that contains FDA requirements for documentation and electronic signatures. Generally, the regulations are intended to assure that a process or product has been designed and implemented in a way that is suitable for its intended purposes.

The FDA also publishes guidance documents. Even though these are not laws, you can be sure that an FDA inspector will want to know why you aren't following the guidance (i.e., guidelines), if that is the case. Alternative approaches may be used if the inspector agrees. Good Manufacturing Practice (GMP) and current GMP (cGMP) regulations are guidelines for designing, testing, and maintaining the integrity of manufacturing processes, so that the quality of the product does not deteriorate over the life of the process. The World Health Organization (WHO) and many countries have their own versions of GMP. See http://en.wikipedia.org/wiki/Good_manufacturing_practice for more information.

Good Automation Manufacturing Practice (GAMP) is specific to automated processes. Practitioners using GMP began saying that the acronym meant "Great Mounds of Paper" because automation, with its many computer systems and many, many subsystems, is complex. The FDA responded by finding ways to make the quality system less "burdensome." An analysis of the quality risk of each subsystem has been introduced that reduces the amount of paper work required if the subsystem is not critical to the quality of the product. See Chapters 16 and 17 in this book; see also chapter 17 in *ISA-95 Implementation Experiences*, volume 3 in the WBF series, for a brief discussion of the usefulness of GAMP in any automated process.

The FDA quality system depends on verification and validation documents. 820.30(f) gives the example of an architect specifying an air-conditioning unit for a building. When the air-conditioner is installed, the manufacturer's specifications for the machine may be verified with appropriate tests. These tests do not validate that the building occupants will be comfortable as the seasons change.

In 820.3(aa), *verification* means confirmation by examination and provision of objective evidence that specified requirements have been fulfilled. In 820.3(z), *validation* means confirmation by examination and provision of objective evidence that specified that the particular requirements for a specific intended use can be consistently fulfilled.

The FDA requires documents that define the objective evidence. A User Requirements Specification (URS) defines the specific intended use. This is translated to one or more Functional Specifications (FS) that define the technical requirements to be verified. These documents form the left leg of the "V-model" that is discussed in Chapters 11 and 14.

The V-model starts with the URS and descends the left leg of the "V" as specifications until software development occurs at the bottom of the "V." Then the right leg is ascended as software is verified, the applications are verified against the various FS, and finally, the project is validated against the URS. The tests performed are grouped into Installation Qualification (IQ), Operational Qualification (OQ), and Performance Qualification (PQ).

Building a regulated process is incredibly more complicated than building a chlorine bleach process, especially if the process is automated.

Bill Hawkins
November, 2010

CHAPTER 1

ISA-88 Provides a Framework for a Pharmaceutical Process Module Library

Presented at the WBF North American Conference, April 13–16, 2003, by

Vince Miller
Team Leader, Automation Services
millerv@bek.com
BE&K Engineering, 2450 Perimeter Park Drive, Morrisville, NC 27560

Abstract

Companies face tough economic realities in the business of discovering and bringing new pharmaceutical products to market. Only one in ten thousand new compounds discovered in the lab will survive to warrant commercial production. When a new product is approved, it is imperative to maximize the return on investment by scaling up to commercial production quickly.

Modular construction and design techniques are one answer to this challenge. Although this can save time at the construction site, front-end design time may actually increase. Thus companies are still faced with the dilemma of being late to market or risking investment in new facilities before the new product is approved.

This chapter presents a method of reducing that risk by designing the building blocks of modern production facilities prospectively—well before a new drug's approval is certain. Because many of the compounds currently in the development pipeline can be manufactured using similar core technologies, robust standardized process modules can be designed in advance and taken "off the shelf" when needed. ISA-88.01 provides the required flexibility and scalability within each module, and it provides the methods that allow the modules to be connected in the configuration necessary to meet the needs of any given process.

With this approach, many of the problems associated with transferring technology from the laboratory to the factory can be solved in advance. A library of core process modules can be constructed using the ISA-88.01 process, physical, and procedural models as design guidelines.

Building the Process Module Library

The process module library is built using the following four ISA-88.01 models: the process, physical, procedural, and control activity models. The process model is derived from the General Recipe (GR) for the unique product to be manufactured. The physical and procedural models provide the manufacturing capabilities needed to perform the process operations. ISA-88.01 batch control activities and functions allow the assembled modules to function as one system. The first step in developing the process module library is to review the GRs for both existing products and those in the development pipeline to determine the core process technologies that form a common basis for several products (Fig. 1.1).

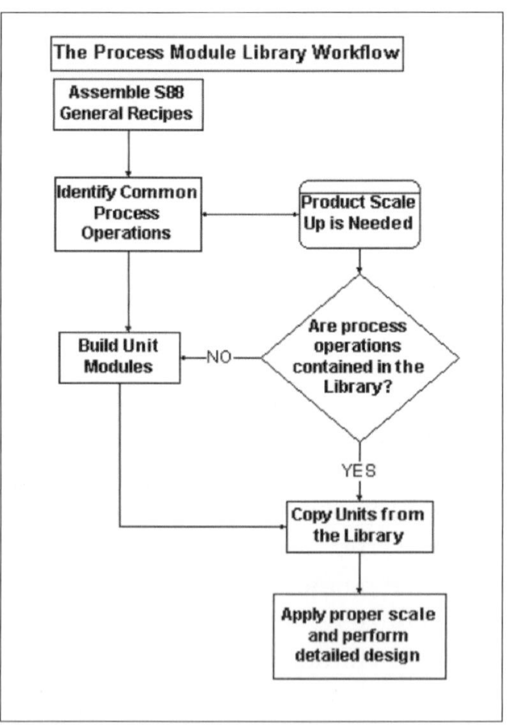

Figure 1.1. The process module library workflow.

Building the GR Library

The GR for each of the products should be organized in an outline that is formatted like the ISA-88.01 process model, showing the process stages, operations, and actions for each. The outlines can be combined and stored in a GR library. Figure 1.2 provides a snapshot of a GR library, showing the partially expanded outline of Product Alpha. Following the ISA-88.01 model, the outline is expandable within each process stage to reveal the process operations and process

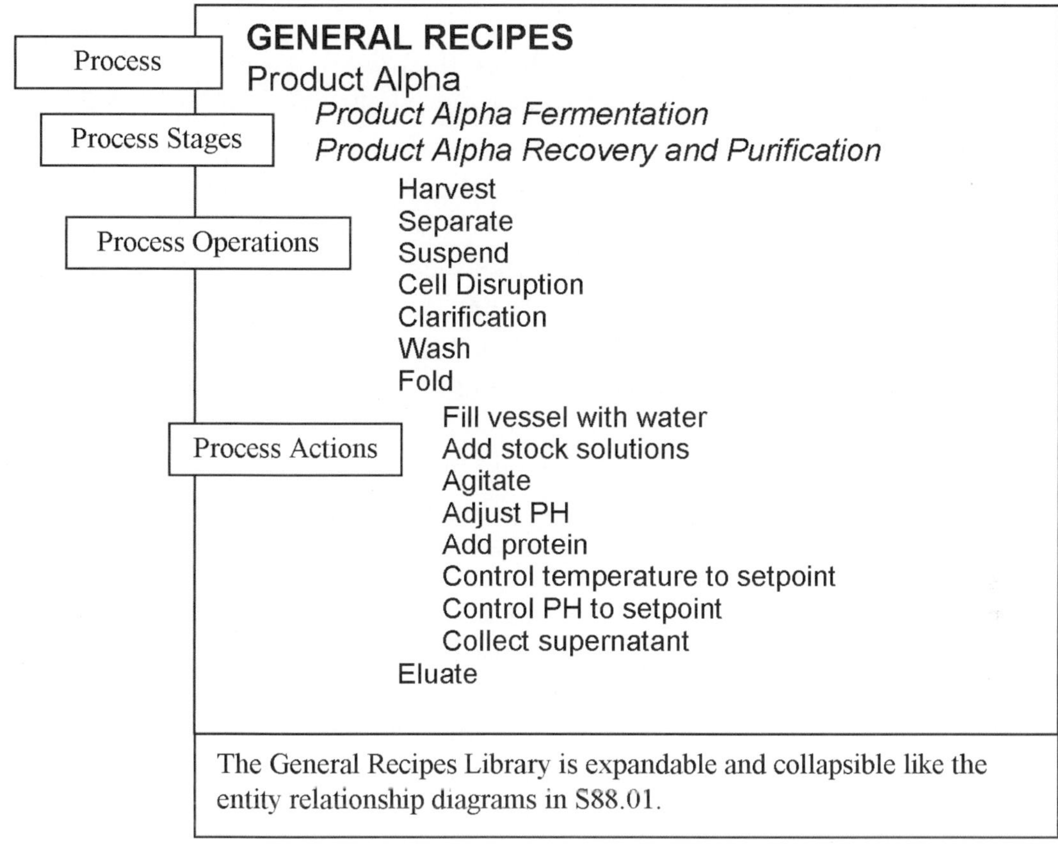

Figure 1.2. GR library.

actions. Looking within each product recipe will identify common process operations that are good candidates for further development as core process modules.

Identifying Core Process Operations

Once the GR library is assembled in an ISA-88.01 format, it can be analyzed to find the equipment needs and constraints that different process operations have in common. Look for common process operations that can employ similar classes of equipment to perform the required process actions. For example, the process operations required to manufacture Product Alpha (Recovery and Purification Stage) can be performed by units such as a harvest vessel, a centrifuge, a homogenizer, and a process vessel. If similar equipment needs were found in other product recipes, then these units would be excellent candidates for further development. The physical model provides the framework to develop and document the required equipment functionality.

Developing the Physical Models

After the core process operations have been identified, the physical model can be built starting at the unit level. In keeping with the library concept, the design specifications for each common unit will be assembled in a standardized design package. Each package contains a Piping and Instrumentation Diagram (P&ID) for the unit, a Functional Specification (FS), and mechanical specifications for the vessel and other equipment. The package must be assembled according to the ISA-88.01 physical model hierarchy to provide maximum flexibility and scalability in support of the library concept. Each documentation package forms a module that is fully self-contained and independent from other units. The ISA-88.01 characteristics of each module make it possible to check the units out of the library and assemble them in the configuration needed to manufacture a product.

P&ID

The P&ID functions as the cornerstone of the design package for each process module. It illustrates the piping, valves, instrumentation, and other equipment needed to perform the process operations required by the GR. Devices that work together to perform finite tasks are grouped into Equipment Modules (EMs) and Control Modules (CM). EM and CM designs should be standardized so that they can be designed once and replicated in other units (Fig. 1.3).

The harvest vessel can be used to perform several process operations. The physical capabilities of the unit are established by the EMs and CMs assigned to it, such as Item 1,"Process Inlet 1"; Item 2,"Clean Air Supply"; and Item 3,"Purified Water Supply." Item 4 is part of "Transfer Unit 1."

In reviewing the GR library, it was determined that a harvest vessel unit was needed to perform several common process operations in conjunction with other units, such as a homogenizer, a centrifuge, a process vessel, and transfer units needed to move product between them. Figure 1.3 is an example P&ID under development for the harvest vessel unit. The P&ID shows the piping, valves, process equipment, and instrumentation needed to establish multiple capabilities for the unit.

The Functional Specification

The FS is a companion document to the P&ID. Using standard word processing software features, it is built as a collapsible outline patterned after the hierarchy of the ISA-88.01 physical model. The major headings of the outline correspond to each of the units depicted on the P&IDs. When collapsed to the highest level, the

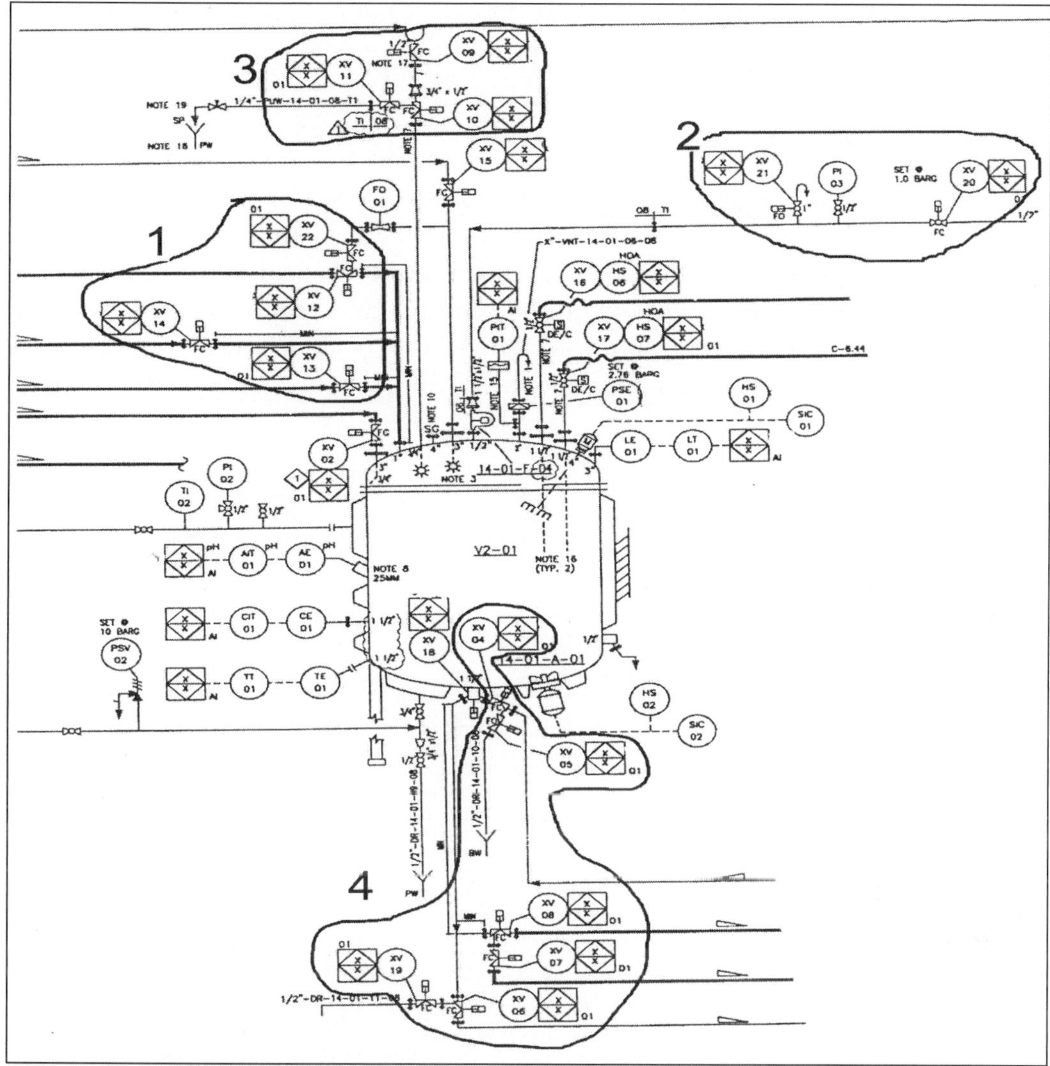

Figure 1.3. Harvest vessel P&ID.

document will show a list of units. Each unit can be expanded to the next level to show EM and again to show CM.

Descriptive generic alias names are assigned to equipment components and instrumentation in order to link the P&ID to the FS. This also makes the model portable. It is important to develop a convention or syntax for alias names and create a concordance of the names as they are assigned to entities depicted on the P&ID. This concordance will provide a guide for the project team to follow and will be used to generate an index of terms for later reference. When the model is checked out of the library for use, the appropriate tag naming convention can be applied and each device can be linked to a unique tag name using its alias name.

Defining the Mechanical Features

Standard equipment data sheet templates are developed for each major piece of equipment. These templates provide a menu of features that the final end user can select. The template serves to guide the end user to achieve a complete and thorough design specification. Unwanted features are simply deleted from the template. The harvest vessel unit library entry is expanded in Figure 1.4 to show the mechanical features of the tank.

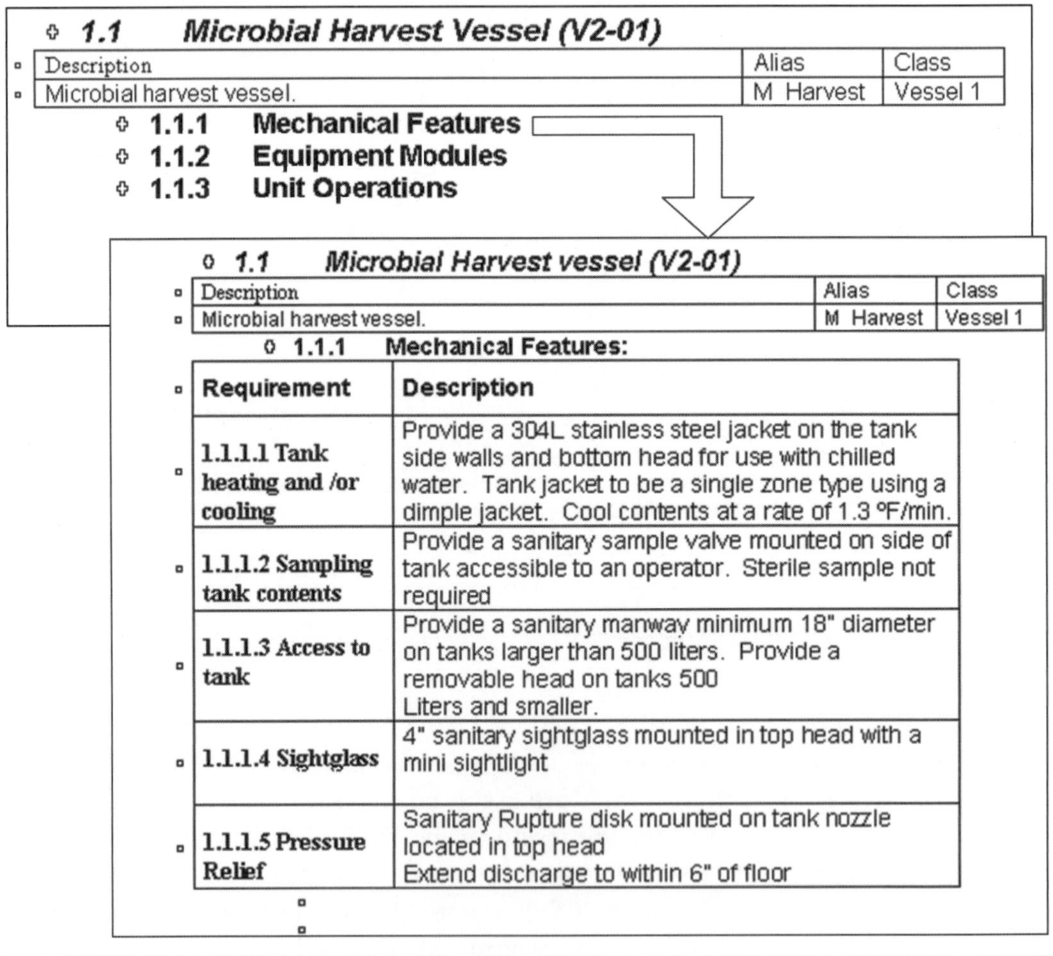

1.1 Microbial Harvest Vessel (V2-01)

Description	Alias	Class
Microbial harvest vessel.	M Harvest	Vessel 1

1.1.1 Mechanical Features
1.1.2 Equipment Modules
1.1.3 Unit Operations

1.1 Microbial Harvest vessel (V2-01)

Description	Alias	Class
Microbial harvest vessel.	M Harvest	Vessel 1

1.1.1 Mechanical Features:

Requirement	Description
1.1.1.1 Tank heating and /or cooling	Provide a 304L stainless steel jacket on the tank side walls and bottom head for use with chilled water. Tank jacket to be a single zone type using a dimple jacket. Cool contents at a rate of 1.3 °F/min.
1.1.1.2 Sampling tank contents	Provide a sanitary sample valve mounted on side of tank accessible to an operator. Sterile sample not required
1.1.1.3 Access to tank	Provide a sanitary manway minimum 18" diameter on tanks larger than 500 liters. Provide a removable head on tanks 500 Liters and smaller.
1.1.1.4 Sightglass	4" sanitary sightglass mounted in top head with a mini sightlight
1.1.1.5 Pressure Relief	Sanitary Rupture disk mounted on tank nozzle located in top head Extend discharge to within 6" of floor

Figure 1.4. Mechanical features of the harvest vessel.

Identifying and Developing the EMs

As the P&ID is being further developed, EMs and CMs are added to the physical model hierarchy in the FS. Several EMs have been circled on the harvest vessel P&ID (Fig. 1.2). Figure 1.5 shows how the "Process Inlet 2" EM can be expanded to define its CM and the phases it can perform.

Similar specifications exist within each EM entity in the FS outline. Each EM is completely self-contained and portable so that each can be removed or added to the physical model to achieve the desired process capabilities. Standard EMs can be developed and stored in a library to be replicated in many different units. Figure 1.6 shows an example of an EM library. Here, the general attributes for all EMs can be defined and reused throughout the project to ensure consistency and

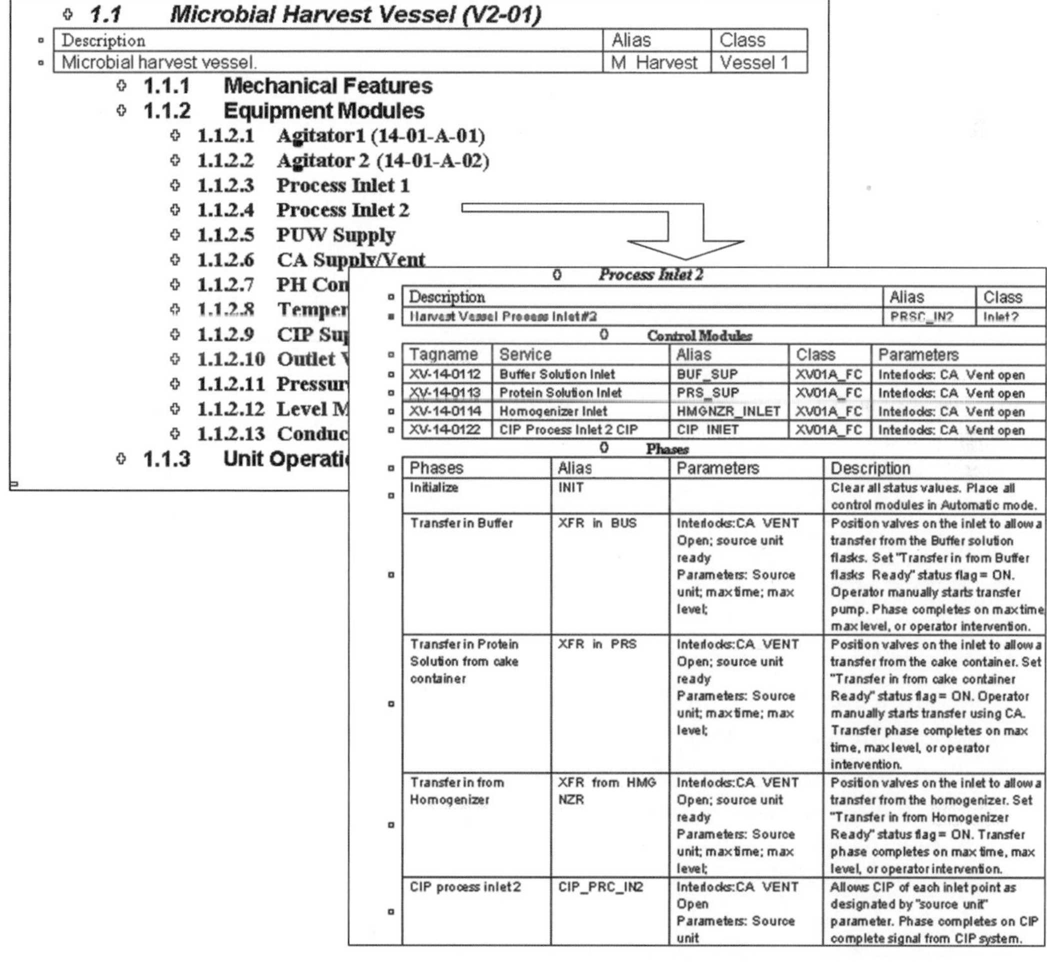

1.1 Microbial Harvest Vessel (V2-01)

Description	Alias	Class
Microbial harvest vessel.	M Harvest	Vessel 1

- 1.1.1 Mechanical Features
- 1.1.2 Equipment Modules
  - 1.1.2.1 Agitator1 (14-01-A-01)
  - 1.1.2.2 Agitator 2 (14-01-A-02)
  - 1.1.2.3 Process Inlet 1
  - 1.1.2.4 Process Inlet 2
  - 1.1.2.5 PUW Supply
  - 1.1.2.6 CA Supply/Vent
  - 1.1.2.7 PH Con
  - 1.1.2.8 Temper
  - 1.1.2.9 CIP Su
  - 1.1.2.10 Outlet
  - 1.1.2.11 Pressur
  - 1.1.2.12 Level M
  - 1.1.2.13 Conduc
- 1.1.3 Unit Operati

Process Inlet 2

Description	Alias	Class
Harvest Vessel Process Inlet#2	PRSC_IN2	Inlet2

Control Modules

Tagname	Service	Alias	Class	Parameters
XV-14-0112	Buffer Solution Inlet	BUF_SUP	XV01A_FC	Interlocks: CA Vent open
XV-14-0113	Protein Solution Inlet	PRS_SUP	XV01A_FC	Interlocks: CA Vent open
XV-14-0114	Homogenizer Inlet	HMGNZR_INLET	XV01A_FC	Interlocks: CA Vent open
XV-14-0122	CIP Process Inlet 2 CIP	CIP INIET	XV01A_FC	Interlocks: CA Vent open

Phases

Phases	Alias	Parameters	Description
Initialize	INIT		Clear all status values. Place all control modules in Automatic mode.
Transfer in Buffer	XFR in BUS	Interlocks:CA VENT Open; source unit ready Parameters: Source unit; max time; max level;	Position valves on the inlet to allow a transfer from the Buffer solution flasks. Set "Transfer in from Buffer flasks Ready" status flag = ON. Operator manually starts transfer pump. Phase completes on max time max level, or operator intervention.
Transfer in Protein Solution from cake container	XFR in PRS	Interlocks:CA VENT Open; source unit ready Parameters: Source unit; max time; max level;	Position valves on the inlet to allow a transfer from the cake container. Set "Transfer in from cake container Ready" status flag = ON. Operator manually starts transfer using CA. Transfer phase completes on max time, max level, or operator intervention.
Transfer in from Homogenizer	XFR from HMG NZR	Interlocks:CA VENT Open; source unit ready Parameters: Source unit; max time; max level;	Position valves on the inlet to allow a transfer from the homogenizer. Set "Transfer in from Homogenizer Ready" status flag = ON. Transfer phase completes on max time, max level, or operator intervention.
CIP process inlet2	CIP_PRC_IN2	Interlocks:CA VENT Open Parameters: Source unit	Allows CIP of each inlet point as designated by "source unit" parameter. Phase completes on CIP complete signal from CIP system.

Figure 1.5. Harvest unit EMs.

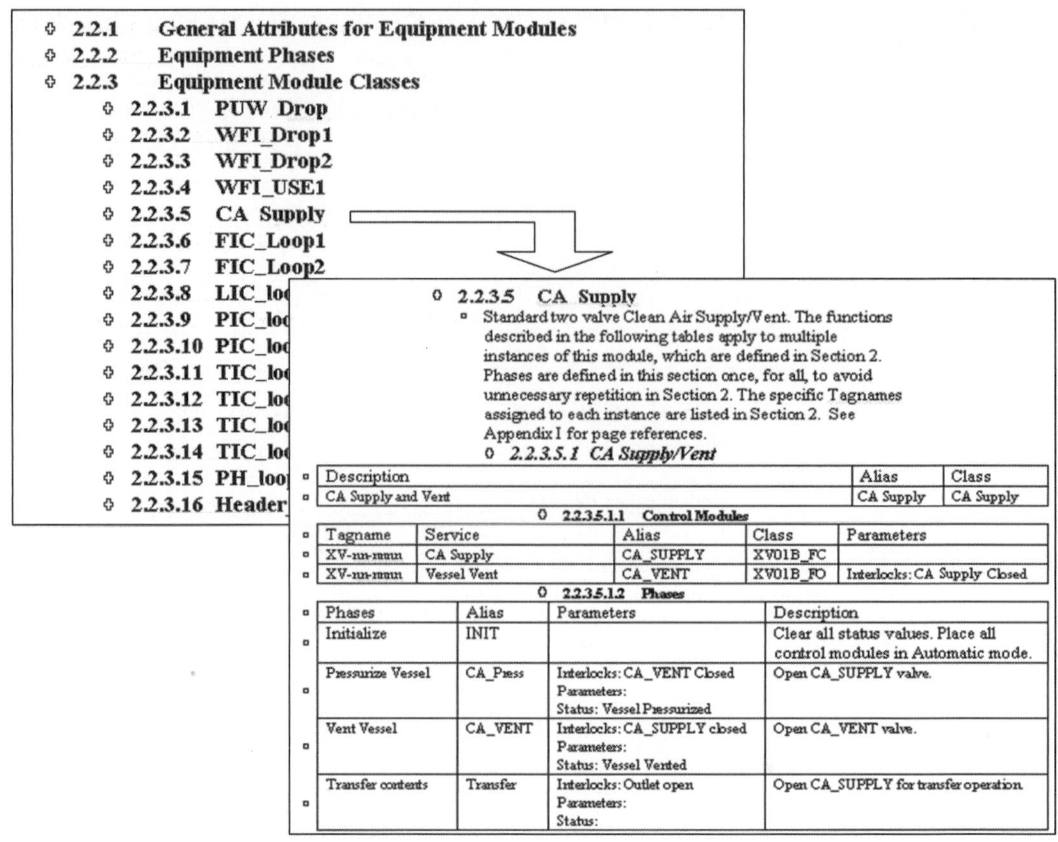

- 2.2.1 General Attributes for Equipment Modules
- 2.2.2 Equipment Phases
- 2.2.3 Equipment Module Classes
  - 2.2.3.1 PUW Drop
  - 2.2.3.2 WFI_Drop1
  - 2.2.3.3 WFI_Drop2
  - 2.2.3.4 WFI_USE1
  - 2.2.3.5 CA Supply
  - 2.2.3.6 FIC_Loop1
  - 2.2.3.7 FIC_Loop2
  - 2.2.3.8 LIC_lo
  - 2.2.3.9 PIC_lo
  - 2.2.3.10 PIC_lo
  - 2.2.3.11 TIC_lo
  - 2.2.3.12 TIC_lo
  - 2.2.3.13 TIC_lo
  - 2.2.3.14 TIC_lo
  - 2.2.3.15 PH_loo
  - 2.2.3.16 Header

2.2.3.5 CA Supply

- Standard two valve Clean Air Supply/Vent. The functions described in the following tables apply to multiple instances of this module, which are defined in Section 2. Phases are defined in this section once, for all, to avoid unnecessary repetition in Section 2. The specific Tagnames assigned to each instance are listed in Section 2. See Appendix I for page references.

2.2.3.5.1 *CA Supply/Vent*

Description	Alias	Class
CA Supply and Vent	CA Supply	CA Supply

2.2.3.5.1.1 **Control Modules**

Tagname	Service	Alias	Class	Parameters
XV-nn-mmm	CA Supply	CA_SUPPLY	XV01B_FC	
XV-nn-mmm	Vessel Vent	CA_VENT	XV01B_FO	Interlocks: CA Supply Closed

2.2.3.5.1.2 **Phases**

Phases	Alias	Parameters	Description
Initialize	INIT		Clear all status values. Place all control modules in Automatic mode.
Pressurize Vessel	CA_Press	Interlocks: CA_VENT Closed Parameters: Status: Vessel Pressurized	Open CA_SUPPLY valve.
Vent Vessel	CA_VENT	Interlocks: CA_SUPPLY closed Parameters: Status: Vessel Vented	Open CA_VENT valve.
Transfer contents	Transfer	Interlocks: Outlet open Parameters: Status:	Open CA_SUPPLY for transfer operation.

Figure 1.6. EM library with expansion of "Clean Air Supply."

efficiency. The "Clean Air Supply" EM is expanded in Figure 1.6 to show its specifications, CMs, and phases of operation.

Transfer Units

Transfer units are critical components that are needed to move product from one process operation to the next. They contain valves, pumps, and instrumentation to connect process modules in either a network or train configuration, as required by the GR. When the transfer unit is copied from the library for an actual application, it can be classified as a stand-alone unit or an EM belonging to the upstream or downstream unit, or it can be shared as an EM, depending on the physical capabilities required by the application. Cleaning and sanitation requirements are an important consideration when making this choice. If the transfer unit needs to be cleaned independently of the upstream and downstream units, then it probably should be classified as a unit, so that it can have modes of operation and status values independent of other units. An ISA-88.01 EM can also have these properties, although not all control systems allow this. Figure 1.7 provides an expanded view of a typical transfer unit that contains CMs and performs phases.

Developing the Procedural Models

With the physical model fully developed within each process module section, the procedural model can be added to the outline, with the physical capabilities

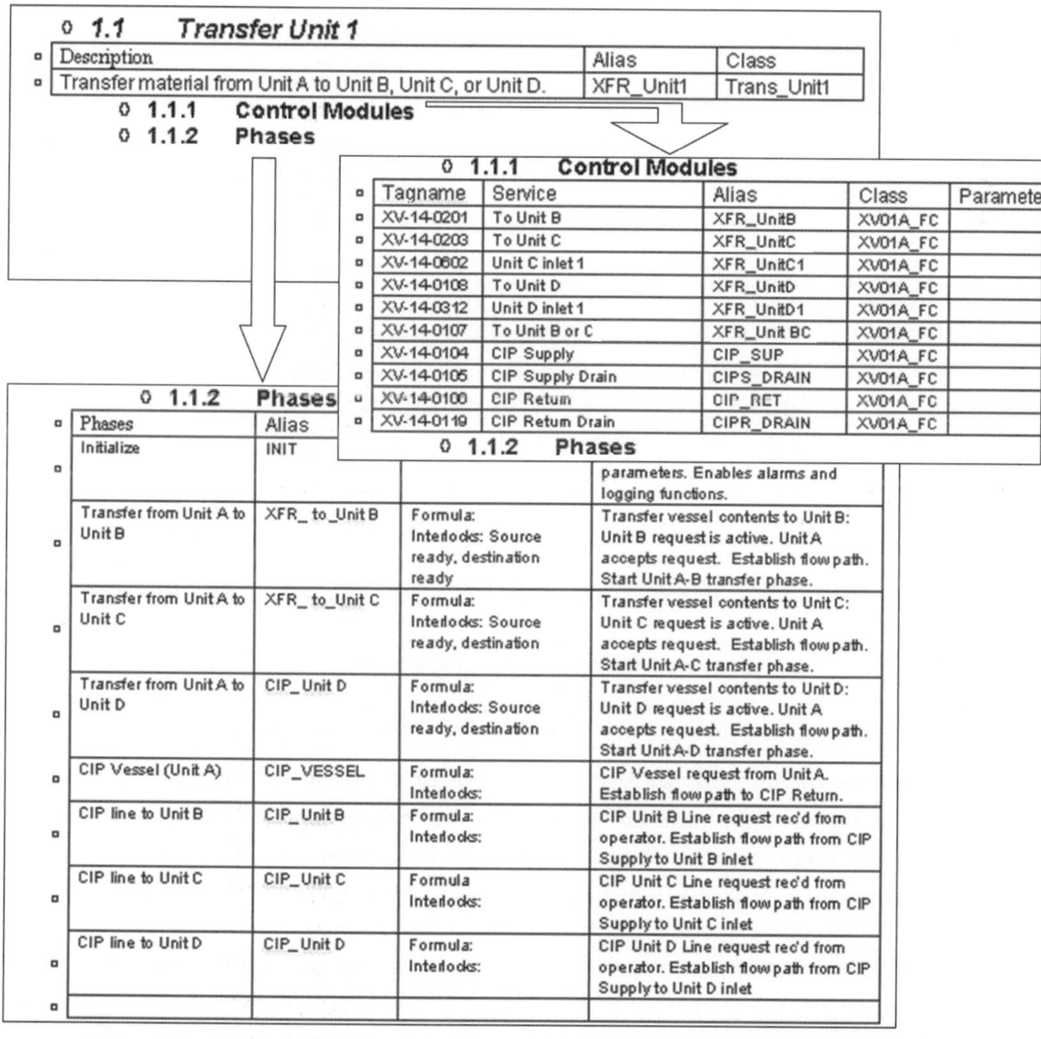

o **1.1 *Transfer Unit 1***

Description	Alias	Class
Transfer material from Unit A to Unit B, Unit C, or Unit D.	XFR_Unit1	Trans_Unit1

o **1.1.1 Control Modules**

o **1.1.2 Phases**

o **1.1.1 Control Modules**

Tagname	Service	Alias	Class	Paramete
XV-14-0201	To Unit B	XFR_UnitB	XV01A_FC	
XV-14-0203	To Unit C	XFR_UnitC	XV01A_FC	
XV-14-0602	Unit C inlet 1	XFR_UnitC1	XV01A_FC	
XV-14-0108	To Unit D	XFR_UnitD	XV01A_FC	
XV-14-0312	Unit D inlet 1	XFR_UnitD1	XV01A_FC	
XV-14-0107	To Unit B or C	XFR_Unit BC	XV01A_FC	
XV-14-0104	CIP Supply	CIP_SUP	XV01A_FC	
XV-14-0105	CIP Supply Drain	CIPS_DRAIN	XV01A_FC	
XV-14-0100	CIP Return	CIP_RET	XV01A_FC	
XV-14-0119	CIP Return Drain	CIPR_DRAIN	XV01A_FC	

o **1.1.2 Phases**

Phases	Alias		
Initialize	INIT		parameters. Enables alarms and logging functions.
Transfer from Unit A to Unit B	XFR_to_Unit B	Formula: Interlocks: Source ready, destination ready	Transfer vessel contents to Unit B: Unit B request is active. Unit A accepts request. Establish flow path. Start Unit A-B transfer phase.
Transfer from Unit A to Unit C	XFR_to_Unit C	Formula: Interlocks: Source ready, destination	Transfer vessel contents to Unit C: Unit C request is active. Unit A accepts request. Establish flow path. Start Unit A-C transfer phase.
Transfer from Unit A to Unit D	CIP_Unit D	Formula: Interlocks: Source ready, destination	Transfer vessel contents to Unit D: Unit D request is active. Unit A accepts request. Establish flow path. Start Unit A-D transfer phase.
CIP Vessel (Unit A)	CIP_VESSEL	Formula: Interlocks:	CIP Vessel request from Unit A. Establish flow path to CIP Return.
CIP line to Unit B	CIP_Unit B	Formula: Interlocks:	CIP Unit B Line request rec'd from operator. Establish flow path from CIP Supply to Unit B inlet
CIP line to Unit C	CIP_Unit C	Formula Interlocks:	CIP Unit C Line request rec'd from operator. Establish flow path from CIP Supply to Unit C inlet
CIP line to Unit D	CIP_Unit D	Formula: Interlocks:	CIP Unit D Line request rec'd from operator. Establish flow path from CIP Supply to Unit D inlet

Figure 1.7. Transfer unit example.

defined in the EM phases. Using the functionality of the EMs and CMs as basic building blocks, process operations are built to perform all possible functions of the unit. It is important to build transfer and cleaning operations as well. Each process operation must stand alone, so that each can be deleted or moved to meet the GR requirements of a specific product in the future.

The level of granularity is also important. Process operations perfom major processing activities, so each must match a requirement for its generic function. Operations are built from process actions that will become equipment phases. It is better at this stage to have many small single-function actions rather than a few large multifunction actions. Process actions should be structured so that they can be combined with other actions to form many different operations. This will provide the flexibility needed later when the unit is checked out of the library and configured as part of a complete process stage.

Figure 1.8 provides an expanded view of the harvest vessel process module, showing how an operation is constructed using EM phases. The "Transfer in Buffer Solutions" operation is expanded to show the EM phases that are used.

Developing Other Components

Additional components can be added to the process modules depending on the needs of the targeted industry. For example, commissioning and validation protocols can be developed for CMs, EMs, or entire units. Other attributes such as budgetary equipment costs can also be added to the modules so that cost estimates for a given configuration of modules can be generated quickly.

Using the Process Module Library

With a library of core process modules on the shelf, designed, and ready to use, a project can move into a detailed design phase very quickly once product approval is assured. The required process operations for the unit are selected and assembled as per the process model for the successful product. Transfer units are then selected to provide the appropriate connections. EMs and operations can be removed from each module to meet the specific needs of a particular product. Vessels, pipes, and valves are sized to provide the appropriate scale. Units are copied from the library to create the physical and procedural models for Product Alpha, as shown in Figure 1.9.

The units needed to manufacture Product Alpha are copied from the library to create a FS for the Recovery and Purification Stage required by the GR. Each unit copied from the library contains the complete physical and procedural model and is capable of performing a predefined set of process operations. The P&IDs

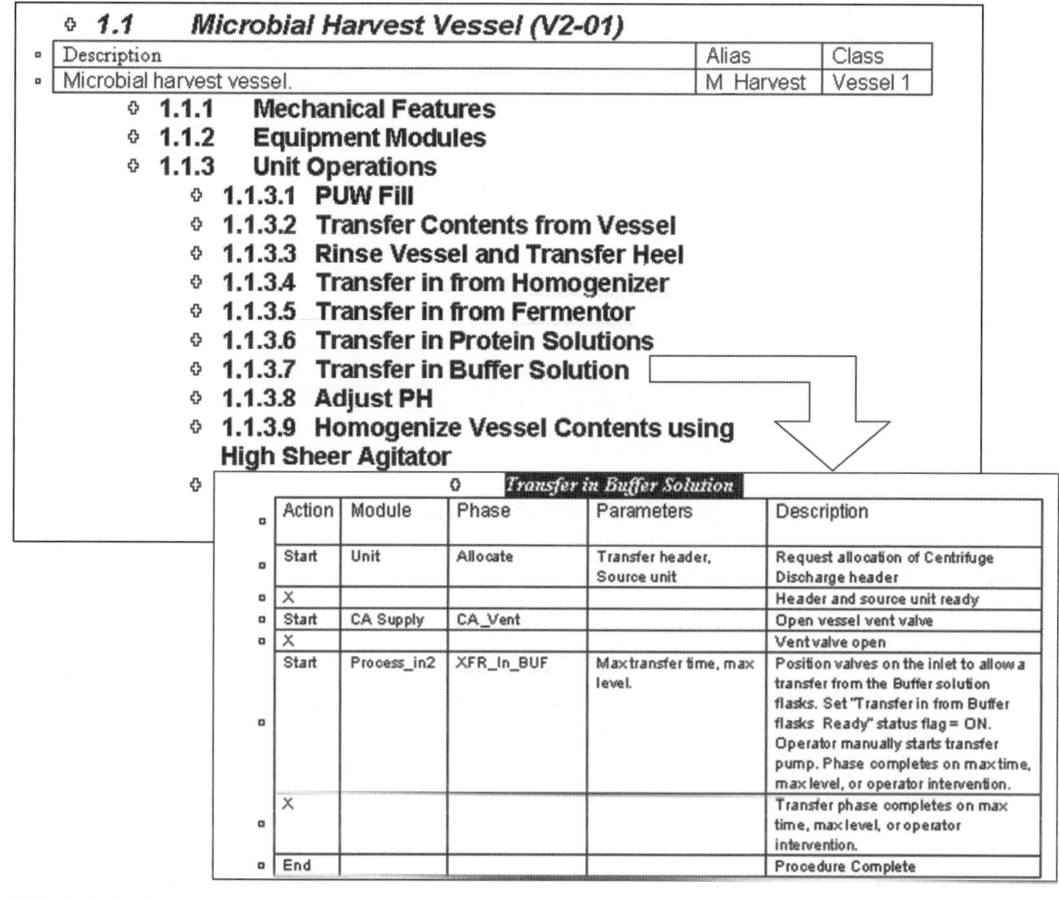

1.1 Microbial Harvest Vessel (V2-01)

Description	Alias	Class
Microbial harvest vessel.	M Harvest	Vessel 1

- 1.1.1 Mechanical Features
- 1.1.2 Equipment Modules
- 1.1.3 Unit Operations
  - 1.1.3.1 PUW Fill
  - 1.1.3.2 Transfer Contents from Vessel
  - 1.1.3.3 Rinse Vessel and Transfer Heel
  - 1.1.3.4 Transfer in from Homogenizer
  - 1.1.3.5 Transfer in from Fermentor
  - 1.1.3.6 Transfer in Protein Solutions
  - 1.1.3.7 Transfer in Buffer Solution
  - 1.1.3.8 Adjust PH
  - 1.1.3.9 Homogenize Vessel Contents using High Sheer Agitator

Transfer in Buffer Solution

Action	Module	Phase	Parameters	Description
Start	Unit	Allocate	Transfer header, Source unit	Request allocation of Centrifuge Discharge header
X				Header and source unit ready
Start	CA Supply	CA_Vent		Open vessel vent valve
X				Vent valve open
Start	Process_in2	XFR_In_BUF	Max transfer time, max level.	Position valves on the inlet to allow a transfer from the Buffer solution flasks. Set "Transfer in from Buffer flasks Ready" status flag = ON. Operator manually starts transfer pump. Phase completes on max time, max level, or operator intervention.
X				Transfer phase completes on max time, max level, or operator intervention.
End				Procedure Complete

Figure 1.8. Harvest vessel unit process operations.

for each of these units are also copied from the library, so that the combination of the P&IDs and the FS form a complete package ready for detailed design to begin.

Benefits of an ISA-88.01 Based Process Module Library

Up-front investments in engineering time to create a library of core process modules can shorten the time required for the design, construction, and commissioning phases of a project. The design phase is shortened because basic design is essentially completed when the appropriate units are selected from the library, assembled, and scaled to establish the physical capabilities needed to produce a product. Since each process module is fully self-contained and inherently compatible with each of the other units, the library naturally supports modular design

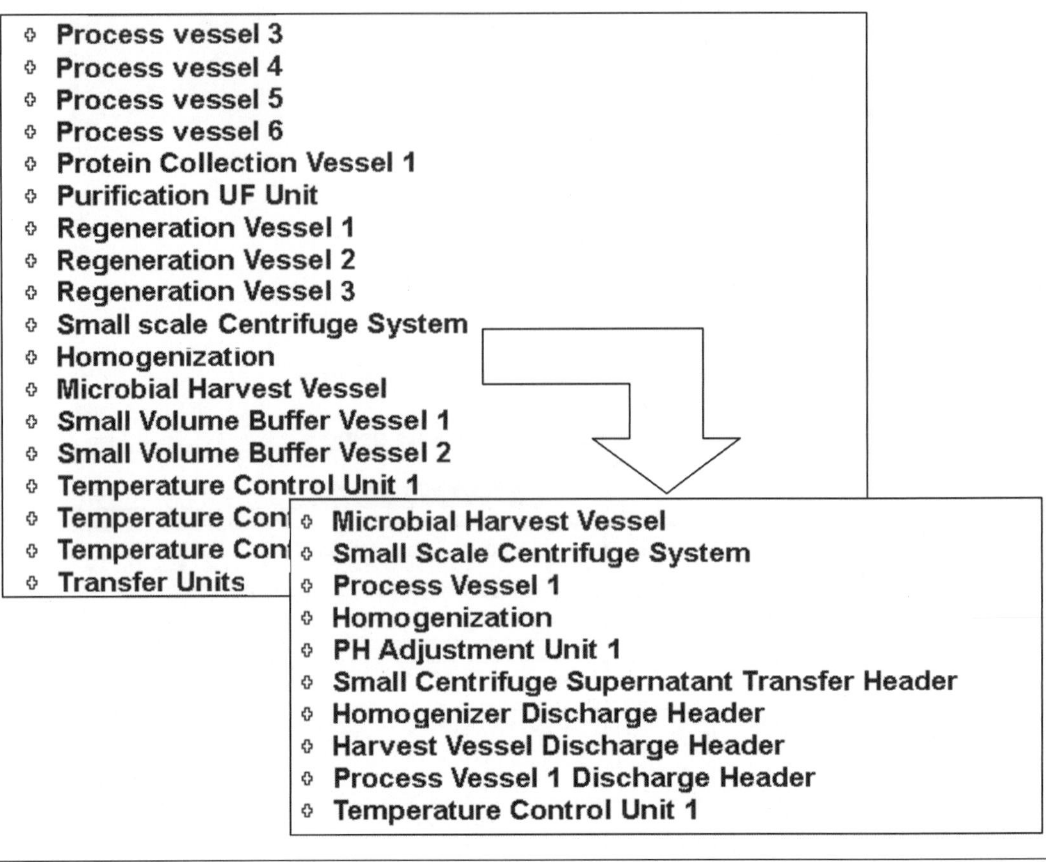

Figure 1.9. The process module library, with units for Product Alpha.

and construction techniques that can shorten the construction and commissioning timeline. The construction timeline can be further shortened because equipment can be ordered sooner, which keeps long lead-time items off the critical path. The modules contain sufficient detail regarding control functions, so a capable control system and systems integrator can be selected earlier. Also, validation protocols can be developed earlier thanks to the details contained in the physical and procedural models for each unit. This helps to keep the software development and validation off the critical path.

All of these factors tend to shorten the timeline from project approval to commercial production, which allows for a quicker scale-up of both new and existing products. The shortened project timeline can allow capital investments in facilities to be delayed until after late-stage clinical trials are completed, when the risk of product failure is lower. Thus, successful products emerging from the development pipeline can be brought to commercial-scale production more quickly and with less financial risk.

CHAPTER 2

ISA-88 Design and Implementation Case Study for a Complex Bulk Pharmaceutical Batch Process

Presented at the WBF North American Conference, April 13–16, 2003, by

Scott C. Clark
Sr. Engineer
scott_clark@merck.com
Merck & Co., Inc., 2778 South East Side Highway, Elkton, VA 22827, USA

Abstract

This chapter presents an ISA-88.01 implementation case study derived from a complex, multistep bulk pharmaceutical process. It methodically steps through the design process used to build the application software components that would ultimately control a campaign style of processing, consisting of five distinct trains and a representative cross-section of unit operations seen in Active Pharmaceutical Ingredient (API) bulk processing. Software reliability and flexibility were of equal importance, and this chapter will review the ISA-88.01 approach, including model development, object-oriented Equipment Modules (EMs), common and unique phases, recipe-based unit-to-unit coordination, batch reporting, and automated documentation routines. The end product was a fully automated, flexible, ISA-88.01 based system consisting of reliable and reusable phases, where the automation of processing changes was not resource intensive and batch reporting was highly integrated.

Case Study Attributes

Process Attributes

This case study is based on a multicampaign bulk pharmaceutical process where the product is campaigned through the same processing equipment during each process step. After each process step, the material is placed in storage and the equipment is mechanically reconfigured. The processing vessels ranged from 100 gallons to 18,000 gallons, and a process manifold was used extensively to move material from one vessel to another. There were many shared EMs, including flow meters, flow totalizers, condensers, vacuum pumps, and utilities. The list of unit operations included distillations, organic reactions, quenches, chromatography, filtering, drying, countercurrent extraction, nanofiltration, diafiltration, centrifugation, and other separation processes.

Control System Attributes

The automation systems included a Distributed Control System (DCS) and five ancillary Programmable Logic Controllers (PLCs) providing machine- and phase-level skid control with approximately 2000 total I/O. The PLCs were connected with the DCS via networked links. The automation systems were designed for fully automated continuous control and batch execution and included an integrated continuous historian, batch historian, and event chronicle.

Model Development

ISA-88.01 contains three models: the process model, the physical model, and the procedural model. The process model organizes the process into stages, operations, and actions. The physical model defines units, EMs, and Control Modules (CMs). The procedural model organizes the process into procedures, unit procedures, operations, and phases. The procedural model defines the order in which the process model is run on the physical model. ISA-88.01 also supports current Good Manufacturing Practices (cGMPs), in that it emphasizes good practices for the design and operation of batch processes. The following sections will discuss the process of developing the ISA-88.01 models and each model's deliverables.

Process Model

The process model provides the first opportunity to apply ISA-88.01 principles and the idea of common and unique objects (e.g., operations) that can be reused to accomplish the process requirements. This model's deliverables can help bridge the gap between classic process descriptions and ISA-88.01 and is useful to the process-development team members, regardless of their experience with ISA-88.01.

Process Model Diagram

Process model diagrams were developed from process descriptions, Piping and Instrumentation Diagrams (P&IDs), and Process Flow Diagrams (PFDs) to align the process descriptions with ISA-88.01 and provide the basis for subsequent models. These diagrams listed the process steps, stages, and operations in a graphical format.

As the design matured, vessel IDs and material transfer lines (dotted) were added to further define and clarify the process model. These material transfer lines were added, in part, to clarify the unit-to-unit coordination that would be required by the automation system. Figure 2.1 provides a simplified example of a process model diagram.

Developing an accurate and useful process model requires experienced personnel with a thorough understanding of the process. A well-developed process model will translate directly into the procedural model.

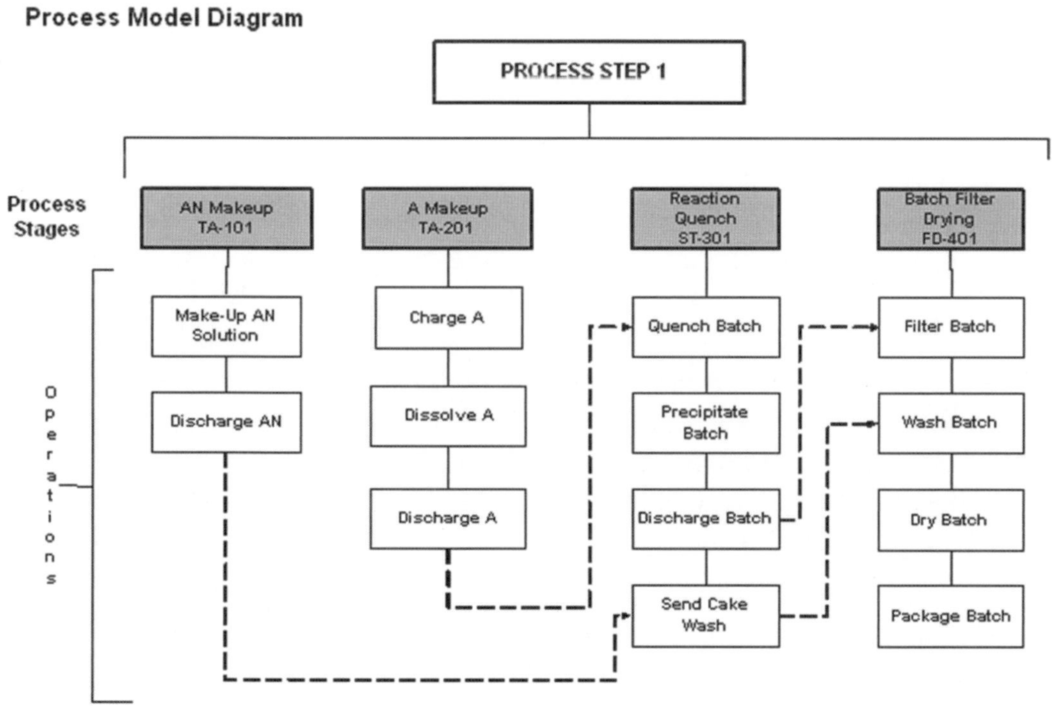

Figure 2.1. Example process model diagram.

Physical Model

There was a need to define the usage of CMs, EMs, and units for each mechanical group of the processing equipment. This was defined in the physical model. This model and its documentation were instrumental in helping the process engineers and configuration engineers visualize the equipment setup and control strategies for each processing step. The physical model deliverables also provide information that is not typically organized on the P&IDs and PFDs.

Quite often, the P&IDs would not include all the EMs and CMs associated with a unit, so there was a need to develop a drawing that would define all these components. After they were identified, EMs also needed to be defined, and this was done through the development of detailed requirement documents. This combination of drawings and requirement documents became the physical model.

Physical Model Drawings

Physical model drawings were typically based on units and all the resources (i.e., shared or nonshared EMs or CMs) that were necessary for each process step. They clarified how the equipment was physically configured and subsequently controlled. The drawings identified the applicable CMs and EMs and helped define the procedural control equipment-mapping model. Figure 2.2 provides an example physical model drawing with typical EMs identified. Typical EMs and their functionalities included the following:

- *Agitator*. Auto on or off, speed control
- *Inlet/Outlet/Nozzle/Vent*. Multiple discrete valve sequencing
- *Condensers*. Coolant circulation control, distillate, or coolant temperature control
- *Vacuum Pumps*. Seal water control, pressure and temperature control, booster control, interface to selector switch EM
- *Vessel Jacket*. Multiple discrete valve sequencing, multiple cooling and heating services, sequencing of fill and blow, vessel and jacket fluid temperature control
- *Selector Switch*. Switching between multiple control scenarios
- *Utilities*. Fluid service controls to multiple EMs or multiple units

The physical model drawings were developed so that shared EMs could be encapsulated as drawing objects and then pasted onto additional physical model

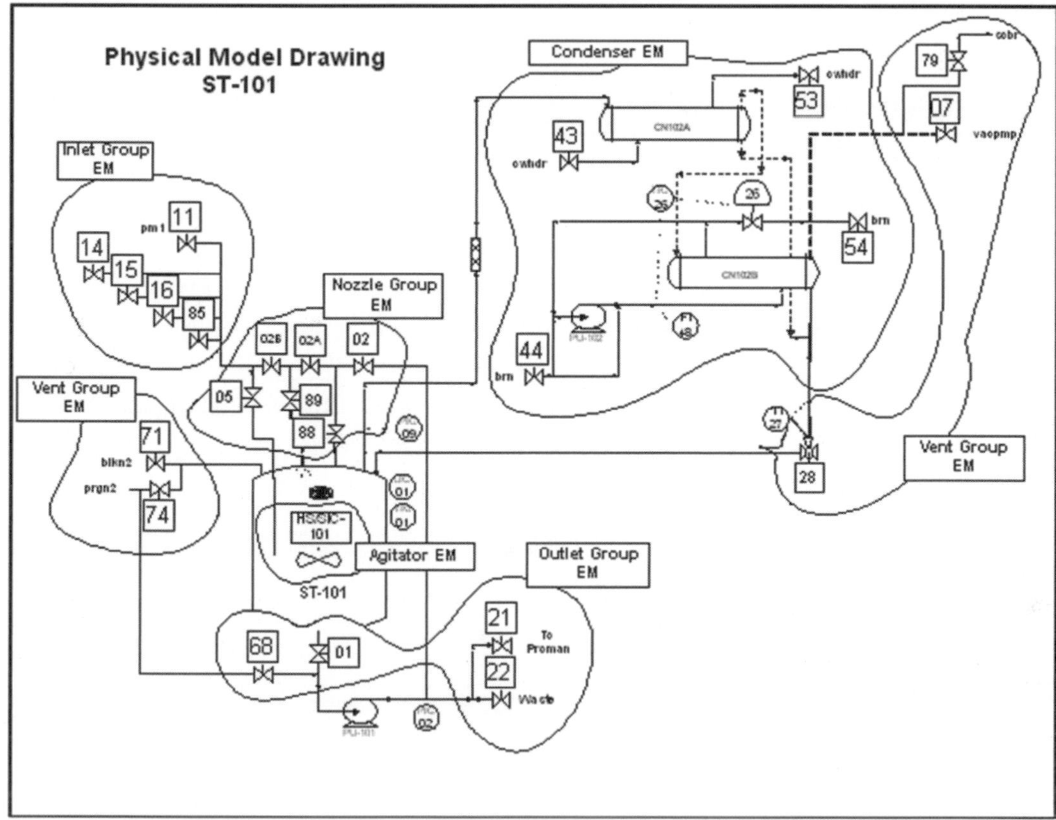

Figure 2.2. Example physical model drawing.

drawings as needed. For example, since the condenser EM was used on multiple units, it was created as a library object and pasted into the physical model drawings that used this condenser EM.

EM Detailed Requirements

After EMs were identified, a requirement specification was developed to define their functionality as object-oriented entities. The EM requirements that were specified were as follows:

- Internal and external parameters
- Setpoints
- Modes
- Actions and sequence of actions

- Alarms
- Interlocks
- Failure processing
- Human interface
- Batch execution interface
- Reporting

Figure 2.3 provides a requirements outline for a condenser EM.

It is important that the team developing the physical model can effectively translate the equipment's control needs based on the process requirements. This may include defining advanced control strategies, cascades, overrides,

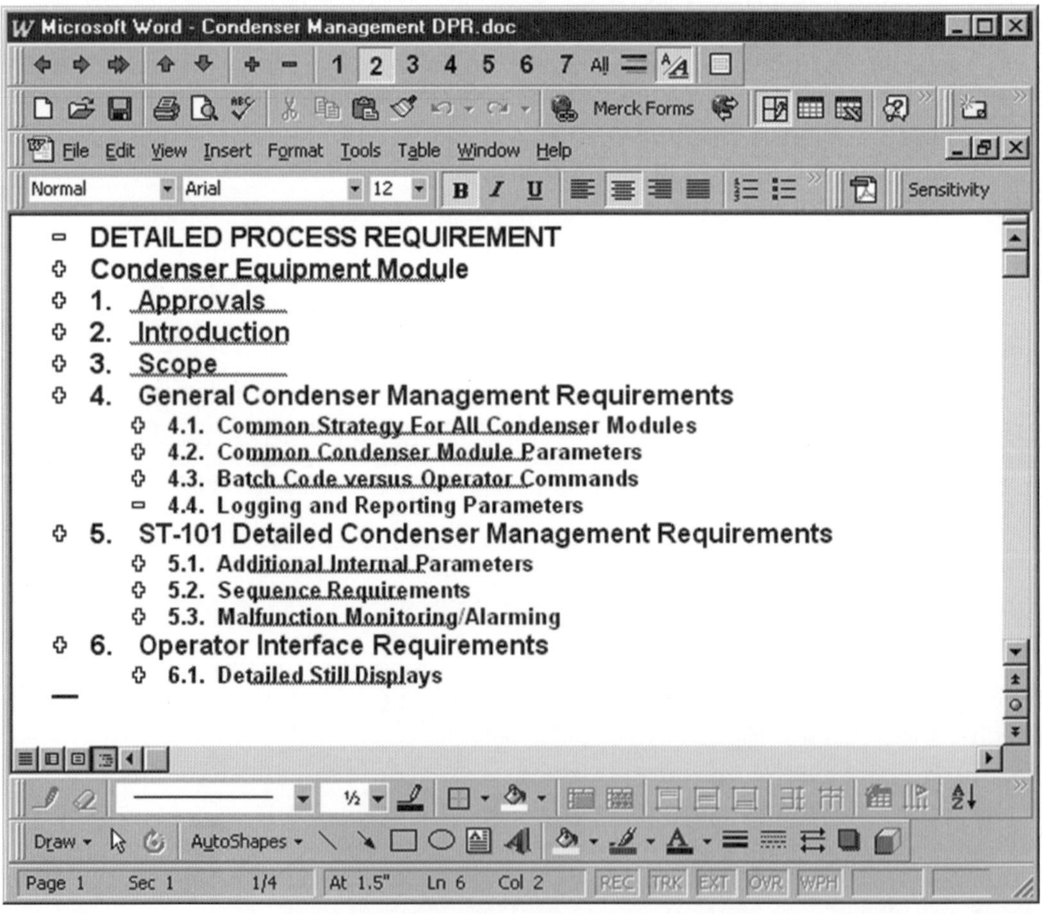

Figure 2.3. Example requirements outline.

split ranging, and EM setpoints, to name just a few. This team must have a firm understanding of object-oriented programming and possess extensive processing experience and control experience to ensure that the physical model will be flexible and integrate well into the procedural model.

Procedural Model

The procedural model provides the link between the physical model and the process model. A flexible design tool was needed to organize this link and document the requirements.

Procedural Model Database

The procedural model was developed as a relational database that specified the order in which the process model was run on the physical model and provided a flexible solution to deal with a maturing process design. The first step in its development was to import the process and physical model data. Tables were specified so that the procedural model data could be easily reconfigured as the process design matured with minimal repetition. Figure 2.4 provides an example database structure to store this information.

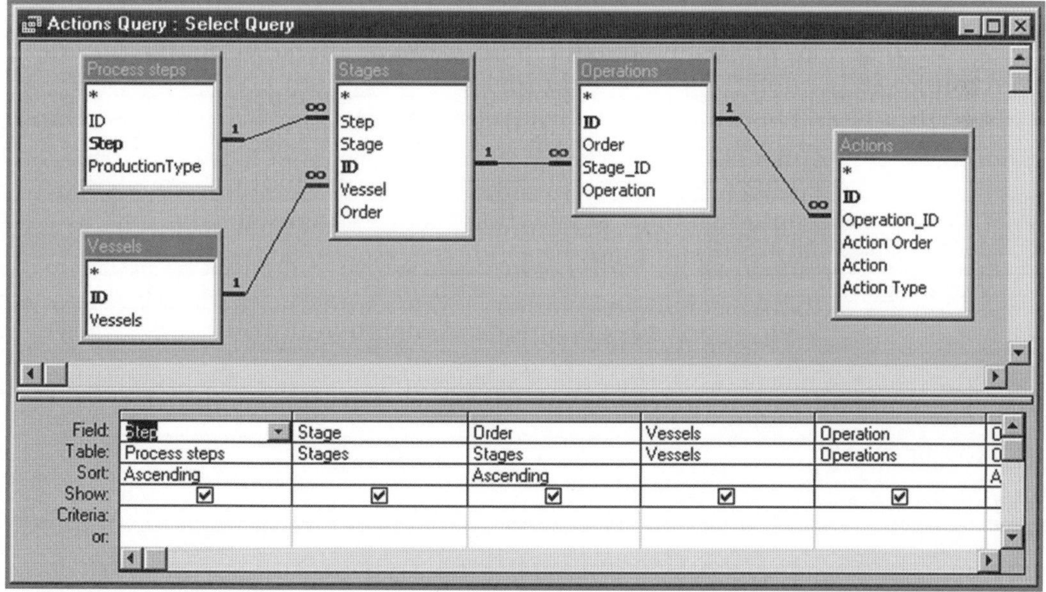

Figure 2.4. Example procedural model database.

Reports could be generated from this database in an outline format to form the procedural requirements. Typical reports included the following:

- Process stages and operations (i.e., the sequence of execution)
- Action (phase) types and descriptions
- Operation recipe and report parameters list
- EM list

Common versus Unique Phases

Reusability is a core benefit to ISA-88.01 and was taken advantage of whenever possible. As the process model was developed, the number of process operations that were reused became well defined. Since the process and procedural models were developed using a database, queries were run to help define their requirements (or at least the instances) of each of the phases. Phases were then categorized as the following:

- Common (i.e., executable on all units in a class)
- Unique (i.e., executable on a subset of units in a unit class or an alternate unit class)

Table 2.1 lists some of the common and unique phases.

Table 2.1. Example list of common and unique phases

Common phases	*Unique phases*
Age	Harvest
Temp adjust	Filter
Constant volume distill	Dry
Concentration distill	Quench
Vent setup	Decant receive
Receive	Continuous extraction
Discharge	Nanofilter
Manual add	Skid specific
Line blow	
Decide path	
Continuous level	
Alarm set	

Phase Requirements

A requirement specification was written for each phase to define the following phase attributes and requirements:

- Recipe parameters
- Run/Hold/Restart/Stop/Abort sequence of actions
- Method of completion
- Failure monitoring and recovery
- Batch execution interface
- Operator interface
- EM control and interfaces
- CM control and interfaces
- Alarm management
- Reporting parameters
- Unit-to-unit coordination

Phase Development

The granularity of phase control was an important issue during phase development. This included the extent of EM and CM control embedded within each phase. Programming strategy (state versus sequential), failure-handling strategy, recovery and restart strategy, complexity, and coordination of EMs and CMs during phase execution are all factors in determining this granularity. An example of this might include determining whether the Age phase, which provides a batch age function, should control the inlet and outlet EM groups or whether there should be a separate parallel phase that controls the inlet and outlet groups. As control of the EMs and CMs became more complex, it led to integrating more EM and CM control into one phase, thereby reducing or removing phase-to-phase coordination issues.

Unit-to-Unit Coordination

During the development of application software for a flexible facility, a recipe-driven methodology to coordinate phases on different units was required. A typical example of this is the requirement to coordinate the opening of outlet valves and the starting of a pump from a sending vessel with the opening of the inlet valves on the receiving vessel. The two vessels also needed to notify each other if there was a failure or hold condition or if the ending condition of the transfer had occurred.

Requirements

The specific requirements for unit-to-unit coordination were as follows:

- Specify partner by recipe (i.e., "Recipe Parameter 'THEY1' = TA-101").
- Dynamically build and release unit-to-unit links.
- Require more than one partner, if necessary (i.e., ratio two flow streams into a receiver).
- Allow a unit to notify others of a HOLD or FAILURE condition.
- Allow the Totalizer EM to indicate the NEAR_COMPLETE or COMPLETE condition.
- Require coordination messages to be used between units, such as:
    - SYNCHRONIZE (Partner is ready to communicate.)
    - READY_TO_TRANSFER (Partner is ready to begin transfer.)
    - TRANSFER (Both partners are transferring.)
    - NEAR_COMPLETE (Transfer is nearly complete.)
    - COMPLETE (Transfer is complete.)
    - BREAK (Sender/receivers are to break/maintain communications.)
    - SILENT (Channel is available for use.)

Example

A novel approach was taken where a 32-channel communication dispatch module, or "CB radio," provided this dynamic unit-to-unit coordination. Phases had linked, embedded logic to find an open channel, post and read messages on the radio, and send the appropriate message (such as TRANSFER) to the dispatch module at the appropriate time during phase execution. The radio unit-to-unit link message was built at run time and recipe based. Figure 2.5 provides an example of this design at work:

- CH 4 (i.e., channel 4) indicates a transfer has just completed between two vessels.

- On CH 5, the phases are determining whether the channel should be maintained or released.
- CH 6 through CH 8 indicate a constant volume distillation in progress, where two tanks are providing a ratio feed, and there is one distillate receiver.
- CH 10 through CH 16 are currently not in use and are available to any unit.

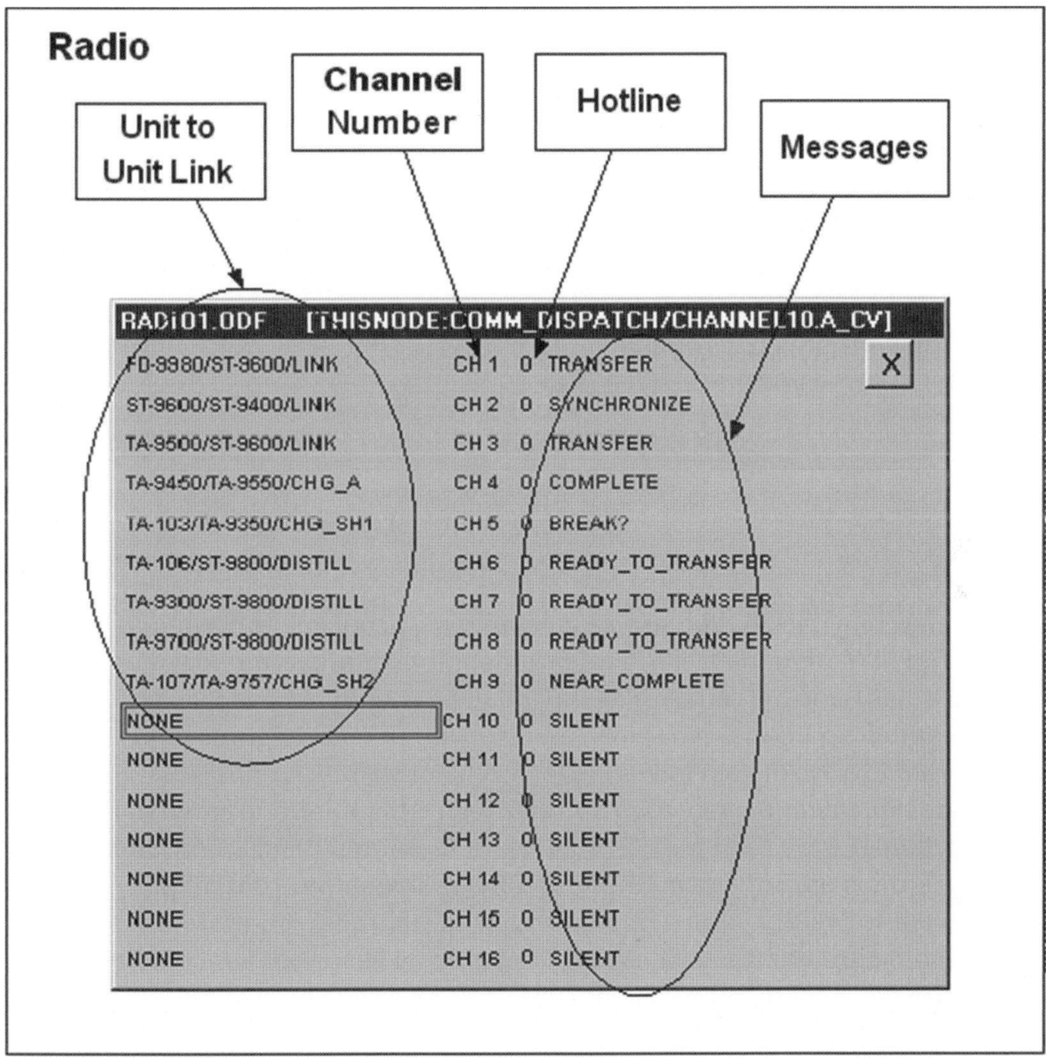

Figure 2.5. Example of radio channels for coordination control.

Skid Control

Communications between a centralized control system and skids were handled by creating "wrapper" phases that would provide batch execution control of the skid phases. Recipe downloads and Start, Stop, Restart, and Abort commands were sent to the skid. The skid phase would then proceed based on these instructions and report status information back to the centralized control system (e.g., Running, Holding, general status messages) Unit-to-unit coordination, as described previously, would be handled on the centralized control system, within the wrapper phase, thus providing integrated batch control regardless of where the phase control logic resided.

Batch Reporting

The development of a standardized methodology for batch reporting can be somewhat elusive. Common issues that may arise during the requirement definition and development are as follows:

- The report may be difficult to define. (What and how much information should be included?)
- Events or data to be reported may originate from multiple databases and other sources.
- The redundancy or recovery of the data and reporting system may be complex.
- The procedural model may exist in two locations, thereby increasing the complexity of changes.
- Reporting triggers may be dynamic within a critical process stage or operation.

Requirements

The requirements of the batch-reporting system included the following:

- Specify critical parameters by recipe.
- Monitor and report critical process parameters and calculate statistics.
- Monitor and report critical alarm events.

- Have tightly integrated process event and batch event data.
- Have strict synchronization with a centralized clock.
- Be robust—monitor during Run, Hold, Restart, and Stop or Abort states.
- Integrate the reporting model into the procedural model.

Example

A design approach was implemented where the Critical Process Parameter (CPP) monitoring (of minimum, maximum, and average calculations) and critical alarm monitoring were done in the respective CMs executing in a redundant controller. A reporting phase was developed to turn monitoring on or off at the critical stage within the procedural model, as specified by the recipe. Once the monitoring logic was turned on in the CM, it would continue to execute even during a batch Hold, Restart, or other state change. The calculated values and triggers (i.e., stop and start events) were written directly to the batch historian as standard report parameters at the start and end of the critical process stage or operation. Figure 2.6 provides an example of the reporting model within the procedural model.

For critical alarms, the batch monitor phase changes the module alarm priorities at the start of a critical stage to a priority reserved for critical alarms. The batch report was designed to query specific report and event parameters from the batch historian, as logged by the batch monitor phase and the event journal subsystem (for alarms). Since the batch monitor phase was placed in the procedural model, all batch reporting events were logged in context with the procedural model.

Documentation

Application software documentation typically falls into two categories: (1) documentation developed during the system design or commissioning phase or (2) documentation developed or revised during the system maintenance phase. The goal was to provide documentation from both categories that required the least amount of resources to maintain in an as-built state. Various software tools were developed to extract requirements and design documentation directly from the control system. The information contained on the originally developed model drawings and documents was moved into the control system. Software tools were used to extract these into an acceptable documentation format, as needed for

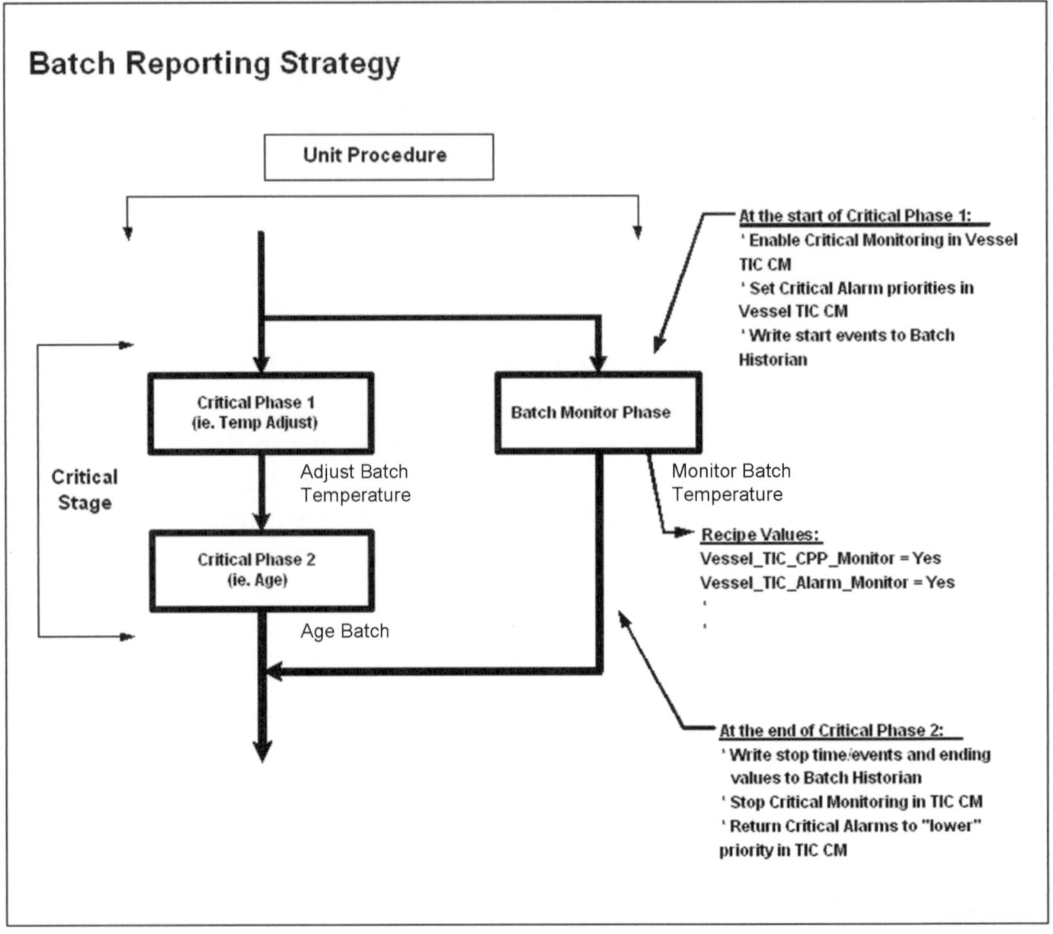

Figure 2.6. Example batch reporting strategy.

reviews and approvals. Table 2.2 provides a list of documents and how they could be maintained during the maintenance phase.

When detailed requirements were necessary, an approach was adopted where the requirements were documented in the application software design environment by developing a design shell that was, in essence, the detailed requirements. Coding was then done within this shell to become the design. This approach combined the detailed requirements and designs, which minimized documentation redundancy and reduced the resources needed to maintain them in an as-built state. The documentation can be exported to allow for review and approvals outside the application software environment.

Table 2.2. Example documentation list		
Document	*Description*	*Maintenance*
Process model	Process descriptions organized into stages and operations	Maintained in area and unit procedures with a tool to extract and present them in a commented sequential function chart format. A tool was developed to list the mapping of phases to units.
Requirements	EM, phase, and CM general requirements	Maintained as text or references in the application software function blocks with a tool to extract and present them for review or approval.
Designs	EM, phase, and CM designs	Maintained in application software with a tool to extract and present them.
Physical model	Units, EM, and CM	Maintained in application software. For EMs the control system provides a list of referenced CMs that are controlled by the EM. See the "Requirements" row.
Procedural model	Unit procedures, operations, and phases	Maintained in the same fashion as the process model.
Formula	Listing of recipe values used to process a batch	Tool developed to extract formulas in a compact tabular format.
General design specification	Design guidelines for software development	Maintained outside the control system.

Conclusion

This chapter presents an ISA-88.01 implementation case study derived from a complex, multistep bulk pharmaceutical process. It reviews the processes that are involved with building application software to control a campaign style of processing, consisting of five distinct trains and a representative cross-section of unit operations, seen in API bulk processing. The chapter also reviews model development; use of object-oriented EMs; common and unique phases; recipe-based, unit-to-unit coordination; batch reporting; and automated documentation routines. The end product of this case study was a fully automated, flexible, ISA-88.01 based system consisting of reliable and reusable phases, where automating processing changes was not resource intensive and batch reporting was highly integrated.

CHAPTER 3

Managing Complex Equipment Status in a GMP Environment

Presented at the WBF North American Conference, March 5–8, 2006, by

Mark Albano
Senior Marketing Specialist
mark.albano@honeywell.com
Honeywell, 2500 W. Union Hills Drive, Phoenix, AZ 85027 USA

Tom Farenholtz
Principal Consultant
tom.farenholtz@honeywell.com
Honeywell, 13655 Dulles Technology Drive, Herndon, VA 20171 USA

Steve Zarichniak
Applications Consultant
steve.zarichniak@honeywell.com
Honeywell, 1100 Virginia Drive, Fort Washington, PA 19034 USA

Abstract

Batch manufacturing systems must maintain the status of equipment and identify which ones are available, dirty, out of service, and so on. In industries that enforce Good Manufacturing Practices (GMPs), additional status types and values for process equipment must be maintained. Rules associated with the equipment status can be complex and involve the equipment class, current equipment state, and time in the current state. This chapter will cover the operator interface used to review the equipment status and the interlocking requirements between the recipe and batch phases based on multiple status types. Coordination, timing, and reporting of the equipment status transitions, both internal and external to the batch system, are discussed. Sample system architectures that illustrate how GMP status can be applied to current batch implementations will be presented.

Introduction

For manufacturers who are subject to regulations enforced by the Food and Drug Administration (FDA) in the United States and other agencies worldwide, it is critical that processes and systems are validated. By definition, equipment, units, and devices are not compliant by themselves. The combination of the equipment, process definition, operating practices, and maintenance policies are all subject to the validation criterion established by a manufacturer. This chapter provides an overview of equipment status in a regulated environment and explains its application in a complex batch environment.

Validation

Validation requires the establishment of documented evidence that provides a high degree of assurance that a specific process will consistently produce a product meeting its predetermined specifications and quality attributes. The purpose of validating automation systems is to prove with a high degree of assurance that the system will work correctly and consistently. For FDA 21 CFR Part 11 compliance, validation should also prove the accuracy and reliability of the data collection and management system.

Equipment Status Management Issues

Methods for managing equipment status vary from paper-based Standard Operating Procedures (SOPs) to computer-based Manufacturing Execution Systems (MESs), Enterprise Asset Management (EAM) systems, and process automation systems. These methods have evolved over time and tend to be specific to each manufacturer. Mergers and acquisitions of companies have led to multiple systems and methods across their corporate enterprises.

Maintaining and harmonizing computer-based systems that meet GMP regulations and internal corporate standards is an expensive proposition. Internally developed systems promise maximum functionality and flexibility, but the reality is that these systems rarely get completed with all the expected features. When a new operating system platform emerges, support for older versions is dropped. Compliant systems must be on a supported platform so, eventually, the custom system must be migrated to a new platform. The revalidation cost is often as much as the original system implementation.

Commercial Off The Shelf (COTS) applications, such as maintenance management systems, provide affordable solutions but are generally stand-alone and not integrated with batch or process control systems. Before defining a solution, a discussion of some equipment management requirements is in order.

Equipment Definitions

ISA-88.01 defines the following terms:

- *Unit.* "A collection of associated control modules and/or equipment modules and other process equipment in which one or more major processing activities can be conducted" (3.60)
- *Equipment Module.* "A functional group of equipment that can carry out a finite number of specific minor processing activities" (3.16)

Presently, multiple-batch automation systems may be needed to manage units, Equipment Modules (EMs), and other resources. The primary management functions conducted by the batch automation systems are availability (i.e., ownership) and process capabilities (i.e., class parameters). An additional requirement for GMP processes is the management of equipment unit status.

This chapter defines a common equipment status management scheme for all equipment, including units, EMs, and other process equipment. Process capability is not considered part of equipment status. It is important to manage all these equipment types, not just the units, in the same fashion as in a GMP plant. Figure 3.1 contains four units (T1000, T1001, T2000, and T2001), one Control Module (CM; TP100X200X), and four other pieces of process equipment (TL1000200X, TL1001200X, TL100X2000, and TL100X2001).

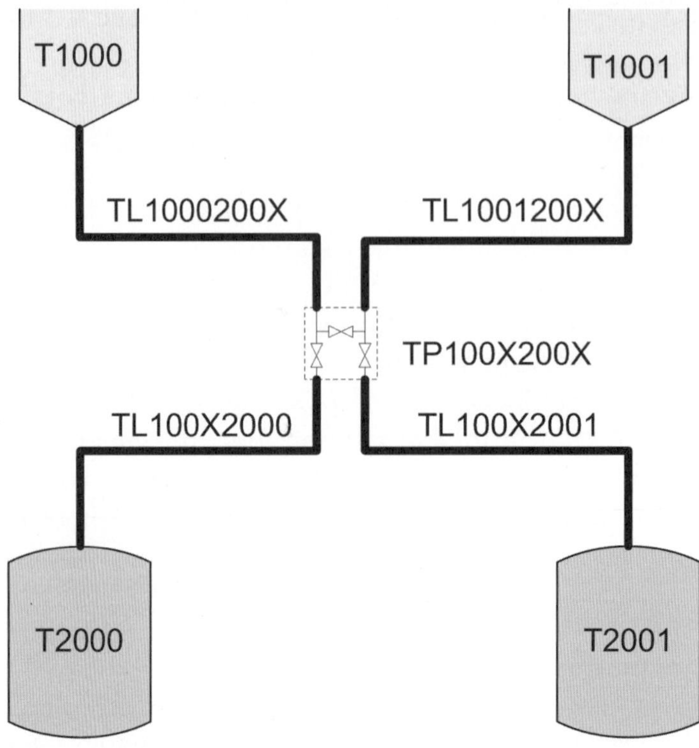

Figure 3.1.: Equipment model.

Equipment Status Data Management

In addition to managing ownership and capacity, additional status types need to be defined for the equipment in a GMP plant. Status types are class based. By looking at the information in Figure 3.1, the status types described in Table 3.1 were defined.

Table 3.1 shows how different classes of equipment have their own requirements for the management of several different types of status values. Batch automation systems usually have a single status type for equipment. If a single status type is all that is available, then there are a great number of unique status values to cover all possible combinations. In Table 3.1, there would need to be six (3 * 2) values for the prep tank, transfer line, and transfer panel and 27 (3 * 3 * 3) values for the mix tank. Transition logic would use enumerations of these combinations. Adding status values or types multiplies the number of combinations. These large enumeration sets for each unit class are difficult to maintain and make it difficult to configure control strategies based on status. A better approach is to allow equipment to have multiple status types defined and to enable the control strategies to set and check them in any combination.

Table 3.1. Class-based status types and values

Equipment class	*Members*	*Status type*	*Status values*
Prep tank	T1000, T1001	Availability	Available, In-use, Out-of-service
			Dirty, Clean
Mix tank	T2000, T2001	Availability	Available, In-use, Out-of-service
		Cleaning	Dirty, Clean, Sterile
		Process	Empty, Mixing, Mixed
Transfer line	TL1000200X	Availability	Available, In-use, Out-of-service
	TL1001200X		
	TL100X2000	Cleaning	Dirty, Clean
	TL100X2001		
Transfer panel	TP100X200X	Availability	Available, In-use, Out-of-service
		Cleaning	Dirty, Clean

Batch Independence and Genealogy

GMP equipment status is persistent across batches and exists even if no batch is active on a unit. Batch systems are designed to process batches and do not carry status from batch to batch. Historically, the process control system has been the usual place to maintain persistent data for the process. GMP requires the maintenance of an equipment log, showing historical status change events and genealogy of the equipment and batches. The process control system maintains current status and relies on a data historian or process logs for any type of event data based on time but not by batch. Batch association is important for genealogy (i.e., association of equipment with a batch).

Persistence of the equipment status must also be maintained if the batch automation system fails or undergoes a warm or cold restart. Storing status values in process control registers requires careful design and analysis of failure modes.

Unit Class Status for Process State

A status type ("Process") has been defined for the mix tank in Table 3.1. This is a very useful status type that can be used to assist in the coordination of processes. For example, consider a mix tank that holds an intermediate product for use in the next process step. Once the mixing is done and the batch is complete, the mixer availability status is set to "Available," so that it can be acquired by a process that needs to transfer material out of the mixer. What is to prevent a cleaning procedure from being run on the unit? Without maintaining a process status, the unit is available and can be acquired by a cleaning batch. This could unintentionally ruin the intermediate material in the mixer.

Currently, this problem is solved by the use of either an expensive, finite scheduling system or by communicating the current status of a unit to operators on a whiteboard mounted on the unit. Manual log sheets for the equipment are used to verify that the equipment has followed the correct sequence of events. SOPs require the operator to check the logs and whiteboards before starting a batch and then sign the batch record to indicate that the check was performed. As part of the batch release process, manual verification of equipment logs is required. This process is manual and as such is prone to human error.

Association of Other Non-class-based Equipment

Management of other process equipment is left up to the control-system vendor or process engineer, who must determine how to add these to the overall strategy for a project. It falls somewhere between a batch and process control system. Referring to Figure 3.1, the transfer lines and transfer panels are shared between different units at different times. Batch systems allow these types of equipment to be managed as resources. One of the drawbacks to batch systems is a built-in feature that automatically releases resources at the end of some process steps. The ability to associate other process equipment with a unit and have the association persist across operations, unit procedures, and batches would remove unnecessary logic from CMs and phases.

An example that uses the model in Figure 3.1 would be a cleaning procedure using a reusable operation to clean two lines and a transfer panel. A Clean-In-Place (CIP) skid would be required to support each cleaning cycle. If the CIP skid was released between CIP cycles, then this would prevent the orderly completion of all cleaning cycles.

Equipment Status Rules

Up to this point, our discussion has assumed that the status values are static and only change when commanded by a process request or event. There has been no restriction on legal transitions from one status value to another. Also, several of the status values may require an expiration to be associated with them.

Table 3.2 illustrates the addition of transition and timing rules for the cleaning status type for the mix tank. One additional status value, "Clean-expired," was

Table 3.2. Status value rules					
Equipment class	*Members*	*Status type*	*Status values*	*Timeout and next value*	*Allowed transitions*
Mix tank	T2000, T2001	Cleaning	Dirty	No timeout	Clean
			Clean	14 days Clean-expired	Dirty, Sterile
			Clean-expired	No timeout	Clean, Dirty
			Sterile	10 hours Clean	Clean, Dirty

added to the list of status values. This value is set if the "Clean" value expires after the timeout period.

An independent process that monitors equipment rules and changes the status when a timeout occurs needs to be part of the equipment management solution. In order for this to work, a time stamp needs to be associated with each status change.

Equipment Status Solution

A solution for equipment status management in a GMP environment requires significant additions to the standard capabilities of batch and process control systems. Status management also involves and impacts other plant systems including MES, EAM, ERP, finite scheduling, and the data historian. Automatic timeout monitoring requires a component that is not currently present in any plant system.

Given the global impact of equipment status management, there is sufficient functionality, commonality, and complexity to define a new class of system (Fig. 3.2) that includes the following:

- *Equipment status database.* Relational database to support the equipment status model that includes a configuration utility to manage the model and status rules
- *Equipment transaction processor.* Standard to high-performance (depending on the plant size), scalable, real-time transaction processor
- *Common set of transactions.* Well-defined and documented set of transactions to support equipment status change requirements
- *Equipment status monitor.* Processes automatic status changes based on rules

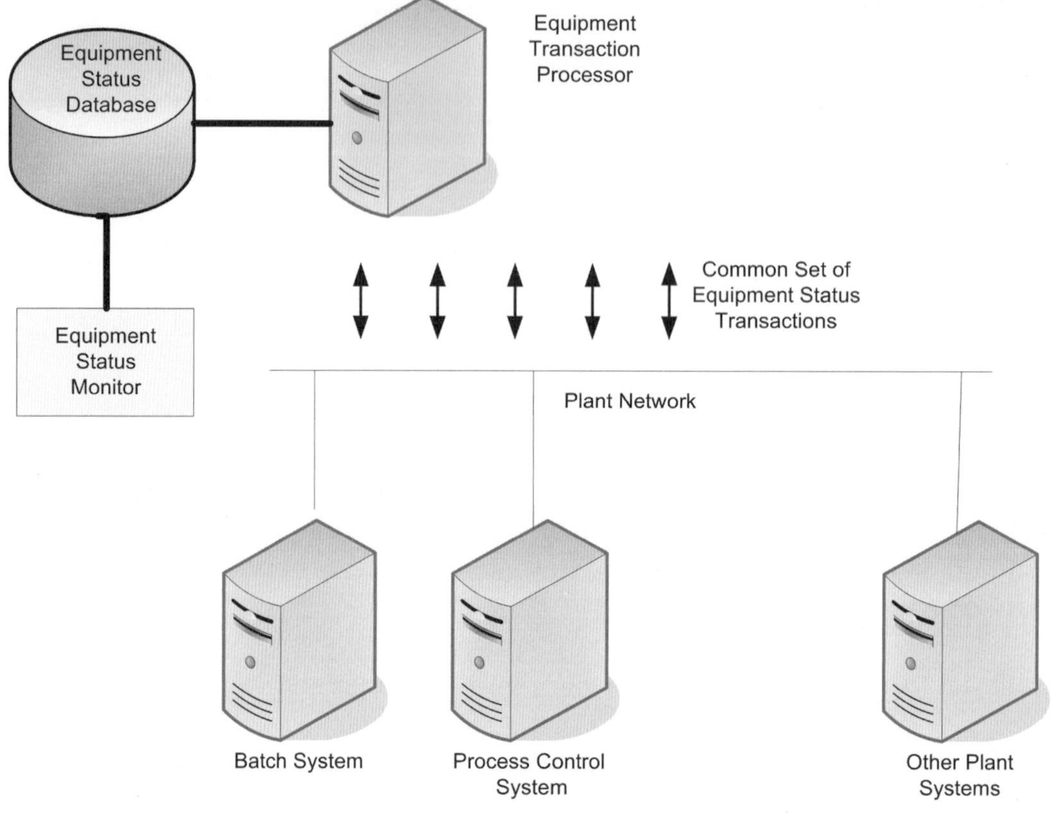

Figure 3.2. Equipment status architecture.

Characteristics of Equipment Status Management

Some important characteristics of status management include the following:

- *Real time*. Needs to support process control systems
- *Highly available*. Must be as reliable and as available as the process control system
- *Secure*. Must keep data on a secure server that is configured with data redundancy and backup capabilities
- *Scalable*. Must be able to expand to accommodate growing plant needs
- *Configurable*. Must be able to be configured on-line without a system restart for changes to take effect

Conclusion

We can conclude the following from the topics discussed in this chapter:

- Today's systems are too simplistic for complex equipment status management.
- Complexity results in the tendency to go back to paper, which has many disadvantages:
  - Limited production yield
  - Operating in a non-real-time environment
  - After-the-fact checking
- Some solutions to these issues include the following:
  - Status management application component
  - Need to develop industry standards

Definitions

Auxiliary equipment—Equipment that is not a unit but can be used in the execution of a batch.

Class-based equipment—Equipment whose attributes, properties, and status types are defined in an equipment class, whose values are maintained for individual equipment in the class.

Electronic record—Any combination of text, graphics, data, audio, pictorial, or other information in digital form that is created, modified, maintained, archived, retrieved, or distributed by a computer system.

Electronic signature—Computer data compilation of any symbol or series of symbols executed, adopted, or authorized by an individual to be the legally binding equivalent of that individual's handwritten signature.

Equipment genealogy—Record of all activities that occur in a batch pertaining to equipment. The genealogy records are constructed in a manner that allows for tracing both forward and backward through subsequent and prior batches.

Equipment status type—An attribute of equipment relating to its status. Status types are assigned a unique status value from a list of allowable status values and their associated rules.

Equipment status value—State assigned to a status type for a piece of equipment such as "clean" or "dirty."

Non-unit-based equipment—Equipment that requires status management where processing activities do not occur.

Status repository—The repository with all current and historical equipment statuses in the enterprise.

Further Reading

Instrumentation, Systems, and Automation Society. 1995. *ANSI/ISA-ISA-88.01-1995: Batch control part 1: Models and terminology*. Research Triangle Park, NC: ISA.

U.S. Federal Government. *Code of federal regulations, section 21, part 11*. Washington, DC: Government Printing Office.

———. *Code of federal regulations, section 21, parts 210–11*. Washington, DC: Government Printing Office.

U.S. Food and Drug Administration. 1987. *FDA guidelines on general principles of process validation*. Washington, DC: Government Printing Office.

———. 1998. *FDA guide to validation of automated systems in pharmaceutical manufacturing*. Washington, DC: Government Printing Office.

CHAPTER 4

Impact of Batch Software Upgrades on Validated Batch Applications

Presented at the WBF North American Conference, April 13–16, 2003, by

Shri Mariyala
Senior Batch Support Engineer
srinivas.mariyala@wonderware.com
Wonderware Invensys, 26561 Rancho Pkwy South, Lake Forest, CA 92630 USA

Abstract

Batch software upgrades for validated batch applications raise many issues. This chapter describes guidelines for how much revalidation evidence is needed to meet regulatory requirements for validated batch applications, if changes are made to software that is utilized in developing these automated applications. In addition, this chapter talks about the benefits and liabilities expected after making upgrades to batch software.

Often, it is difficult to estimate the level of testing effort that is needed to perform on a production system after implementing software upgrades. The outcome of this testing and documentation should provide sufficient data to demonstrate that software upgrades have no negative impact on equipment, product, or process performance and that the system has been restored to its validated status. This is an absolute requirement for pharmaceutical companies and should meet the guidelines required by Food and Drug Administration (FDA) and internal company standards.

The FDA's analysis of 3140 medical device recalls conducted between 1992 and 1998 reveals that 242 of them (7.7%) can be attributed to software failures. Of those software-related recalls, 192 (6.1%) were caused by software defects that were introduced when changes were made to the software after its initial production and distribution.

Introduction

This chapter provides some of the guidelines and recommendations for evaluating the validation efforts needed after making software upgrades. It does not list all the activities and tasks that must, in all instances, be used to comply with the regulatory requirements. The discussion begins with a brief explanation of some of the key terms that are utilized in validation studies:

- *Software validation*. This refers to the confirmation by examination and provision of objective evidence that software specifications conform to user needs and intended users and that the particular requirements implemented through software can be consistently fulfilled.
- *Requirement*. This refers to any need or expectation for a system or for its software. Requirements reflect the stated or implied needs of the customer and may be market based, contractual, or statutory, as well as based on an organization's internal requirements. There can be many different kinds of requirements (e.g., design, functional, implementation, interface, performance, physical). Software requirements are typically derived from the system requirements for those aspects of system functionality that have been allocated to software. Success in accurately and completely documenting software requirements is a crucial factor in successful validation of the resulting software.
- *Specification*. This refers to a document that states requirements. It may refer to or include drawings, patterns, or other relevant documents and usually indicates the means and the criteria whereby conformity with the requirements can be checked. There are many different kinds of written specifications (e.g., system requirements, software requirements, software design, software test, software integration). All these documents establish "specific requirements" and are design outputs for which various forms of verification are necessary.
- *Functional requirement*. This is a requirement that specifies a function that a system or system component must be able to perform.

- *Physical requirement*. This is a requirement that specifies a physical characteristic that a system or system component must possess (e.g., material, shape, size, weight).
- *Revalidation*. Relative to software changes, revalidation means validating the change itself, assessing the nature of the change to determine potential ripple effects, and performing the necessary regression testing.
- *Quality assurance*. This refers to the planned systematic activities necessary to ensure that a component, module, or system conforms to established technical requirements and all actions that are taken to ensure that a development organization delivers products that meet performance requirements and adhere to standards and procedures. It also refers to the policy, procedures, and systematic actions established in an enterprise for the purpose of providing and maintaining some degree of confidence in data integrity and accuracy throughout the life cycle of the data, which includes input, update, manipulation, and output. It includes the actions, planned and performed, to provide confidence that all systems and components that influence the quality of the product are working as expected, both individually and collectively.
- *Quality control*. This refers to the operational techniques and procedures used to achieve quality requirements.
- *Testing, regression*. This involves rerunning test cases that a program has previously executed correctly, in order to detect errors spawned by changes or corrections made during software development and maintenance.
- *Traceability*. This refers to the degree to which a relationship can be established between two or more products of the development process—especially products having a predecessor-successor or master-subordinate relationship with each other. It determines the degree to which the requirements and design of a given software component match.
- *Installation qualification*. Installation qualification studies establish confidence that the process equipment and ancillary systems are capable of consistently operating within established limits and tolerances. This phase of validation includes the examination of equipment design (e.g., determining the calibration, maintenance, and adjustment requirements and identifying critical equipment

features that could affect the process and product). Examples of equipment performance characteristics that can be measured include uniformity of speed for mixers; temperature, speed, and pressure for packaging machines; and temperature and pressure of sterilization chambers.

Documentation

Once the equipment configuration and performance characteristics are established and qualified, they should be documented with the following items:

- *Installation qualification.* This should include a review of pertinent maintenance procedures, repair parts lists, and calibration methods for each piece of equipment.
- *Process performance qualification.* Establish confidence that the process is effective and reproducible. The performance qualification should include a minimum of three successfully planned qualification runs, in which all the acceptance criteria are met. For example, in the production of parenteral solutions by aseptic filling, the significant aseptic filling process steps to define and challenge should include the sterilization and depyrogenation of containers and closures; the sterilization of solutions, filling equipment, and product contact surfaces; and the filling and sealing of containers.
- *Product performance qualification.* Establish confidence through appropriate testing that the finished product produced by a specified process meets all release requirements for functionality and safety.
- *Process validation.* Establish documented evidence that provides a high degree of assurance that a specific process will consistently produce a product meeting its predetermined specifications and quality attributes.
- *Validation protocol.* This refers to a written plan stating how validation will be conducted, including test parameters, product characteristics, production equipment, and decision points on what constitutes acceptable test results.

Software Validation after a Change

The items described in the following sections need to be verified prior to making any software upgrades on a validated batch automated system.

Evaluate Software Changes

Depending upon the complexity of software, a seemingly small local change may have a significant global system impact. A thorough evaluation of each software change is required, which includes the following questions:

1. What changes were made to the batch software and why?
2. Are there release notes from the software vendor?
3. Is there documentation from the software vendor that confirms that the changes went through Quality Assurance (QA)? Does it provide a brief explanation of the effect of changes on the entire software package?
4. Do vendor documents provide enough evidence to show that the software developer conducted the appropriate level of software regression testing to show that unchanged but vulnerable portions of the system have not been adversely affected?
5. Do vendor documents provide enough evidence to show that the validation analysis was conducted not just for validation of the individual change, but also to show the extent and impact of that change on the entire software system?
6. Do design controls and appropriate regression testing provide enough confidence that the software is validated after a software change?

In addition, the following items should also be considered:

1. Depending on the risk involved, the device or product manufacturer should consider auditing the vendor's software change-control methodologies used in software changes. Assess the development and validation documents generated for the software changes.
2. Most of the automated equipment and systems used by device or product manufacturers are supplied by third-party vendors and are

Off The Shelf (OTS) purchases. The device manufacturer is responsible for ensuring that the product-development methodologies used by the OTS software developer are appropriate and sufficient for the device or product manufacturer's intended use of that OTS software.

3. Some vendors who are not accustomed to operating in a regulated environment may not have a documented life-cycle process that can support the device or product manufacturer's validation requirement. Other vendors may not permit an audit. Where necessary validation information is not available from the vendor, the device or product manufacturer will need to perform sufficient system-level "black box" testing to establish that the software meets their defined "user needs and intended uses."

Evaluate How Much Revalidation Evidence Is Required

The level of revalidation effort should be commensurate with the risk posed by the software complexity, changes implemented, and safety risk—not on firm size or resource constraints. The selection of revalidation activities, tasks, and work items should be commensurate with the complexity of the automation process and the risk associated with the effect of software changes for the specified, intended use.

For lower-risk devices, only baseline validation activities may be conducted. As the risk increases, additional validation activities should be added to cover the additional risk. Revalidation documentation should be sufficient to demonstrate that the system has been restored to its validated status.

In addition to analyzing how much revalidation evidence is required, one should also determine the nature and extent of testing needed as part of the revalidation effort and evaluate how software upgrades will affect equipment, product, and process performance.

Demonstration of Some of the Revalidation Guidelines

A steam sterilizer utilizes different sterilization cycles for the sterilization of stoppers, seals, and fill machine equipment in a pharmaceutical environment. Steam-sterilization cycles were automated utilizing OTS batch software. This automated sterilization application was validated by demonstrating the following studies in the steam sterilizer:

1. Three empty cycles were performed in the chamber with a temperature setpoint of 122°C. In each of the three cycles, temperature was maintained between 122°C and 123°C across the chamber. This demonstrates that temperature is distributed uniformly across the chamber and that the equipment is performing successfully without any negative impact on equipment performance.
2. Three loaded chamber cycles were performed at 122°C for 45 minutes in a chamber with biological indicators placed in each of the load packages. The load contained packages of stoppers and seals that were utilized for capping and sealing filled bottles. The biological indicators containing microorganisms were sent to a QA lab for testing after sterilization. Each of the indicators contained less than 10^{-6} microorganisms after 7 days. All physical characteristics for stoppers and seals were within the specified limits after sterilization. This demonstrates that there was no negative impact on product or process performance.
3. The software vendor made changes to the batch reports. These changes were cosmetic and provided additional information about process variables such as average, min, max, and other statistical data related to sterilization cycles. These changes only added calculations to the report; therefore, reviewing vendor documentation on report updates and functionality for these updates will be verified. No process validation is required because these changes have no negative impact on equipment, product, or process performance.
4. The software vendor made significant changes to the batch manager and other communication modules. The changes improved the communications between the batch manager and its clients, such as the batch scheduler, batch display, and third party clients that request information from batch manager. The vendor made changes to the way that the batch manger processed all tasks; instead of doing things synchronously, it now does thing asynchronously for performance improvement. A new version will be released because of these software updates. If the device or product manufacturer is willing to update existing software with a new version, then they need to perform revalidation on product and process performance and evaluate if there is any effect on equipment performance.

Conclusion

A thorough evaluation of software change; vendor documents; risk posed by the software complexity; and effects on equipment, product, and process performance will help formulate appropriate revalidation test procedures and sufficient documentation to demonstrate that the system has no negative impact on equipment, product, and process performance after software upgrades. The studies described in this chapter will help lower costs and the amount of resources required for testing and documentation; they may also prevent having insufficient data during FDA inspections and thus help to avoid warnings.

CHAPTER 5

Is It Possible to Build a Pharmaceutical Plant in 18 Months or Less Using ISA-88?

Presented at the WBF North American Conference, April 13–16, 2003, by

Lars Petersen
Manager, Automation
LarP@nne.dk
Novo Nordisk Engineering, Krogshojvej 55, 2880 Bagsvaerd, Denmark

Abstract

ISA-88 is an important factor when reducing time in both design and construction phases and in commissioning and qualification. But is it enough? Will ISA-88 give an answer to the question presented in the title of this chapter, or is there a need for modular engineering unseen to the business? This chapter will give different perspectives on this subject and try to give an overview of the life cycle of the modular approach in terms of thinking the methods into early development and ending up with a plant producing the product.

This chapter also outlines some of the requirements for the control systems, so that they fit into the modular approach and give an overview of the different disciplines and how they need to comply with ISA-88 and standardization to reach the goal.

Drivers in the Pharmaceutical Market When Introducing a Product

Setting up the success criteria of a project is essential. Success criteria often involve a trade-off between economies, time, and quality. If you are dealing with pharmaceutical projects, then quality is invariably a fixed parameter. Every company has to deal with the regulations set by the authorities, and in this field, Food and Drug Administration (FDA) regulations are dominating, even for products introduced outside the United States.

This means that projects are often driven by economy and time, and these parameters are connected. Dealing with time is about postponing the investment until the latest moment because a pharmaceutical development project is not safe until finished. Therefore, pharmaceutical projects are economically optimal if the production plant construction time is reduced to a minimum level. At the same time, and for the same reasons, the development time of new products is reduced dramatically within all competitors. This highlights the need for a reduced time from approved product until product launch. Establishing the production plant is one of the longest lead time items.

In recent years, this subject has even forced a number of companies to reduce or withhold their automation strategies because these are often introducing extensive regulation issues. What project group has not been faced with the fact that they started with a Manufacturing Execution System (MES) solution during basic design that has been postponed or dropped before going into detail design?

Time to Market as a Driver

Many methods have been introduced for reducing time schedules in projects (e.g., optimizing technologies, resources, paper workflow). In the last few years, completely new methods for reducing project time have been developed. Engineering and contractor companies have been focusing on executing parallel activities. This has introduced a number of new companies that specialize in producing or executing fast engineering and implementations. At the same time, a number of standards like ISA-88 and ISA-95 have evolved. For the pharmaceutical market, Good Automation Manufacturing Practices (GAMP) and a number of other International Society for Pharmaceutical Engineers (ISPE) guidelines have been introduced.

These standards and the need for parallel engineering have introduced a need for higher modularization. Modularization is a tool that enables an engineering job to be broken down into logical modules with identified interfaces. These modules can then be engineered separately and connected at a later stage (Fig. 5.1).

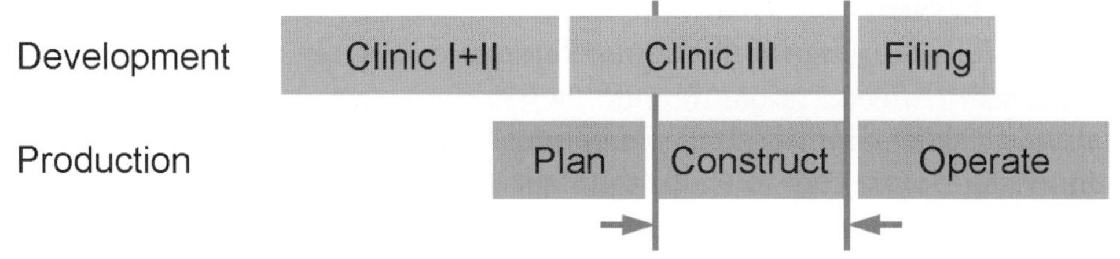

Figure 5.1. The construction period.

Another important time-reducing factor that has been evolving is off-site construction. Off-site construction is used in different construction types, and when used in conjunction with parallel engineering and modularization, the result can influence fast-track projects tremendously (Fig 5.2). Plants constructed in modules using parallel engineering and off-site construction in different ways are a key issue in today's fast-track projects.

This chapter is about different methods of modularization and how it is done on different levels where ISA-88 is very important. Automation is time-consuming, especially when qualifying and documenting the test. ISA-88 can be implemented without using modularization at all levels, but the real gain is reached if this is done.

This chapter will also introduce a term referred to as Process Module (PM). The acronym is only used in this chapter. A PM is a level where all parts of modularization are implemented for all disciplines. The PM idea is to isolate a complete function into a module. This introduces a number of advantages.

A number of engineering companies and contractors are using modularization and modular engineering. But it is important to note that there is a difference between building in standard modules and building in modules that can be reused and replaced in order and then brought together again to function. This can also be expressed as the difference between a puzzle that is required to be put together in one way and LEGO bricks. LEGO bricks can connect to each other no matter what the form and shape. The point is the interface. This is exactly the defining characteristic of PMs.

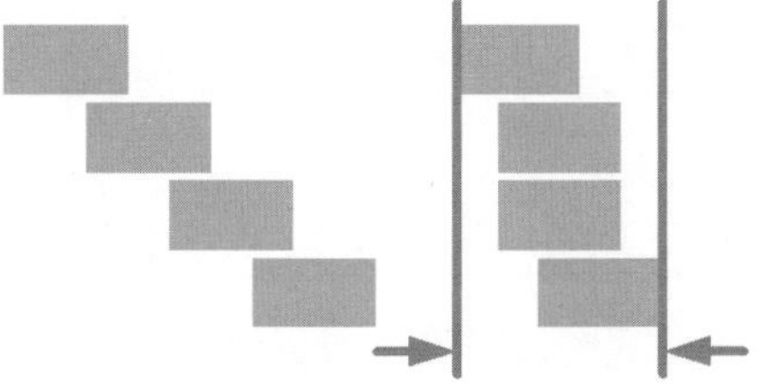

Figure 5.2. Parallel project phases in a construction period.

Modularization

Today, there are three dominating modular solutions in the pharmaceutical industry: preassembled facilities (Fig 5.3), skid mount units (Fig 5.4), and preassembled units (Fig 5.5).

A preassembled facility is a part or section of a plant engineered and constructed off-site. The modules are most likely done as building blocks or modules where all modules are custom designed. The modules are often tested together in the same way that the finished plant will function. There are different challenges of integration, especially regarding automation systems, depending on how many sections of preassembled parts are required to be integrated.

Skid mount units are fully functional modules, standardized with a complete control system (Fig 5.4). These modules differ from the preassembled facility by being Off The Shelf (OTS) units. This means that the modules can be delivered very quickly, if and only if there are no changes to the design. The challenge is integration to the rest of the plant. 21 CFR Part 11 is also a challenge because often the plant will end up with a number of different automation systems.

For both preassembled facilities and skid mount units, it is vital that ISA-88 develops standard recipe-to-equipment interfaces. Presently, automation vendors are handling this very differently. We thought OPC gave us a lot of solutions, but

Figure 5.3. Preassembled facilities.

Figure 5.4. Skid mount units.

it gives only a technology. Although this technology is important, OPC can be implemented in many different ways.

Preassembled units are equipment engineered and constructed off-site as modules but without control systems (Fig 5.5). These modules are evident for all types of modules, but more as part of the fully off-site delivered module.

Another method that is being used today is the PM, defined as a fully functional module with standard interfaces (Fig 5.6). As mentioned previously, the best way of describing it is to use a LEGO brick. The module can consist of different types and with different forms, but it can always connect to another module. This kind of module is not really used in the engineering industry today.

If a module type like this is constructed, then all the existing issues could be addressed (Fig 5.7). Different vendors could construct the modules, although this process requires multiple skills, especially when both engineering and constructing. The module could then be tested off-site and be integrated, even though it comes from different vendors. The real challenge here would be to integrate the knowledge from process, mechanical, automation, and validation skills.

The PM is handled in between area and units in ISA-88.01 terms, but it is a challenge for ISA-88.01 implementation (Fig 5.8). The reason for this is that a PM needs to be of a certain size. It has to be small enough to be powerful as a building block, but it also has to be able to handle several batches inside at the same time; otherwise it cannot be fully functional. A module should be able to handle Clean

Figure 5.5. Preassembled units.

In Place (CIP) and production at the same time, and that will require two batches to be active at the same time, if CIP is treated as a batch. The module needs to be handled as a small ISA-88.01 process cell, but a process cell is often treated as a section of a plant. The interface between two process cells is not really addressed in ISA-88, and the PM needs to be seen as a stand-alone module, while still interfacing other parts of the batch process in a process cell. This is similar to the problem of shared equipment that lies between a unit and an EM.

Optimizing the Use of Modularization

The full effect of the modularization requires that the engineering organization and methods be organized around the modules. A structure that is built around the project activity model needs to be adapted to the modules. The module has a given shape in each phase.

Figure 5.6. PM.

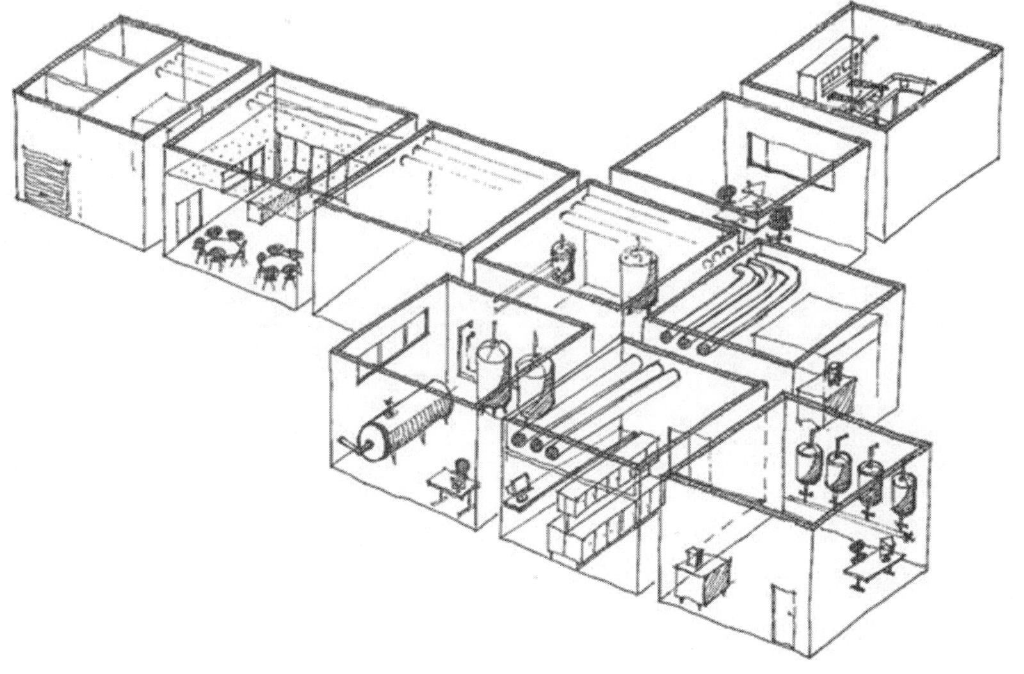

Figure 5.7. A set of PMs connected as LEGO bricks.

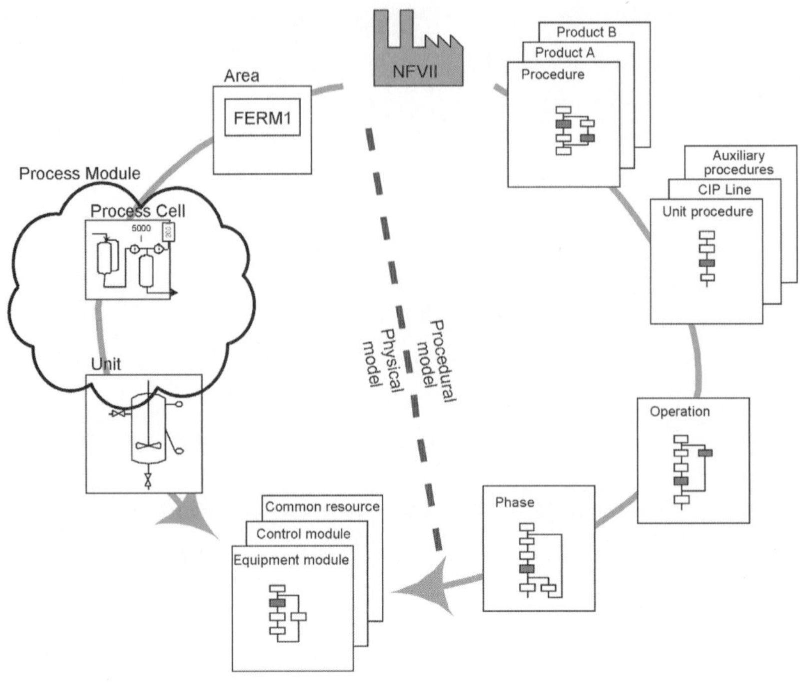

Figure 5.8. The PM adapted into the ISA-88 structure.

The qualification concept of the plant also needs to be adapted to the modules to gain optimal efficiency and flexibility when testing and integrating on-site in parallel phases (Fig 5.9).

Figure 5.10 illustrates the many test phases that are used when parallel activities are required. This concept is used when testing is done at the same system in several different physical locations at the same time.

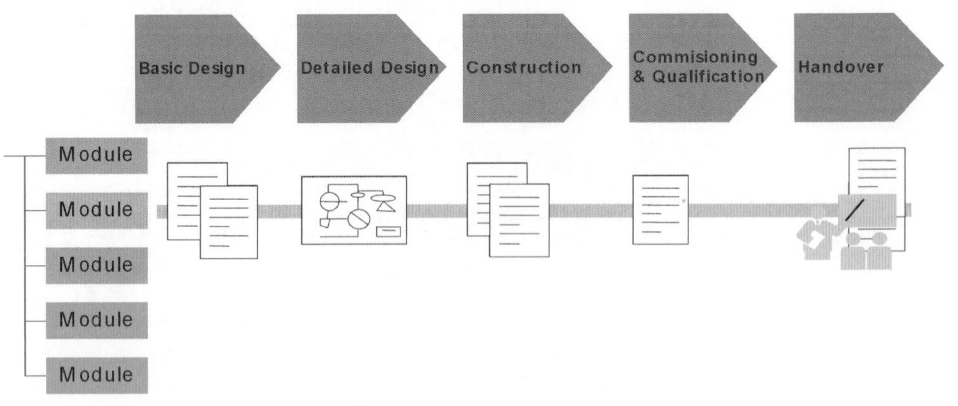

Figure 5.9. Project activity model and the modules for each phase.

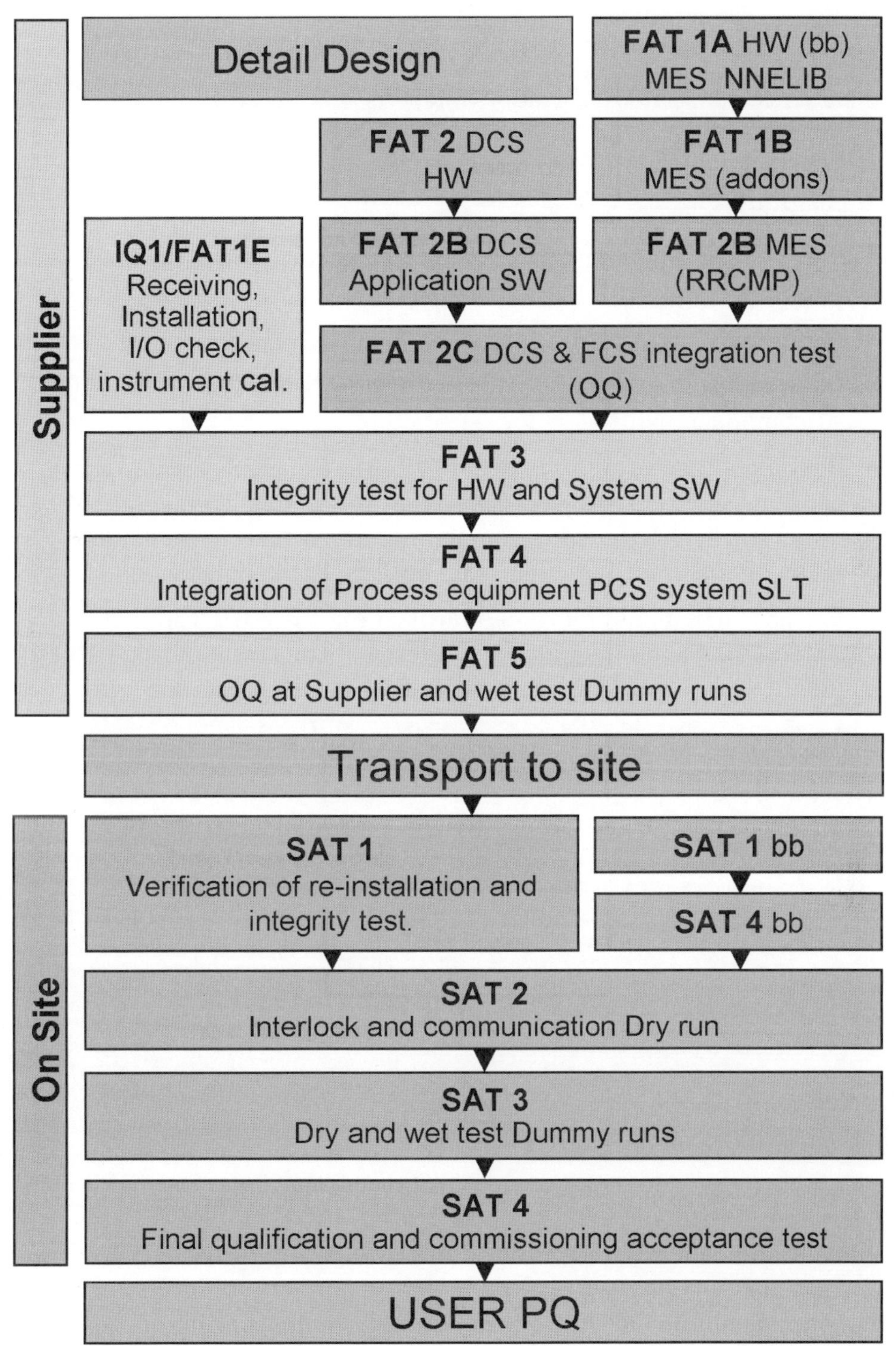

Figure 5.10. Test phases for parallel activities.

Conclusion

Our experience using modularization and PMs revealed the following:

- This is not done without commitment and common goals in the project group.
- It is a new process for vendors.
- It requires high skills in project management.
- It requires very strict discipline and loyalty to the strategies.
- It is a tough process for Quality Assurance people, and it requires strict management.
- It requires control systems with flexibility, especially around the handling of databases, because configuration management is very difficult when performing engineering in parallel. There are too many people modifying (and deleting) the same database.
- ISA-88 is a help, but more standardization is needed, especially around the recipe-to-equipment interface. The structure above the unit level also needs to be addressed.

CHAPTER 6

Batch On-line Analytics: A Solution Beyond Six Sigma

Presented at the WBF European Conference, November 10–12, 2008, by

Willy Wojsznis
willy.wojsznis@emerson.com
Emerson Process Management, 12301 Research Blvd., Austin, TX 78759

Randy Reiss
Emerson Process Management, 12301 Research Blvd., Austin, TX 78759

Terry Blevins
Emerson Process Management, 12301 Research Blvd., Austin, TX 78759

Chris Worek
Emerson Process Management, 12301 Research Blvd., Austin, TX 78759

Abstract

The Process Analytical Technology (PAT) guidelines established by the U.S. Food and Drug Administration (FDA) have sparked renewed interest in analytic tools for multivariate analysis. In particular, Principal Component Analysis (PCA) for fault detection and Projection to Latent Structures (PLS) for end-of-batch prediction of quality parameters are seen as pivotal techniques for improving process operation. A number of software packages are currently available for off-line analysis. The difficulty for an average control or production engineer is that these tools are not designed for on-line operation. Many PAT issues may be addressed by tightly integrating analytic tools with the production and control system. This chapter examines basic design requirements, associated with batch analytics, that can be applied for on-line operation. This chapter delivers an in-depth look at the data processing requirements, calculations, and limitations for the application of on-line analytics to a batch process. It explains how to achieve proper data alignment for different batches—a key requirement for building good statistical models—and how to use the model for on-line analysis. A mammalian cell simulated bioreactor is used to illustrate the advantages of on-line process analytics over traditional monitoring and control techniques. The process simulation and PAT application runs on a commercial control system and uses production operation screens.

Introduction

The PAT initiative launched and supervised by the FDA boosted research and applications of analytical techniques. It is important to note that the term *analytical* in PAT is viewed broadly and includes chemical, physical, microbiological, mathematical, and risk analysis conducted in an integrated manner. The goal of PAT is to understand and control the manufacturing process, which is consistent with the current drug quality system: quality cannot be tested into products; it should be built-in or should be by design.[9]

This chapter presents methodology and tools for batch multivariate statistical monitoring and control as essential PAT components. Multivariate statistics is based on PCA for fault detection and PLS for end-of-batch prediction of quality parameters. These techniques are seen as pivotal to improving process operation. Many PAT issues may be addressed by tightly integrating analytic tools with the production and control system. The chapter focuses on batch analytics applied to on-line operation, off-line modeling tools, and PAT integration into control infrastructure. The distinctive features of the presented solution are recent research advances in batch data unfolding and modeling,[3] data alignment by using Dynamic Time Warping (DTW) for off-line modeling,[2] and on-line operation. Ease of use and interpretation of the results from an integrated PAT user interface are an important part of the solution discussed in this chapter.

It is also important to incorporate Six Sigma into our discussion, as it provides a successful and business-driven approach to process improvement, reduced costs, and increased profits.[8,7,10] Six Sigma is rooted in fundamental statistical and business theory; consequently, the concepts and philosophy are very mature. A few critics claim Six Sigma is merely a basic version of quality improvement.[6] It is evident that Six Sigma is not adequately supported by the tools for multivariate statistics, cross-correlated data analysis, and process modeling. A simple illustration of how univariate statistics can fail is found in the works of Boudreau and McMillan[1] and Mason and Young.[4] The application of PAT within the framework of Six Sigma will show the benefits of the extension into multivariate and on-line batch processes. The presented batch data analysis involves the following steps:

- Data extraction and preprocessing
- Data alignment
- Process modeling and model validation
- On-line process monitoring

Details about how the PAT tools are configured and each of the steps' functionality are given in the sections that follow, along with our conclusions.

PAT Configuration and Data Preprocessing

Batch data modeling requires huge amounts of data from various sources: transmitters, control loops, analyzers, virtual sensors, calculation blocks, and manual entries. Most of the data is stored in the Distributed Control System (DCS) continuous data historians. However, significant amounts of data (particularly manual entries) are associated with process management systems. Data extraction from both systems must be merged to satisfy model-building requirements. Since this can be extremely time-consuming, it is recommended that the procedure be split into two phases:

1. Merging data from the data historian and other sources in a common, readable format for review and preliminary validation
2. Selecting and extracting data subsets for the predefined model configuration and phases of batches for the developing PAT model

A workstation dedicated to PAT tasks is an effective solution, serving as a data concentrator and on-line operation platform. An example of such a solution with a DeltaV system for a laboratory bioreactor is shown in Figure 6.1.

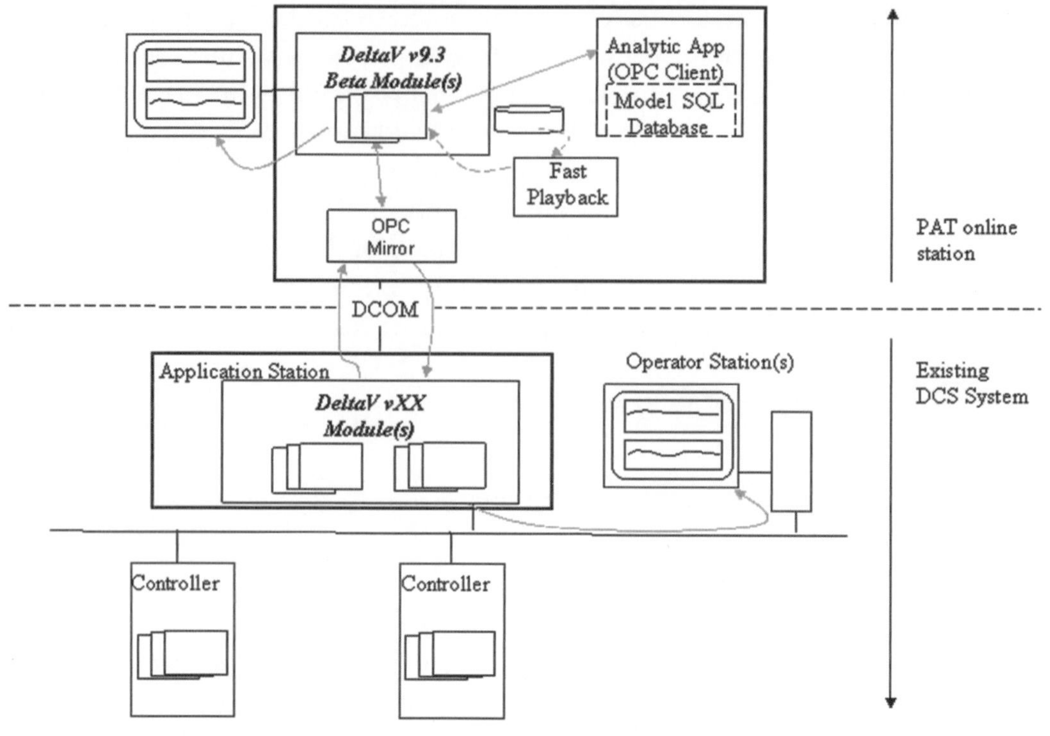

Figure 6.1. PAT workstation used for a bioreactor application.

The proposed architecture of Figure 6.1 utilizes a layered approach to integrating PAT on-line functionality. The solution has minimal impact on the existing system. Using established standards for communication (in this case OPC), it allows for vendor and version independence between the existing DCS system and the PAT on-line system.

Batch operation normally undergoes several significantly different stages from a processing technology and modeling standpoint. Therefore, a batch period is subdivided on the phases and the model built for each phase. Data for the same phase, from many batches, are grouped to develop a model for that phase. The purpose of this data arrangement is to remove or alleviate process nonlinearity and typical-for-batch nonstationary behavior.

An equally important procedure for model development is properly aligning data from various batches. It is normal that data trends for different batches have different length- and time-shifted locations of the important process landmarks. Data length and process landmarks should be aligned to get a legitimate model (Fig. 6.2).

The traditional technique used for aligning batch data is an indicator variable for representation of the progress of the batch. DTW, a technology borrowed from

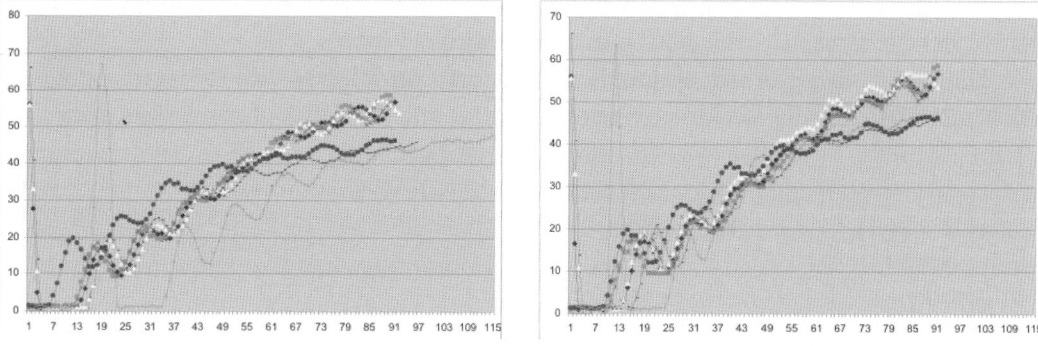

Figure 6.2. Data trends: original (left) and aligned (right).

speech recognition, was selected for the implementation. DTW minimizes the distance between respective process trajectories for different batches. In doing so, DTW takes into account all the variables in the analysis and is an effective approach for batch data alignment. In addition, an indicator variable was created and defined as a fraction of the batch phase completion time. This indicator variable is added to the original process variables set in order to improve the robustness of the DTW calculation and to prevent convergence to local minima over an excessive period of time.

The DTW algorithm applied prior to the PCA/PLS model development can be sketched as the following: from a number of process trajectories for every parameter, we select a reference trajectory, **R**, and define the distance, **d**, between the reference trajectory and any remaining trajectory, **X**, as

$$\mathbf{d(i(k),j(k)) = \{R[i(k)] - X[j(k)]\} \cdot W \cdot \{R[i(k)] - X[j(k)]\}^T} \qquad (6.1)$$

W is a positive weight matrix that reflects the importance of each measured variable. k is a grid point along the path ($k = 1,2, \ldots, K$). The lengths of **X** and **R** are t and r. The total distance **D(t,r)** between the two trajectories is

$$\mathbf{D(t,r)} = \sum_{k=1}^{K} \mathbf{d(i(k),j(k))} \qquad (6.2)$$

The optimal path that minimizes **D(t,r)** is

$$\mathbf{D^{opt}(t,r)} = \min\{\sum_{k=1}^{K} \mathbf{d(i(k),j(k))}\} \qquad (6.3)$$

A path can be denoted as

$$\mathbf{f = \{c(1),c(2), \ldots, c(K)\}} \qquad (6.4)$$

where **c(k) = [i(k),j(k)]** represents a grid point connecting *i* and *j*. A simple illustration of developing optimal path can be found in Figure 6.3.

Usually, minimization (Eq. 6.3) is subject to the global (Eq. 6.5) and local (Eq. 6.6) constraints, which enforce matching initial and end trajectory points and limit for the point (*i,j*) the choice of the predecessor point.

$$\mathbf{c(1) = [1,1]} \qquad \mathbf{c(K) = [t,r]} \tag{6.5}$$

$$\mathbf{c(k - 1) = (i - 1,j), (i - 1,j - 1) \text{ or } (i,j - 1)} \tag{6.6}$$

PCA Batch Process Modeling Basics

After DTW, the aligned model data file is a three-dimensional array: ***I*** batches, ***J*** variables, and ***K*** scan periods. Prior to the model development, the data file is unfolded into two dimensions: ***IK*** × ***J***. So-called hybrid unfolding,[3] as in Figure 6.4, was applied because of its advantages over commonly used batch-wise and variable-wise unfolding.

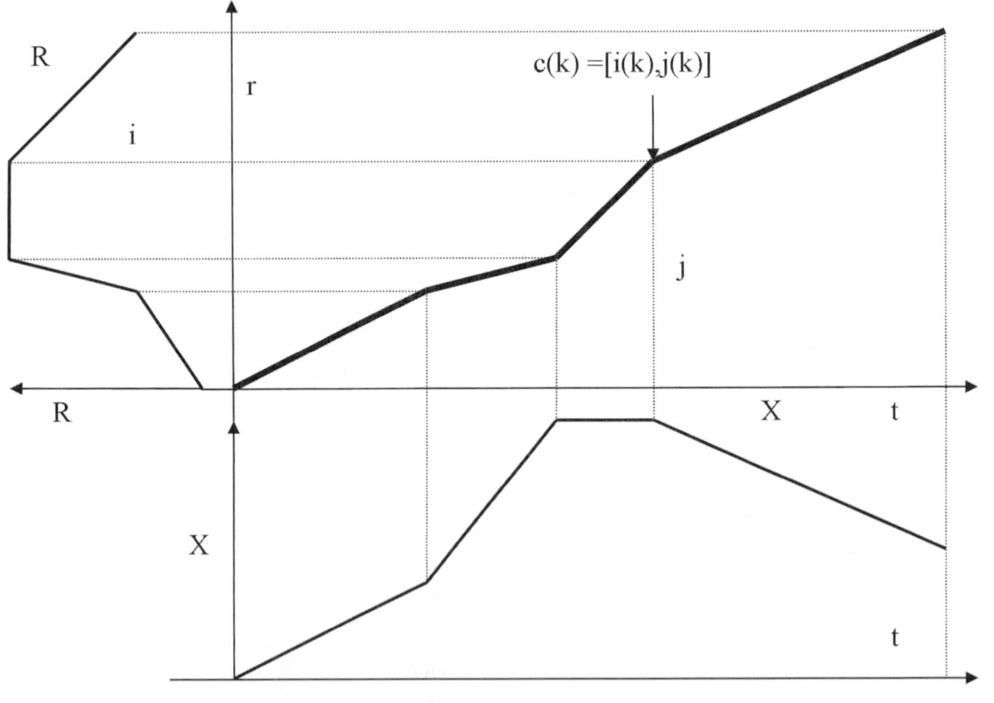

Figure 6.3. Illustration of the raw trajectories R and X and the aligning optimal path c(k).

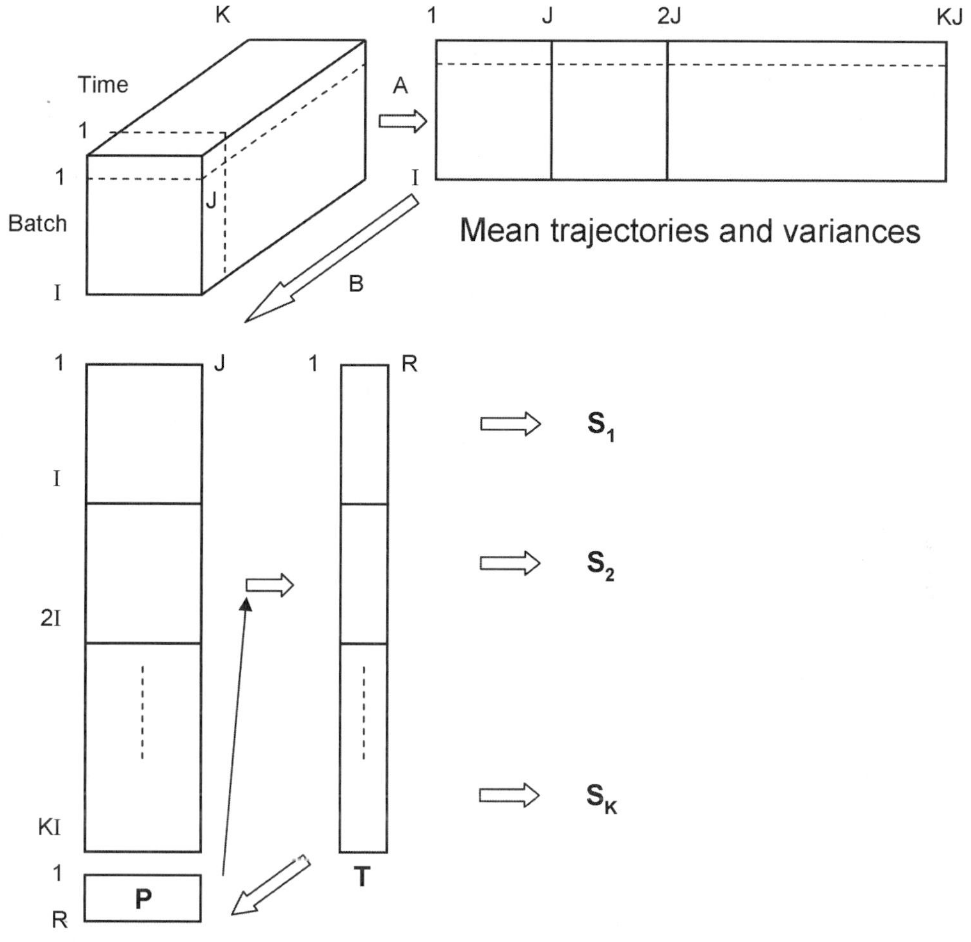

Figure 6.4. Hybrid data unfolding and PCA model development.

With hybrid unfolding, mean values and variances are calculated for every time period with data in batch-wise unfolded states (arrow A in Fig. 6.4), thus diminishing the effect of nonlinearity and dynamics on the model. The data is then rearranged as variable-wise unfolded (arrow B in Fig. 6.4); therefore there is no need to assume arbitrary trajectories from the current time until the end of the phase, as in the original batch-wise unfolding.

For model development, the Nonlinear Iterative Partial Least Squares (NIPALS) algorithm was applied because it requires fewer calculations than alternatives like Singular Value Decomposition (SVD). The model consists of the loading matrix, **P**, and cross-covariance matrices, $\mathbf{S}_k$, for the principal components score matrix. A graphical illustration of the PCA modeling is shown in Figure 6.5.

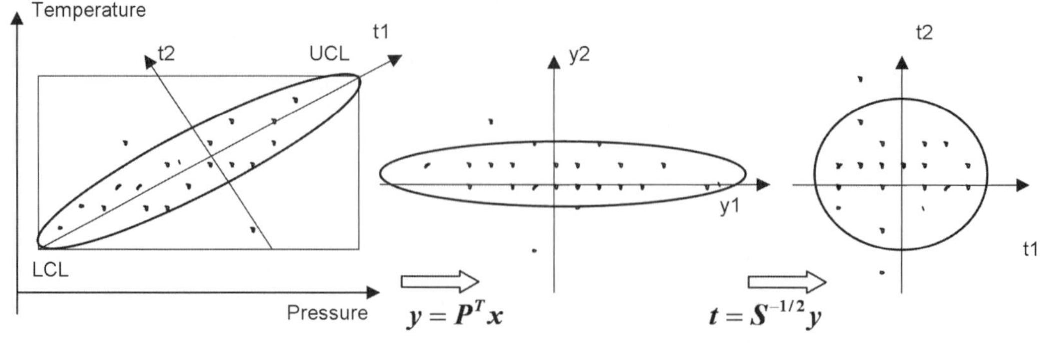

Figure 6.5. Data transformation from the original space to the normalized principal component space.

The model projects the space defined by the original process measurements, **X**, to the orthogonal Principal Component (PC) space, **T**, of a lesser number of coordinates aligned to the maximal process variance directions:

$$y = P^T x \qquad t = S^{-1/2} y \tag{6.7}$$

Total variances for a PC at every scan, k, are calculated as a $\mathbf{T}^2$ statistic:

$$T_k^2 = t_k^T S_k^{-1} t_k \tag{6.8}$$

Upper Control Limit (UCL), T_α^2, for the $\mathbf{T}^2$ with an $\boldsymbol{R}$ principal component model developed from $\boldsymbol{I}$ batches is defined by F-distribution with $\boldsymbol{R}$ and $\boldsymbol{I} - \boldsymbol{R}$ degrees of freedom:

$$T_\alpha^2 = \frac{R(I^2 - 1)}{I(I - R)} F_\alpha(R, I - R) \tag{6.9}$$

Unmodeled errors are expressed by the Square Prediction Error (SPE), also known as the Q statistic:

$$SPE_k = e_k^T e_k \tag{6.10}$$

The UCL for Q statistic is defined by χ^2 distribution:

$$SPE_k^\alpha = \frac{v_k}{2m_k} \chi_\alpha^2 \left(\frac{2m_k^2}{\sigma_k^2} \right) \tag{6.11}$$

where m_k and v_k are mean and variance of the SPE statistic at time period, k.

An extension of PC modeling concept is PLS, where original process measurements are projected to the PC space in such a way as to maximize variances of both the operational data, **X**, and quality data, **Y**. In this way, a better prediction is achieved for the batch process quality parameters, which is usually only available from the lab analysis some time after the end of the batch. The PLS principle component decomposition is illustrated in Figure 6.6.

On-line Process Monitoring

The objective of on-line process monitoring is to detect abnormal operations and identify the sources of abnormality. An equally important task is predicting the end-of-batch quality. To satisfy these objectives, PAT on-line application performs the following functions:

- Automatically collecting, filtering, and preprocessing process measurements, lab analyses, and manual entry data
- Aligning current samples with the model
- Batch operation monitoring by using PCA
- Batch quality prediction by using PLS

On line measurement contains the same variables as are used for model development. An on-line DTW procedure is applied to fit the on-line data to the most similar point on the model trajectories. Then the data is projected to the PC space,

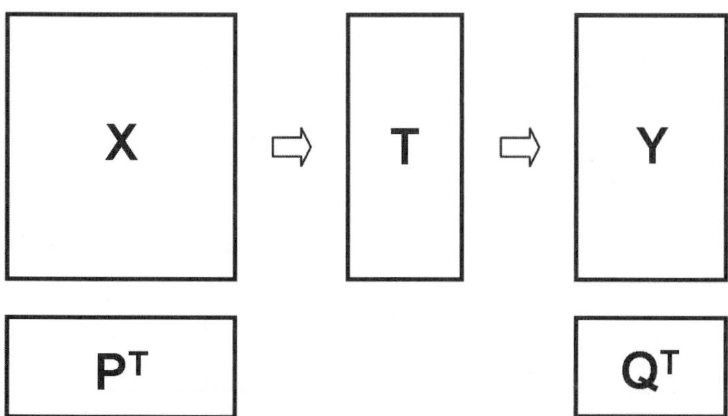

Figure 6.6. PLS, T, maximizing variances both from operational data, X, and quality data, Y. P and Q are PLS model matrices.

UCL for $\mathbf{T}^2$ and **Q** is tested, and process variable contribution to the total variance is defined (Fig. 6.7).

Tests Results and Discussion

Using a first-principles mammalian cell simulated bioreactor,[5] data was collected from twenty batches, and statistical models were generated. The model included twenty parameters measured continuously and one quality parameter known at the end of the batch from the lab analysis. Figure 6.8 shows the T^2 and Q statistical trends for a PCA analysis that captures (1) a univariate deviation and (2) a multivariate deviation.

The chart in Figure 6.9 shows the contribution plot for the univariate deviation (labeled "1" in Fig. 6.8). Process tags "tic41-7.out" and "tic41-7.pv.," both representing the broth temperature of the reactor, are the primary contributors. Notice that both tag contributions by themselves exceed UCL for the process model. The Q statistic relates variation in the on-line process that is not captured in the principal components of the model. In the case of a mammalian cell culture, temperature is not expected to vary. Looking at the raw data for the bioreactor temperature in Figure 6.10, we see the outlier data circled where the perturbation occurred.

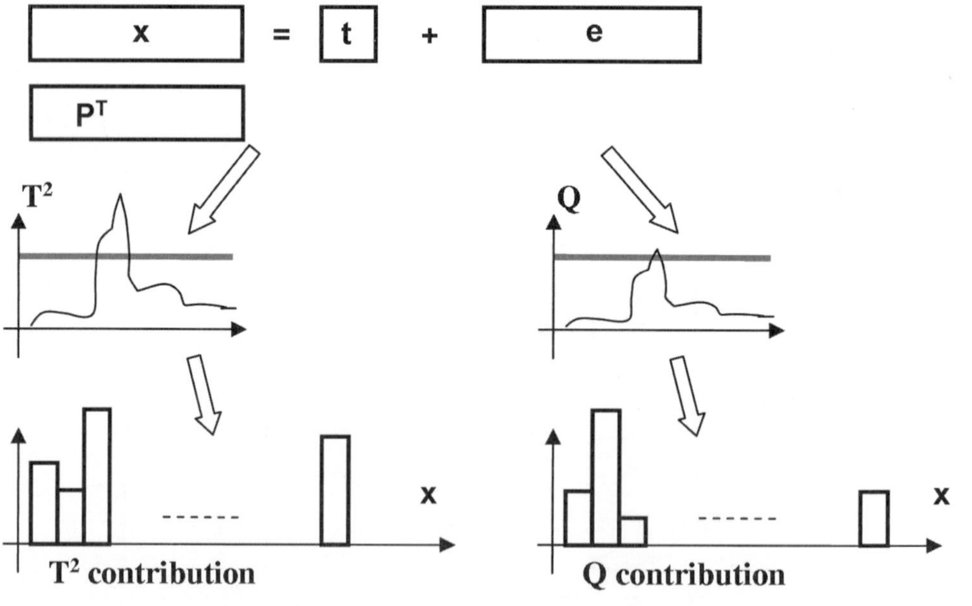

Figure 6.7. PCA on-line functionality.

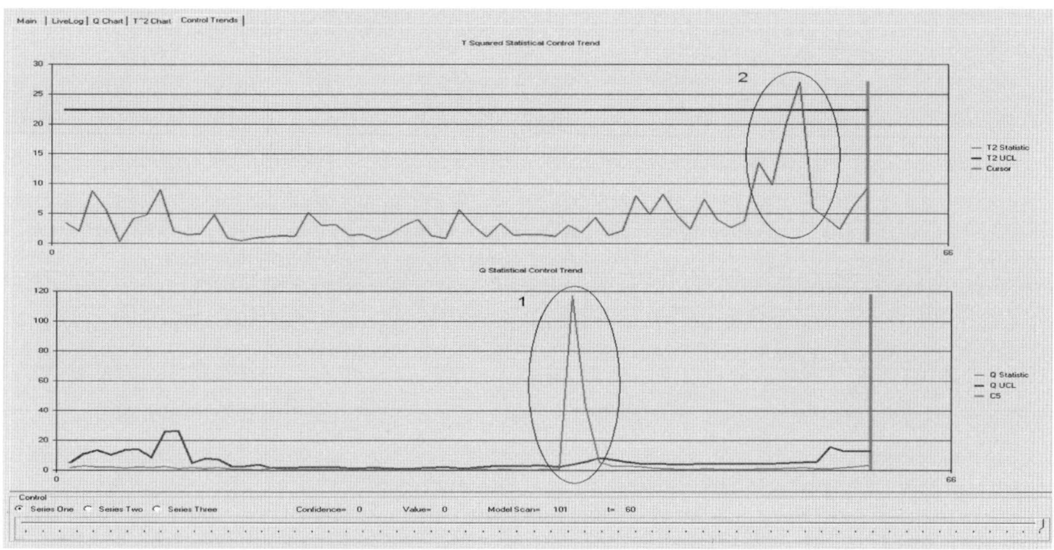

Figure 6.8. PCA on-line statistical trends for a mammalian cell simulated bioreactor.

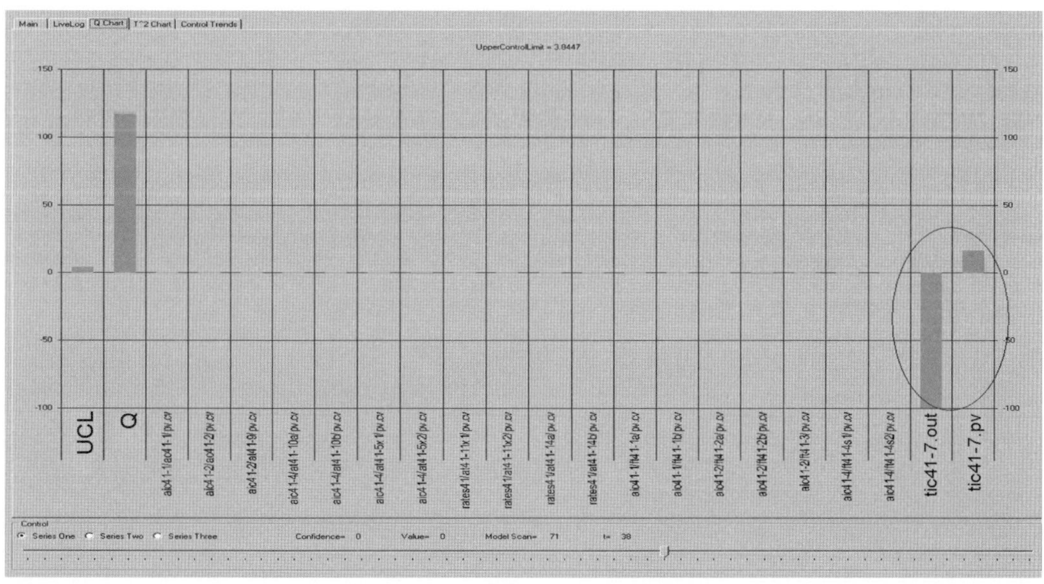

Figure 6.9. Process variables contribution plot.

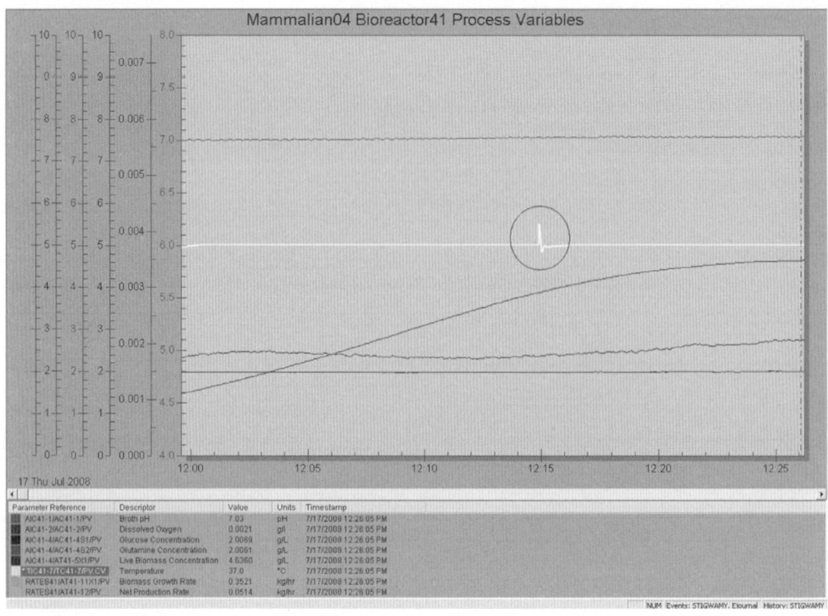

Figure 6.10. Bioreactor temperature perturbation.

The chart in Figure 6.11 shows the related process tag contributions for the multivariate deviation (labeled "2" in Fig. 6.8). Notice that none of the contributions are enough to exceed the UCL, but the compilation of the variation into the T^2 statistics does exceed the UCL. The major contributors are broth pH, pH reagent flow, and dead-cell biomass. Thus the process tag contributions translate the abstract statistics of the analysis into actionable information. The specific scenario that played out in the bioreactor is that an increase in the live cells produced too much lactic acid and ammonia, which changed the pH level and caused an increase in dead cells. Possibly better pH control could prevent the fault in the future.

An examination of the raw data to identify the actual perturbations is often difficult and is a characteristic inherent to multivariate analysis. Thus multivariate analysis adds to the robustness of statistical monitoring by bringing the less obvious into focus for better interpretation.

Conclusion

PCA monitoring applied to the mammalian cell simulated bioreactor detected abnormal parameters, change of pH, regent flow, and dead-cell biomass, as expressed in the T^2 and Q statistics above the UCL. The contribution plot properly identified the sources of abnormality for all tested cases. It should be noted that

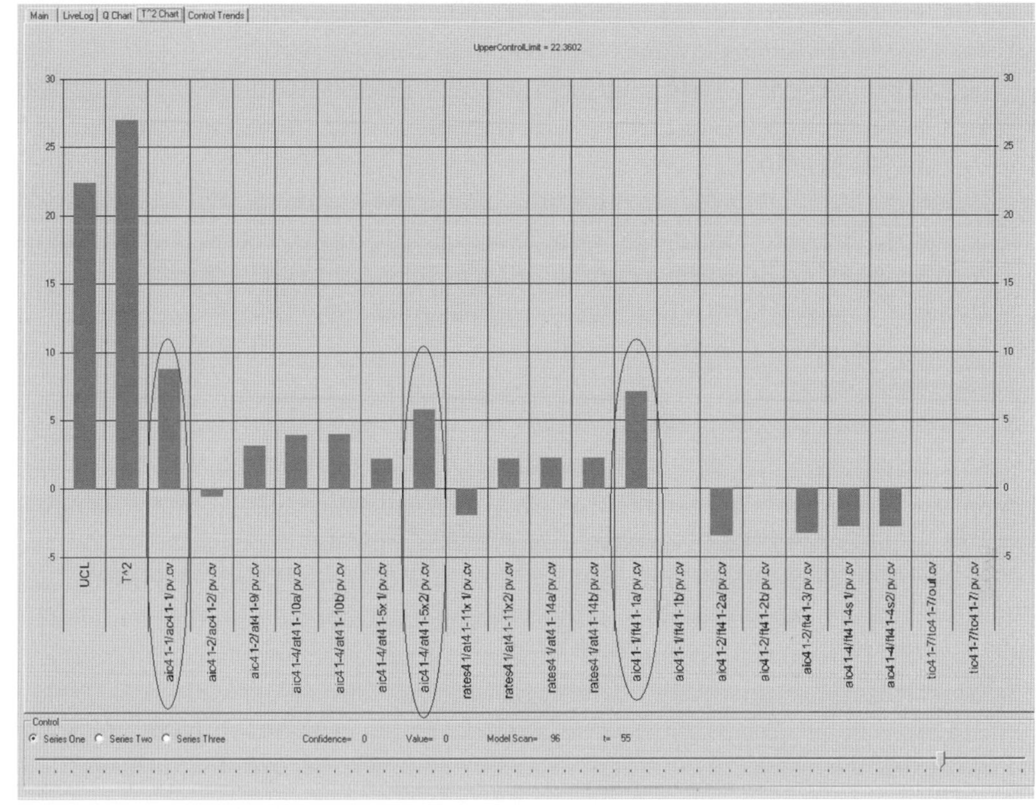

Figure 6.11. PCA on-line process tag contribution chart for related statistical trends.

some of the parameter changes were within acceptable level and would not be detected by the univariate statistics. The quality predicted by PLS matched the end-of-batch quality measurement with expected accuracy.

References

1. Boudreau, M. A., and G. K. McMillan. 2006. *New directions in bioprocess modeling and control*. Research Triangle Park, NC: ISA.
2. Kassidas, A., J. F. MacGregor, and P. A. Taylor. 1998. Synchronization of batch trajectories using dynamic time warping. *American Institute of Chemical Engineers Journal* 44(4).
3. Lee, J. M., C. K. Yoo, and I. B. Lee. 2004. Enhanced process monitoring of fed-batch penicillin cultivation using time-varying and multivariate statistical analysis. *Journal of Biotechnology* 110: 119–36.

4. Mason, Robert L., and John C. Young. 2002. *Multivariate statistical process control with industrial applications.* Philadelphia: ASA-SIAM.
5. McMillan, Gregory, Trish Benton, Yang Zhang, and Michael Boudreau. 2008. PAT tools for accelerated process development and improvement. *BioProcess International* (March).
6. Paton, Scott M. 2002. Juran: A lifetime of quality—an exclusive interview with a quality legend. Interview with Joseph M. Juran. *Quality Digest,* August. http://www.qualitydigest.com/aug02/articles/01_article.shtml.
7. Pyzdek, Thomas. 2001. Six Sigma and beyond: Why Six Sigma is not TQM. *Quality Digest,* February. http://www.qualitydigest.com/feb01/html/sixsigma.html.
8. ———. 2001. *The Six Sigma handbook.* New York: McGraw-Hill.
9. U.S. Food and Drug Administration, Office of Pharmaceutical Science. OPS Process Analytical Technology (PAT) initiative. http://www.fda.gov/AboutFDA/CentersOffices/CDER/ucm088828.htm.
10. Wikipedia. 2010. Six Sigma. http://en.wikipedia.org/wiki/Six_Sigma.

CHAPTER 7

Implementing ISA-88 across Life Science Development Operations

Presented at the WBF North American Conference, March 24–26, 2008, by

Louis Ciabattoni, CPIM
Sr. Product Manager, Life Sciences
lciabattoni@conformia.com
Conformia Software, 150 Mathilda Place,
Suite 202, Sunnyvale, CA 94086 USA

Abstract

Implementing ISA-88 may be perceived as a challenge in most industries. The benefits of the ISA-88 standard test this notion—reducing the time to reach full production levels for new products, making recipe development straightforward, and reducing life-cycle engineering efforts, as well as enabling vendors to supply appropriate tools for implementing batch control. Even with new Food and Drug Administration (FDA) recommendations for pharmaceutical, current Good Manufacturing Practices (cGMP) for the 21st century, the pharmaceutical industry has been slow in accepting the standard. Imagine the added complications in implementing ISA-88 in the pilot plant development area, where recipes are constantly evolving. This chapter will discuss the strategies used in implementing an ISA-88 compliant electronic development record application for small- and large-molecule drug substance and drug product development.

From my perspective as an observing member of the ISA88 committee, combined with my experiences with Quality by Design (QbD), the International

Conference on Harmonization (ICH), Q8 and Q9 from the FDA-Conformia Current Research And Development Agreement (CRADA) project (Docket number 2005N-0353), and the top twenty life science companies, I will describe a strategy to roll out ISA-88 in your company, present new ISA-88 concepts found only in development, and outline a straightforward mechanism for technology transfer of a new product to commercial manufacturing.

Overview of Development Operations

Development is concerned with proof of manufacturability, safety compliance, and determining ideal dosing levels and delivery mechanisms. In the United States, the pharmaceutical development phase can cost $10 million to $500 million, and only approximately one in ten compounds identified by basic research pass all developmental phases and reach the marketplace.[2] Since the product and process is in a state of continual improvements and possibly codeveloped with third-party partners, development operations are constantly finding more efficient ways to organize, manage, and leverage their intellectual capital.

Long overlooked by corporate IT as a sunk cost center, pharmaceutical development operations have struggled with a variety of point solutions. Trying to formulate a reproducible methodology is a constant struggle. In commercial manufacturing, software automation has significantly improved product quality and manufacturing efficiency, but a majority of development operations are still done on paper. Research and Development (R&D) organizations must develop their processes and are continually challenged with the ability to actually make the product and track the related information. Today, time to market is not the only driving force; efficiency is becoming a top priority for all companies.

A variety of efforts are underway to improve efficiency, including internal and FDA initiatives, but ISA-88 has been a quiet rumbling. The struggle to improve internal operation takes various forms. Some companies are forming project teams that are assigned to leading drug candidates. The team is formed at a pivotal point in the maturity life cycle. Other companies move commercial manufacturing closer to development and focus their critical resources to ease the process of technology transfer. As stated earlier, the IT organizations have implemented multiple software solutions, but enhancements such as integration or tailoring are difficult to attain.

With an average of four and a half years to bring a target drug candidate to market—all hinging on the success of the clinical trial—the regulatory group must somehow collate all the significant development information regarding the scale-up of the recipe and present a logical control strategy in their FDA

New Drug Application (NDA) submission. Pharmaceutical development exists across multiple R&D operations (e.g., development laboratories, kilo labs, pilot plant development, late development, Active Pharmaceutical Ingredient [API] manufacturing, preformulation, formulation, and clinical trials manufacturing). Collating and relating all pharmaceutical development information into the Common Technical Document (CTD) can be difficult when you are dealing with different representations of a recipe, from laboratory notebooks to formal, paper batch record manufacturing tickets.

In fact, one of the most overlooked areas of improvement across the pharmaceutical development operations life cycle is the management of the recipe, in a recipe structure from early experiments to formal batch record manufacturing tickets. If a company is able to develop a product and process life-cycle approach that centers around the recipe structure, then they have the possibility of creating a highly significant shift in the way that "tech transfer" occurs. In fact, if done in the right way, we believe that tech transfer could be eliminated—reduced from 8 months to 8 hours—due in large part to the ISA-88 recipe structure, which provides a common language across development and commercial manufacturing. The million dollar question is how to best apply ISA-88 in a development environment where there is a high degree of variability and the environment is far more dynamic than the commercial manufacturing organization.

The FDA is actively promoting the Quality by Design (QbD) initiative. Pharmaceutical companies are trying to understand and map their existing methodologies to the QbD model and, as a whole, are currently following various flavors of QbD. By working with industry, the FDA wants companies to better highlight the current use of their quality designs. Acceptance of ISA-88 can be a significant, positive effect for the pharmaceutical industry. Since Biotech companies may spend over 40% of their revenue in R&D,[5] the pressure to be more efficient is driving their acceptance of ISA-88.

This chapter will discuss several strategies used in implementing an ISA-88 based electronic development record application for small- and large-molecule drug substances and drug product development. This chapter will also describe a strategy to roll out ISA-88 in your company, present new ISA-88 concepts in development, and outline a straightforward mechanism for technology transfer to commercial manufacturing. This chapter will also discuss the importance of standards and how ISA-88 (as IEC 61512) is becoming an international standard across the pharmaceutical industry.

Differences from Commercial Manufacturing

Early development occurs when scientists are utilizing laboratory procedures that are performed at bench scale. It involves a simple list of materials and a simple, small-scale recipe. As the recipe scales up, the process becomes more formal. The lab bench apparatus is replaced with pilot plant, large-scale equipment. The chemistry begins to change as the scale of raw materials increases, but the major impact is variability. Figure 7.1 from ISA-88.03[4] accurately describes the typical development process, but pharmaceutical development is anything but typical.

Pharmaceutical development involves countless iterations of recipes across all its functional areas. Creating the thread that links the development methodology is a group effort to craft the control strategy. The process is further complicated by the fact that it has two major development components—the drug substance and the drug product—as well as the interactions that occur between them.

Looking at the definition of Product and Process Life-cycle Management (PPLM), ISA-88.03 understands the relationship between product and process, as shown in the following quotes:

> 4.4.3.2 Product development
> Product development results in the definition of the product and product specifications. It includes definitions of how to make the product, at least in laboratory scale. It results in basic understanding of the chemistry and processing requirements peculiar to the product. Product development can result in equipment requirements that are described in enough detail to define the type of equipment needed.

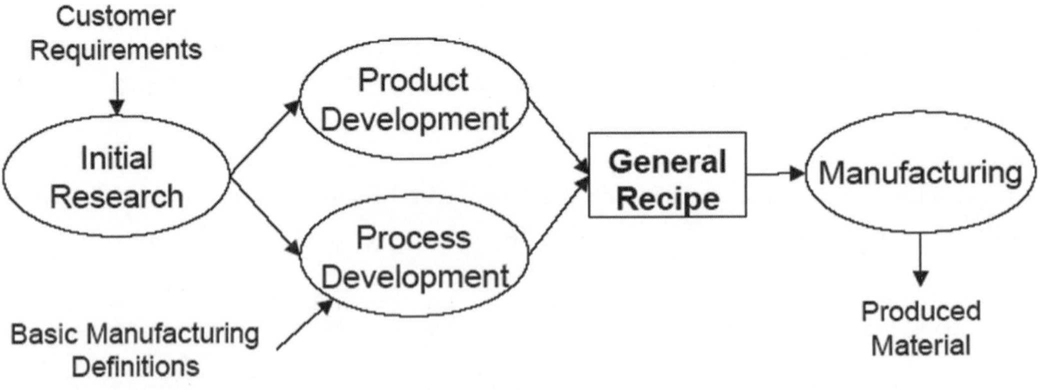

Figure 7.1. General recipes in a typical development function.

4.4.3.3 Process development

Process development results in the definition of the manufacturing processes required to produce the product within product specifications. Process development takes into account the currently defined basic manufacturing process definitions. If additional manufacturing processes are needed, for example because of unique new chemistry that requires new process actions, then process development defines the new process requirements.

Commercial versus development is as different as night and day. GMP for FDA-regulated products takes on all variants. In commercial manufacturing, compliance is enforced by strict GMP adherence to the approved recipe. In life science development, a recipe to create a new chemical or generate a biologic product can be as simple as a laboratory procedure or as detailed as a full batch ticket. GMP becomes significant during clinical trials or whenever humans are dosed to test safety and efficacy. The operating mode characters in development must toggle between GMP and non-GMP. Even in GMP mode, changes may still occur, whether there are known exceptions or major deviations. Throughout development, lessons are captured and organized into structured recipe repositories.

Product and process knowledge can exist in a variety of forms and includes the following information:

- Final Master Batch Record (MBR)
- Phase I, II, III Campaign MBRs
- Campaign history of project
- All associated dispositioned batch records
- Bill of Materials (BOM)—includes supplier qualification, specification plans, and quality history
- Bill of Equipment (BOE)—includes qualifications, calibration histories, and so on
- Bill of Process (BOP)—process operations with process actions, process parameters, process results, assays and results, and so on
- Parameters linked to product quality attributes that link to human benefits
- Scientific factors—design space, design of equipment, development research findings, and scaling observations

This information is packaged in an information dossier that includes the following:

- Development of product specifications
- Flat and hierarchical recipes
- ISA-88 recipe standards
- Customer- and market-specific labeling and packaging, including multilingual support
- Collaboration with external vendors, suppliers, and distributors
- Tracking and registration of product and process changes in multiple markets over product life cycle
- Secured internal and external information exchange
- Risk assessment
- Integration and learning across specialized domains
- A focus on process and design space to identify critical-to-quality parameters

FDA Initiatives Shaping Drug Development

Pharmaceutical guidelines are pointing manufacturers in the life science industry in the direction that other industries have already taken—a more systematic approach to development called *Quality by Design* (QbD). QbD has origins in the discrete manufacturing sectors. Similar logic that applied to cars, semiconductors, planes, trains, and chips is now being introduced and encouraged in the life science industry—that good science leads to better quality products, fewer product rejects or recalls, and enhanced public health protection.

What does the FDA QbD initiative encompass? A QbD system should demonstrate the following characteristics:

- The product is designed to meet patient needs and performance requirements.
- The process is designed to consistently meet product critical quality attributes.
- The impact of initial raw materials and process parameters on product quality is understood.

- Critical sources of process variability are identified and controlled.
- The process is continually monitored and updated to allow for consistent quality over time.

ISA-88 is an up-and-coming standard for the pharmaceutical industry. Unit operations have long dictated the structure of recipes. Pharmaceutical companies use different terminology, even within a single company. Generally, biotechnology companies have advanced more in the acceptance of ISA-88. Mentions of ISA-88 are just starting to make their way to the FDA standards group. Long associated with consumer products and chemicals, ISA-88 has made significant inroads to pharmaceuticals. Looking at the ISA88 committee, the assumption holds true.

QbD references other FDA standards documents. Q8[1] focuses on pharmaceutical development for drug products. The main premise is that quality cannot be tested into products. Information from pharmaceutical development studies is a basis for risk management. Q9 focuses on quality risk management:

> The manufacturing and use of a drug (medicinal) product, including its components, necessarily entail some degree of risk. The risk to its quality is just one component of the overall risk. It is important to understand that product quality should be maintained throughout the product lifecycle such that the attributes that are important to the quality of the drug (medicinal) product remain consistent with those used in the clinical studies. An effective quality risk management approach can further ensure the high quality of the drug (medicinal) product to the patient in providing a proactive means to identify and control potential quality issues during development and manufacturing.[3]

The early concepts of QbD were established with the initiative called "Pharmaceutical cGMP for the 21st Century." Our FDA Cooperative Research And Development Agreement (CRADA) has led pharmaceutical, biotech, and generics industry workshops to a better understanding by FDA of gaps in knowledge about Q8 and Q9 and considerations on how to best fill the gaps. The FDA is seeking improved collaboration with industry on the overall concepts of the new paradigm and how to apply them. The workshops allow companies to discuss the application of the QbD scope and form goals for delivering the best pharmaceutical products.

ISA-88 Concepts for Development

The most important feature, which ISA-88 advocates, is flexibility. Recipe management takes on various forms from simple to complex. Understanding the importance of the equipment and raw materials used with the recipe is critical to the knowledge gained in developing and understanding the process. In early development, the scale is small, as shown in Figure 7.2.

Each step in the procedure described in Figure 7.2 can be represented and stored as ISA-88 phases. As recipes scale-up and become more detailed, ISA-88 is ideal for handling the complexity, as shown in Figure 7.3.

In the pharmaceutical industry, the unit operation usage translates to ISA-88 procedural elements. Traditionally, the use of standard statements (i.e., Microsoft Word templates) serves to hold the recipe elements for reusable, standard instructions. A process engineer or scientist references standard statements and defines the exact details while creating a recipe for a given process. Changes or enhancements

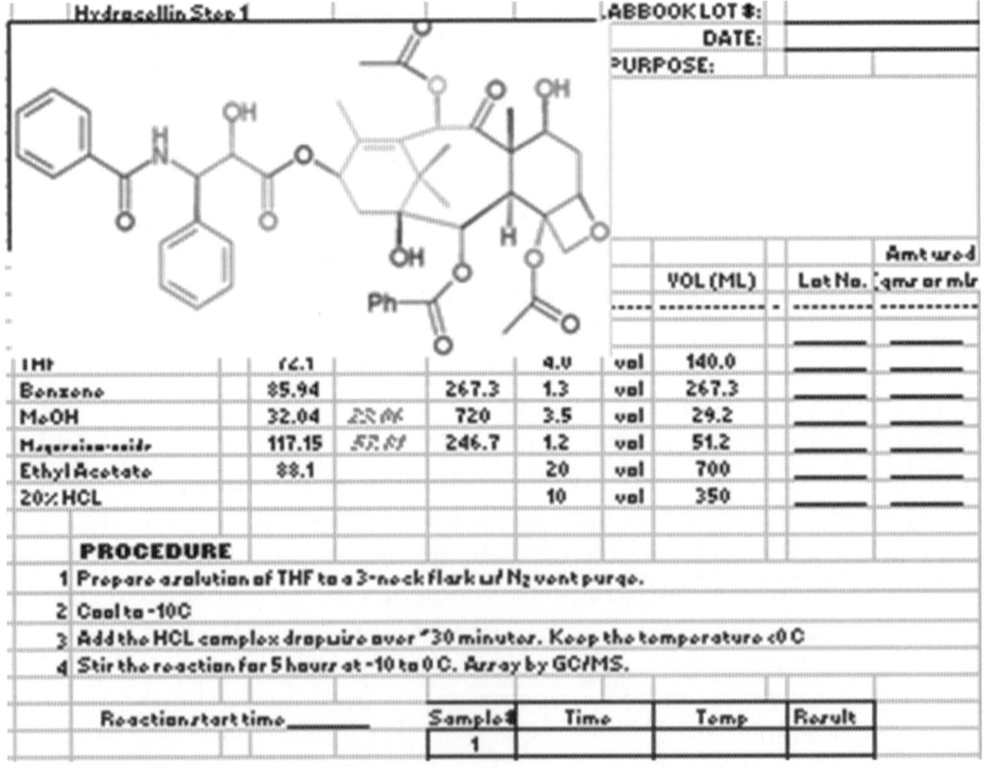

Hydracellin Step 1

LABBOOK LOT #:
DATE:
PURPOSE:

						VOL (ML)	Lot No.	Amt used (gms or mls)
THF	72.1			4.0	vol	140.0	____	____
Benzene	85.94		267.3	1.3	vol	267.3	____	____
MeOH	32.04	[illegible]	720	3.5	vol	29.2	____	____
Magnesium-oxide	117.15	[illegible]	246.7	1.2	vol	51.2	____	____
Ethyl Acetate	88.1			20	vol	700	____	____
20% HCL				10	vol	350	____	____

PROCEDURE

1 Prepare a solution of THF to a 3-neck flask w/ N_2 vent purge.
2 Cool to -10C
3 Add the HCL complex dropwise over ~30 minutes. Keep the temperature <0 C
4 Stir the reaction for 5 hours at -10 to 0 C. Assay by GC/MS.

Reaction start time________

Sample #	Time	Temp	Result
1			

Figure 7.2. Example laboratory procedure.

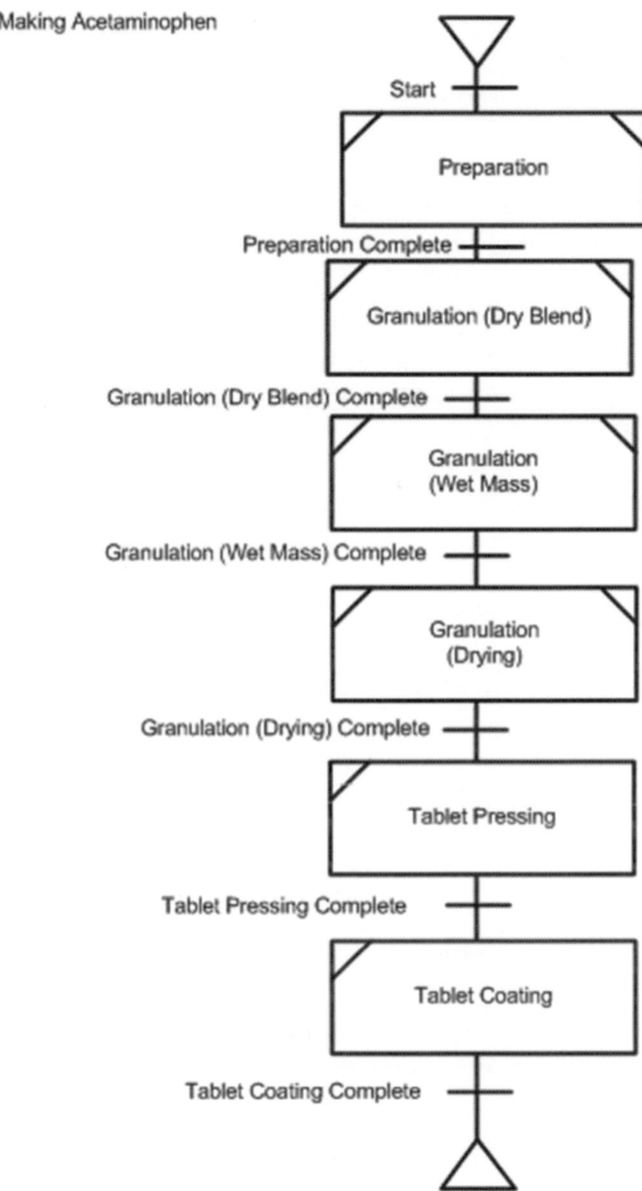

Figure 7.3. Drug product ISA-88 operations.

made to the recipe during execution can be later saved in a standard statement format for later reuse.

The most significant enhancement is in the ISA-88 command functions for development operations. Since a recipe is unproven, the author of the recipe will define conditional logic in the recipe instruction. Depending on the possible outcomes, the process engineer will predefine the expected "if, then" conditions that allow the operator to proceed in a variety of recipe paths. This is clear for anticipated exceptions, but the operator needs to record unexpected results, observations, and even possibly add new recipe elements in order to salvage the Work-In-Progress (WIP) material being manufactured. Depending on each instruction outcome, the operator may need to act using the ISA-88 function commands such as Stop, Hold, Pause, and Abort. Other ISA-88 commands are required when a recipe element must be redone. This includes the use of the ISA-88 Reset and Restart commands.

A new command that is required for development is a Skip command that gives the operator the ability to not perform a recipe element by skipping over it. In a development environment, change and variation are the norm. An added flexibility is required to allow the person executing the process to proceed down a variety of avenues. Supervisory notifications are necessary to ensure the batch success. For known or foreseen process exceptions, the process engineer can outline the possible recipe strategies.

Along with skipping recipe elements, the concept of inserting new elements is common in development operations. Since the recipes are as novel as the drugs coming to market, the scientist, process engineer, and operator are sometimes working side by side. Having an unchanged, approved control recipe only lasts until the execution process has been started. Once execution has started, observations on how well the WIP material is forming will dictate whether the recipe needs to be modified to recover the process.

Strategies for ISA-88 Acceptance

With the FDA QbD initiative and the acceptance of ISA-88, how should the pharmaceutical industry proceed? Here are a few approaches that pharmaceutical companies could try.

Mapping from Unit Operations

In pharmaceuticals today, the diverse use of terminology, paper-based protocols, and isolated point solutions must be overcome. For example, the term *unit operation* is widely referenced in journals, across the industry, and by the FDA. ISA-88.01 defines a *unit procedure* and a *process operation*. The first obstacle to overcome is change. The industry uses paper-based batch records to record a development batch ticket. The first thing you notice in a typical development paper batch record is all the additional instructions added, sections crossed out, and notes in the margin.

Taking the first step would be to leverage the ISA-88 recipe elements. Pharmaceutical unit operations directly translate to ISA-88 procedural elements. The commonly used unit operation standard statement libraries stored in Microsoft Word files could exist as a procedural element library. The basic elements in a laboratory procedure could be simplified to form a phase object of small, lab-scale instructions. Using recipe objects in a structured format allows reusability and assists in migration of recipe elements. Often, a laboratory procedure that is designed to produce laboratory scales of material is repeatedly used as one of the primary assays to evaluate the raw materials of scaled recipes. Simply called a *lab trial*, the laboratory procedure is executed similarly as an analytical assay to test the quality of the raw materials. The specification for the bill of raw materials requires that execution of the associated laboratory procedure results in the successful generation of the target material. The structural form of the laboratory procedure as an ISA-88 recipe object can aid in the automation of this small-scale recipe.

Reducing Process Steps

In early development, a small molecule will have many intermediates before the final product can be generated. Each intermediate is generated in a synthesis called a *process step*. Reducing the number of process steps is important in the overall efficiency of the manufacturing process. Complicated recipes require time and resources and often lead to product-quality issues. Grouping recipe elements in an ISA-88 model begins to refocus the process into recipe management and equipment control areas. Focusing on the recipe itself allows product and process optimization.

Process Optimization

In development, special batches are executed for learning, improving, and optimizing a portion of the recipe; understanding of new equipment or techniques that may improve yields; or experimenting with varying control parameters. ISA-88 isolates critical elements into easily transferable items where development engineers can optimize or resolve any conditions.

Structured Recipe Repository

During the NDA submission process, referencing learnings across development operations helps reduce the process optimization time through better data management. Having a structured methodology to store associated process development controls and associated product quality attributes allows for a better understanding of process knowledge. With ISA-88 structure as the basis for information, the standard allows for better storage and mining of data. Currently, paper batch tickets are scanned and stored in document management systems. Storing the early development recipes in notebooks adds a further complication to understanding. In an unstructured example, the results of early Design Of Experiment (DOE) could have the original hypothesis in a notebook, the data sets in Excel, the analysis in an SAS Institute report, and the conclusion in a management report. ISA-88 provides a basis for a common structure to store product development data.

Enhancing Technology Transfer to Commercial Manufacturing

Using the WBF Batch Markup Language (BatchML) standard is one of the methods by which to transfer development recipes to commercial manufacturing. Another method involves the use of Extensible Markup Language (XML) formatted recipes. Having the recipe in a structured format is essential. The other critical success factor is having development operations and commercial manufacturing both utilize ISA-88.

Conclusion

Pharmaceutical people know of ISA-88, and they often say that they plan to take an ISA-88 course. The adage of "throwing it over the wall" from research to development to commercial manufacturing is alive and well in pharmaceutical companies. There is not much one can do between research and development, but ISA-88 standardizations can accelerate the technology transfer and product launch. Using ISA-88 is an ideal strategy to eliminate the loss of information that often occurs.

ISA-88 is a formal standard, but its flexibility is ideal for development. ISA-88 can be leveraged to represent recipes at various stages in development operations. As process knowledge increases, a development recipe using ISA-88 guidelines easily expands to support the variety of recipe iterations.

Acronyms and Glossary

API—Active Pharmaceutical Ingredient

BOE—Bill Of Equipment

BOM—Bill Of Materials

BOP—Bill Of Process (recipe)

CCP—Critical Control Parameter

cGMP—current Good Manufacturing Practice

CMC—Chemistry Manufacturing and Controls

CRADA—Cooperative Research And Development Agreement

CTD—Common Technical Document

DOE—Design Of Experiment

FDA—Food and Drug Administration

GMP—Good Manufacturing Practice

ICH—International Conference on Harmonization (of Technical Requirements)

MBR—Master Batch Record

NDA—New Drug Application

PPLM—Product and Process Life-cycle Management

PQAS—Pharmaceutical Quality Assessment System

Q8—FDA standard for pharmaceutical development

Q9—FDA standard for pharmaceutical risk management

Q10—FDA standard for pharmaceutical quality system

QbD—Quality by Design

Tech transfer—Technology transfer

Unit operation—This term can be defined as an ISA-88 procedural element; FDA documents specify equipment for the unit operation, which also ties in the ISA-88 physical model. Continuous processes use "unit operation" to mean the equipment and control associated with a chemical unit operation, such as distill, oxidize, crystallize, or centrifuge.

WIP—Work In Progress

XML—Extensible Markup Language

References

1. Berridge, John. 2004. *Q8 pharmaceutical development*. Washington, DC: Government Printing Office.
2. Huff, Bob. 2001. What does R&D really cost? http://thebody.com/content/art13514.html.
3. ICH. *Q9 quality risk management*. Washington, DC: Government Printing Office.
4. Instrumentation, Systems, and Automation Society. 2003. *ANSI/ISA-88.00.03-2003: Batch control part 3: General and site recipe models and representation*. Research Triangle Park, NC: ISA.
5. Wikipedia. 2010. Research and development. http://en.wikipedia.org/wiki/Research_and_development.

Process Definition Management: Using ISA-88 and BatchML as a Basis for Process Definitions and Recipe Normalization

Presented at the WBF European Conference, November 10–12, 2008, by

Mike Power
Business Advisor
mike.power@bearingpoint.com
BearingPoint, Gordonsville, VA 22942 USA

Paul Wlodarczyk
Managing Director
paulw@thecontentguy.net
The Content Guy, Fairport, NY 14450 USA

Abstract

Life science organizations have been late to adopt certain best practices that have long been widely used in other areas of batch manufacturing. In the face of increasing competition and changing regulatory reporting requirements, pharmaceutical and biotech companies are now looking to standard forms of process definition as a means to accelerate time to market, reduce waste, and improve regulatory submissions.

This chapter discusses recipe normalization, Process Definition Management (PrDM), the associated industry standards (e.g., ISA-88 and BatchML), and supporting technologies, which, together, will improve product time to market and embed quality into process design from the earliest stages of product Research and Development (R&D).

The Life Science Industry Today

The average cost of moving a new pharmaceutical or biotech product through the development process in the United States ranges from $300 million to as high as $1.2 billion. Product development takes an average of 12 to 15 years, leaving only 5 to 8 years of patent protection. Only 1 in 10,000 candidate compounds will survive the development process, receiving new product approval and ultimately reaching the marketplace.[2]

Given this reality, life science companies are moving to aggressively focus on process innovations designed to drive down discovery, development, and commercialization costs and to accelerate time to market. However, a surprisingly large portion of the discovery and development life cycle is still based on manual and disconnected process steps. These manual steps lead to delays, inefficiencies, and increased risk during product and process technology transfer.

The ISA-88 Standard for Recipe Representation

ISA-88 is a standard for addressing batch process control and a design philosophy for software, equipment and procedures in batch manufacturing processes. Many manufacturing execution systems use ISA-88 formats for creating control recipes.

Three hierarchical universal models are defined by ISA-88:

1. A **control model** for batch manufacturing, providing a basis for streamlining communications about user requirements, integration among batch automation suppliers and simplifying batch control configuration.
2. A **physical model** that defines "plant-level" equipment in a manufacturing site.
3. A **procedural model** that consists of a hierarchical model for defining recipes in terms of procedures, unit procedures, operations and phases, and defines four types of recipes: general, site, master and control. Procedural models form an excellent representation of all types of recipes from formulation through execution.

Figure 8.1.

As a result of these manual processes and in the face of evolving regulatory requirements, pharmaceutical companies have recognized the critical need for improvements and have been quick to embrace information technologies. In early 2007, Pharmaceutical Commerce encouraged companies to leverage their heavy investments in Information Technology (IT) for maximum value. They suggest that business and IT professionals should search for the intersection of best business practices with the best IT to drive business value.[1] This search includes systems that don't simply improve one element of the process but can be used to align all processes from development to regulatory approval—systems that span the product life cycle while embedding quality into every stage.

Quality by Design

The concept behind the Quality by Design (QbD) initiative is simple: make quality a fundamental part of the development process from the earliest phases of the product life cycle by using automation and standardization to enforce consistency and control. In other words, build quality into the process design, rather than viewing it as a downstream step in manufacturing. QbD principles have been used for decades (with great success) in chemical and high-tech manufacturing. Life science companies have been slower to adopt; however, this is changing as a result of aggressive evangelism from the U.S. and European regulatory agencies (i.e., the Food and Drug Administration [FDA] and the European Medicines Agency [EMEA], respectively).

QbD essentially requires researchers to define and document the Critical To Quality (CTQ) attributes of the product and the corresponding critical and key parameters of the manufacturing process. CTQ attributes include any other variable that is critical to the ultimate quality (e.g., effectiveness, stability, toxicity) of the Active Pharmaceutical Ingredients (API). In QbD, CTQs are identified through rigorous Design of Experiments (DoE) in late-phase discovery and early development. Once identified, the CTQs are refined, documented, reported, and monitored throughout the product life cycle.

Process Definition Management

Process definitions form the foundation for QbD initiatives. PrDM defines detailed process logic as well as parameters for manufacturing phases that include process inputs, process parameters, and process outputs. Process definitions can therefore provide a framework for defining, managing, and submitting key and critical parameters for any regulated manufacturing process. The process parameters can include material parameters (e.g., ingredients and intermediate products, quantities, scale factors), equipment parameters (e.g., speed, duration, and other equipment control settings), operators instructions, Standard Operating Procedures (SOPs), resource requirements (e.g., operator qualifications), control and signature strategies, in-process and postprocess quality attributes, and testing requirements—essentially any aspect of the manufacturing process. Because developers know at design time that certain parameters of the manufacturing process are critical to maintaining the product quality, product developers need a method for capturing and communicating this critical process knowledge throughout the product life cycle.

Room for Improvement

To create process definitions and recipes, product developers typically use flowcharts or documents to map or catalog the steps of the manufacturing process. The quality specifications and parameters behind each step are captured in a separate document that is usually tabular or a spreadsheet. Parameter information can be very complex, often resulting in spreadsheets with thousands of rows. The overall size of some spreadsheets and tables leads to inefficiencies in recipe management, and presents five distinct areas of opportunity for improvement: version control, content reuse, knowledge management, technology transfer, and regulatory submissions.

Version Control

Flowcharts representing process logic and spreadsheets or tables containing process parameters are often managed as separate documents and easily become out of synch. Both version control and synchronization of the spreadsheets or tables are critical to improving process definition management.

Content Reuse

Creating and managing process definitions within unstructured documents like flowcharts and spreadsheets provides no mechanism for reusing process elements. However, recent studies and pilot projects have shown that the reuse of predefined libraries of process segments offers the primary means of codifying knowledge of products and knowledge of manufacturing best practices.

In practice, product development projects proceed independently without sharing insights from other project teams. Process definitions for similar compounds are authored at different times by different process engineers in separate, unstructured documents. As a result, there is no means of sharing common knowledge of product science or manufacturing. Recently, an FDA Cooperative Research And Development Agreement (CRADA) study demonstrated that product developers should be able to share best practices, essentially by sharing "chunks" of recipes or product definitions. By capturing product knowledge (e.g., chemistry, biology, physics) within reusable recipe building blocks,

Average Cost of New Product Development: more than $800 million
Average Time to Market for New Products: 12-15 years
Median Application Time for New Products, 2001: 14 months
Median Application Time for NME, 2001: 20 months

Figure 8.2.

product developers have a method for capturing, sharing, and discovering best methods, which can then be incorporated into new product designs.

In most cases, manufacturing process innovations occur in plant systems (e.g., a Distributed Control System [DCS] or Manufacturing Execution System [MES]), in which case the knowledge is represented in master recipes or master batch records. Technologies and business processes do not exist to apply ongoing manufacturing innovations to earlier process designs and process definitions in development. By capturing best practices in the form of process definitions and codifying them into a library of reusable manufacturing actions, these innovations in manufacturing are available to be used in new product development.

Knowledge Management

Process definitions collectively represent the intellectual property of a life science firm, combining knowledge of product science (e.g., chemistry, biology, physics) and knowledge of manufacturing capability (current and future). Yet because this knowledge is represented in unstructured documents, there are limited effective tools for searching process definitions or recipes to enable the discovery of product or process innovations.

To make matters more difficult, each organization—and, often, each individual within an organization—uses methods and formats that have evolved out of personal habits or "the way things have always been done." From annotations in lab notebooks to operating instructions, rarely are the documentation formats ever consistent. This inconsistency creates additional challenges, making knowledge sharing difficult when attempting to reuse content or transfer technology.

Technology Transfer

The manual nature of technology transfer is perhaps the most significant weakness in the product life cycle. When a process definition (in the form of a document) passes from development to pilot manufacturing, the information from flowcharts and spreadsheets needs to be transferred to the batch control systems. Currently, this process is manual. Re-keying of process logic and parameters is time-consuming and introduces errors in technology transfer, thus putting quality at risk, reducing yield by increasing waste from bad batches, and adding to cycle time. A typical manual technology transfer occurs in 4 to 8 weeks. With the right tools and approach for technology transfer, this process can be completed in days rather than months. For certain biological blockbuster products, the revenue benefit associated with faster time to market may be as high as $1 million per day.

Technology transfer is not just limited to the initial scale-up from pilot to commercial. It is repeated every time the product definition needs to be transferred to a new manufacturing site: when the product enters short-run production for clinical trials; when adding a facility to scale-up; when scaling-down and changing to a smaller plant; and when products go off patent and are moved to lower-cost Contract Manufacturing Organizations (CMOs). Given the addressable impact on time to market, improving technology transfer is likely to be the most significant area of opportunity for most firms implementing PrDM.

Regulatory Submissions

The fifth area of opportunity is regulatory submissions. QbD—as it becomes a regulatory framework—will require reporting on all CTQ attributes and the controls in place to manage them. CTQ attributes and the corresponding critical or key parameters will need to be reported from the earliest preclinical submissions through to large-scale manufacturing. Rather than being part of an entirely separate process, the process definitions themselves can provide a framework for organizing, managing, and reporting on critical attributes and parameters throughout the product life cycle.

Barriers to Effective Process Definition Management

If process definition management can provide such broad benefits to life science firms, then why hasn't it been more widely adopted, and what innovations have occurred to make it easier to implement today?

One barrier to process definition management has been the absence of a standard tool for process specification during development. Another barrier is converting a process definition into an executable recipe. Many batch control systems have recipe editors and use standards-based formats, such as ISA-88, to represent process flows. In batch control, the recipe is authored in a manner that promotes process control and execution rather than process definition, and the master recipes are very equipment specific. Because these master recipes are equipment specific and are not created until the product goes into commercial-scale manufacturing, these recipes cannot provide the same benefits as a process definition generated upstream in the product and process life cycle.

Another barrier is the wide array of standards for representing the process definition itself during development. While this is addressed by ISA-88, even industries that have widely adopted ISA-88 typically use it for master and control recipes during manufacturing but not as a means to represent standard process definitions during development.

A third barrier to effective PrDM is interoperability. At execution, master recipes must be able to be imported by a variety of enterprise systems in the manufacturing process. They often need to share master data from manufacturing, such as references to equipment and materials. Normalization of recipes as ISA-88 procedural models within a single operational language (e.g., an Extensible Markup Language [XML] format such as Batch Markup Language [BatchML]) is a best practice for enabling QbD.

The Rise of XML for Regulatory Submissions

Since the FDA mandated the transition from PDF submissions to XML-based submissions through the electronic Common Technical Document (eCTD) and Structured Product Labeling (SPL) standards, XML tools have changed the way that pharmaceutical companies do business. Corporate adoption of XML continues to accelerate, and this is not merely for the sake of meeting these regulatory mandates; companies are also starting to realize the efficiency that XML-based systems bring to many elements of the life cycle, above and beyond submissions. XML-based systems can help to track, define, and streamline information throughout the product life cycle, from late phase discovery through commercial manufacturing.

Extensible Markup Language (XML) is a key enabler of a QbD initiative, providing a dramatically more efficient way to create, access, manage and reuse information across the pharmaceutical lifecycle. XML helps to unlock the value of unstructured content previously trapped in documents and repositories, and better utilize the structured data in databases, manufacturing execution systems, asset management systems and quality management systems.

Figure 8.3.

XML as a Process Definition Enabler

While QbD brings life science companies the benefits of improved yield and quality, the regulatory costs include increased reporting on critical and key parameters throughout the product life cycle. The current management of process definitions in the pharmaceutical and biotech industries relies heavily on the use of unstructured documents (e.g., flowcharts and spreadsheets) to capture manufacturing process logic and parameters. Spreadsheets can typically grow to be thousands of rows in length; there is no simple way of abstracting only the critical and key process parameters from a recipe for reporting purposes. There is also no method for importing process definition documents into execution systems, so the process of converting process definitions into executable master recipes for batch control is unnecessarily time-consuming, iterative, and prone to errors. XML-based tools and standards can alleviate many of the challenges that life science companies have faced with traditional documentation.

BatchML can address a host of technology transfer and interoperability requirements of recipe management. BatchML—an industry standard that consists of a set of XML schemas written using the World Wide Web Consortium's XML Schema Definition (XSD)—is an emerging best method for representing data for ISA-88 recipes and enabling that information to flow between systems for submissions, recipe and formula management, execution, asset management, and quality control.

BatchML implements the ISA-88 standard and is an excellent tool to use when exchanging ISA-88 based data, providing a set of XML element definitions that may be used in part or as a whole for batch control, master and general recipes, and equipment data.

A Start-to-Finish Solution for Batch Process Definition

With all the possible places where the process could break down in the flowchart and spreadsheet scenarios, documentation of process (flowcharts) and quality parameters (spreadsheets) cannot continue as separate systems. A single solution is needed to create process definitions that combine parameters and process logic, enable reuse of predefined recipe building blocks, and can be transformed into executable master recipes, based on reuseable steps and actions. It is recommended that such a solution would use class-based ISA-88 master recipes as the format for a process definition, with the data stored as BatchML.

ISA-88 compliant sequential function charts may form the basis for capturing procedure, unit procedure, and operation process logic. Parameters may be associated with phases using property sheets and user dialogs. In this way, users may easily navigate the process flow and easily associate parameters with specific recipe elements. Likewise, audit trail (i.e., "header") information must be associated with any recipe element.

Recipe building blocks may be represented as reusable fragments of ISA-88 recipes (visible to users as parts of sequential function charts). BatchML provides a mechanism for specifying recipe building blocks in the form of reusable XML objects at the data layer. Together, this enables content reuse of process logic and standardized parameters through reusable templates. The templates can be stored in libraries using content management technologies.

We have developed a working prototype of a process definition editor. A process definition editor, as described previously, would enable process manufacturers to do the following:

- Improve information transfer between development and manufacturing, including CMOs, through BatchML.
- Improve product quality and overall manufacturing processes through better management of critical and key parameters.
- Increase efficiencies through process reuse across products using recipe building blocks and through recipe normalization.
- Accelerate product time to market through improved technology transfer and faster time to scale from pilot to commercial scale.
- Reduce scale-up cycle time and waste from failed batches through improved technology transfer and better communication of critical and key process parameters.
- Improve submissions to FDA by using class-based ISA-88 master recipes as the basis for capturing and reporting both process logic and critical and key process parameters within the "Description of Manufacturing Process and Process Controls" (section 3.2.S.2.2) of the eCTD.

The cycle time benefits from improved technology transfer through process definition management are significant. Forrester Research reports that a single day of delay on a $1 billion product can cost a life science manufacturer $2.74 million in lost sales.[3] An acceleration and cost savings of just 2% could mean 58 to 73 days in acceleration, a revenue difference of $158 million to $200 million. This level of Return On Investment (ROI) is achievable from a single PrDM pilot project on one product by reducing recipe authoring cycle time and by improving time to scale through more effective technology transfer.

Conclusion

Pharmaceutical companies are investing heavily in R&D to focus on new products, and even incremental improvements can make a huge difference to the bottom line. Yet cost-cutting measures can't come at the expense of quality. Fortunately, companies can greatly reduce time to market, embed quality into their processes, improve productivity, reduce costs, mitigate risks, and simplify regulatory submissions by implementing ISA-88 based process definitions and by representing this information as BatchML. A process definition editor that combines ISA-88 and BatchML can help integrate disconnected business processes formerly locked in static documents, thus resulting in more streamlined internal processes and improved external interactions with contract manufacturers, other supply chain partners, and regulators.

References

1. Chizzo, Scott. 2007. Embracing change. *Pharmaceutical Commerce,* February 11. http://www.pharmaceuticalcommerce.com/frontEnd/main.php?idSeccion=450.
2. Pharmaceutical Research and Manufacturers of America (PhRMA). 2005. *What goes into the cost of prescription drugs?* Washington, DC: PhRMA.
3. UPS Supply Chain Solutions. 2005. Building supply chain capabilities in the pharmaceutical industry. White paper. http://www.ups-scs.com/solutions/white_papers/wp_pharma1.pdf.

CHAPTER 9

ISA-88 Design and Implementation Case Study for a Pharmaceutical Batch Process

Presented at the WBF North American Conference, May 15–18, 2005, by

Richard A. Brill, P.E.
Automation Group Leader
rick.brill@pfizer.com
Pfizer Global Manufacturing, 400 West Lincoln Avenue, Lititz, PA 17543 USA

Gregory M. Skovira
President
gskovira@ezsoft-inc.com
EZSoft, Inc, PMB 304, 3947 West Lincoln Highway, Downingtown, PA 19335 USA

Abstract

This chapter will present an ISA-88 implementation case study of a multistep pharmaceutical process using some leading edge techniques and tools. The chapter details the design process for the application software components that control the batch process. Key objectives were to implement an ISA-88 control strategy that will support process visualization, electronic batch records, work instructions, alarm management, and data collection. The system had to be 21 CFR Part 11 compliant for electronic signatures and electronic records and also adhere to the Good

Automated Manufacturing Practices 4 (GAMP 4) Guide for Validation of Automation Systems. This chapter describes the system design, testing, validation, and implementation methodology employed, including the use of software tools that were used for rapid application development consisting of project design model development, application code and tag generation, as well as automatic test protocol generation. The end product was a fully automated, flexible, ISA-88-based system consisting of reliable and reusable components, where automation of process changes are not resource intensive and batch reporting is highly integrated.

Introduction

There are many reasons to implement process automation. These include accelerated time to market, operational flexibility, process repeatability, and regulatory compliance. The challenge was to meet these objectives and specific user requirements in a cost-effective manner. To meet this challenge, the project team used innovative software development tools and well-established standards, coupled with sound development methodologies like ISA-88.01 and GAMP 4. This case study details how these tools were utilized in the successful implementation of a new, automated Mouthwash Batch System (MBS).

The new batch manufacturing system needed to (1) replace an existing antiquated batch facility, (2) flexibly produce multiple products simultaneously, and (3) include increased responsiveness to the supply chain to quickly launch new mouthwash products. The existing high-volume continuous process did not have the capability to execute the pilot batches required for the development of new formulas, and it was not flexible enough to efficiently change between numerous small volume formulas. Additionally, the small volume batch production facility that did exist was antiquated, with equipment and controls from the 1960s.

The justification for the new flexible and compliant batch system was obvious. There was a need to expand the product portfolio to include new formulas, and additional volume capabilities were required to support future sales growth projections. Additionally, there were regulatory compliance, safety, and ergonomics issues with the existing batch manufacturing facility.

Background

A cross-functional team that included the Manufacturing, Engineering, Quality Assurance, and Validation departments developed the system requirements. These system requirements included the following:

- The ability to produce two products simultaneously
- The need to reduce the area and the amount of equipment required to comply with applicable flammable-material safety codes (alcohol is used in substantial amounts in the formulation)
- The ability to automate bulk material delivery
- The ability to efficiently clean an equipment train for quick formula changeovers

The results of the conceptual design are detailed in the area layout of the process depicted in Figure 9.1. The manufacturing area consisted of the following:

- Staging areas for storing preweighed manual additions
- Two redundant charging suites where the concentrated slurry is mixed

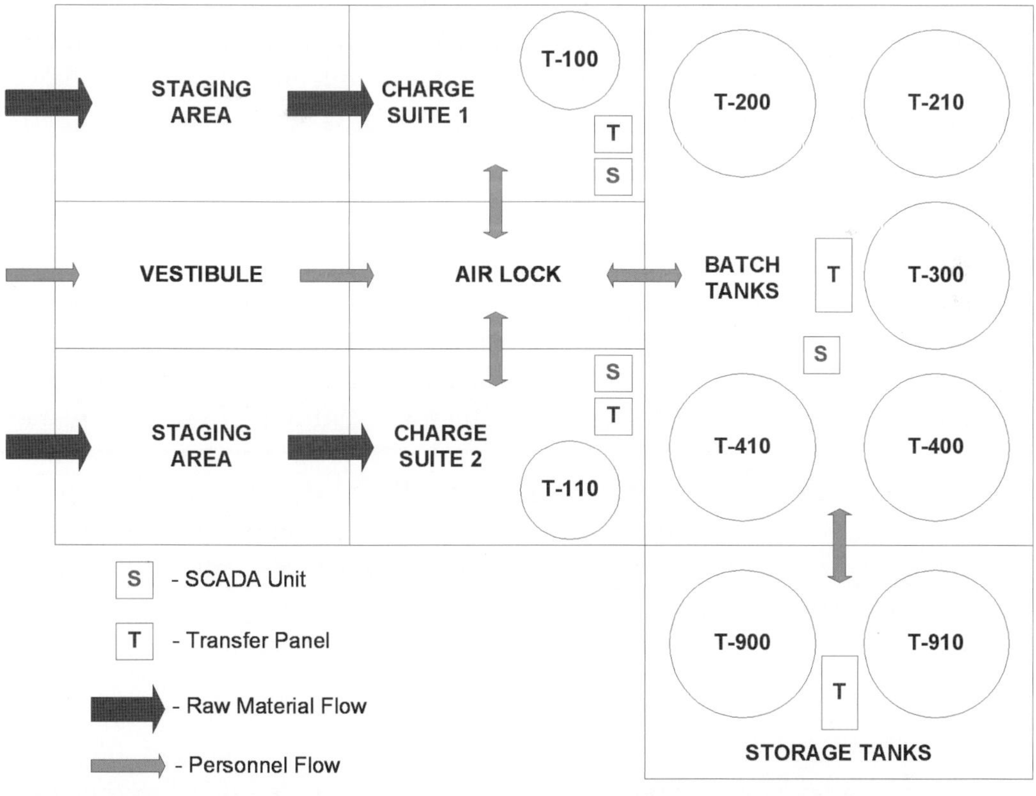

Figure 9.1. Area layout.

- Five batch tanks where the concentrated slurry from either charge suite is combined with bulk materials to produce the product
- Two storage tanks, each capable of being fed from any batch tank and capable of providing finished product to any combination of seven high-speed packaging lines

Transfer panels and associated specialized hard piping were designed to meet the flexibility requirements and allowed the following:

- Any charge tank to tie to any of the five batch tanks
- Any batch tank to tie to any of the two storage tanks
- Either of the two storage tanks to feed either of the two product headers

Once the mechanical conceptual design was completed, focus was placed on developing the requirements of the control system. The control system was created to automate the process as much as possible, while minimizing the reprogramming and revalidation efforts required for process changes and recipe formulation additions. These requirements set the stage for an implementation that followed ISA-88.01 models and terminology.

The Process

The charge tanks are first filled with a water and alcohol solution. After this, liquids and solids from the preweighed kits are manually added to the charge tank. In parallel with the filling and manual additions to the charge tank, bulk raw materials are automatically dispensed to the batch tank selected by the operator during the batch setup procedure. The resultant concentration in the charge tank is agitated and then transferred to the batch tank. The final operation in the charge tank is a water-alcohol rinse that is transferred to the batch tank. The final product in the batch tank is agitated for a predetermined time to complete the process.

The operator initiates product movement from the batch tanks to the storage tanks. The transfer is automated so that, once initiated, the batch tank contents are only transferred as required to maintain a level in the storage tanks.

Regulatory Requirements

All manual charging was to be done via closed systems provided in the charging suites. All charge and storage tanks included nitrogen blanketing, due to the flammability of the raw materials. The charge suite was designed to meet National Electric Code (NEC) Classification I, Division I requirements, and the batch area was designed to meet Classification I, Division II requirements.

In addition to process visualization, electronic work instructions, and the alarm management requirements common to many control systems, the generation of an Electronic Batch Record (EBR) and compliance to 21 CFR Part 11 for electronic records and electronic signatures were critical requirements of the system.

Interface Requirements

The new batch control system required integration with several other systems in the plant. Integration with the site tank farm control system and the site purified water control system were required to coordinate the delivery of bulk, raw materials. Integration with the site's weighing system was required to import the preweighed raw material information for confirmation of manual additions to the batch and to include the information on the EBR. The system requirement to automate confirmation of manual additions to the batch (by comparing scanned information to the data transferred from the weighing system) allowed for the elimination of a second operator to check the addition, thus reducing personnel requirements. Minimal manual data input from the operator was an additional requirement to reduce sources of human error.

Technologies Utilized

The requirements of the system demanded an architecture that was open and flexible to accommodate multiple interfaces. The controllers were selected to match the bulk of the existing installed base. The process visualization, alarm management, and work instruction requirements, among others, were satisfied by designing three Supervisory Control And Data Acquisition (SCADA) servers for the system architecture. The selection of the software was influenced by corporate standards, the existing installed base at the site, and the ability to satisfy the additional system requirements. The system was required to automate the confirmation of the manual material additions, which created the need for scanning bar code information generated by the weighing system. Lack of available scanning devices compliant to NEC Classification I, Division I regulations forced the utilization of pocket personal computing devices.

The system was also required to manage batch production and recipes, create audit trails, provide electronic work instructions, and log and report electronic signatures compliant with 21 CFR Part 11 regulations. A dedicated database was chosen as the secure, compliant data warehouse to store batch data, audit trail information, and alarm logs. Reporting requirements were satisfied using Web-based software accessible from anywhere on the corporate network via a Web application. The overall system architecture is depicted in Figure 9.2.

Design and Development

The Approach

The challenge was to design, implement, and install a large-scale sophisticated system in a regulated industry under a tight schedule and budget without compromising high corporate standards. The design and development process created a project design model that closely followed the GAMP 4 guidelines, relied heavily on the ISA-88.01 models, and used a software development toolset. The project design model applied

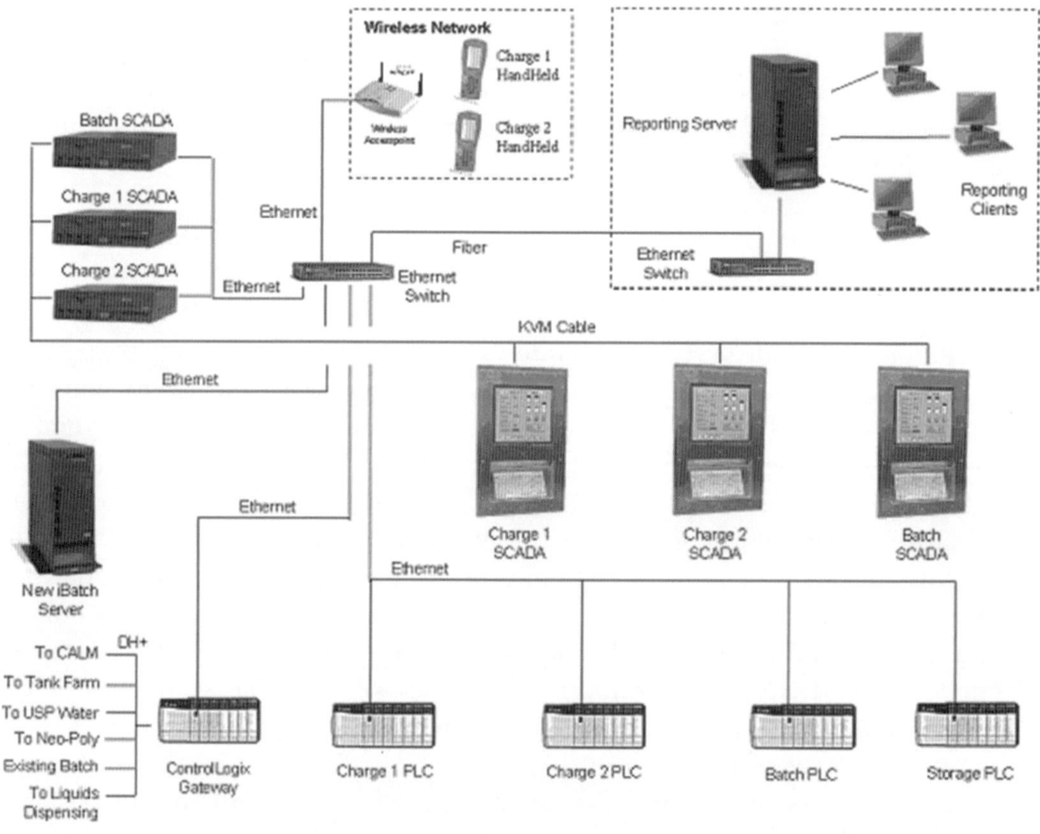

Figure 9.2. System architecture.

to all aspects of the project, including gathering the information for the specifications, system design, documentation, testing, commissioning, and validation. Key to the success of the project was a data-centric software toolset that facilitated the collection of information gathered during each stage of design and development. This information was stored in the toolset's development database. The data collected during the design phase was used for the implementation and documentation of the system.

The impact of the project design model on the project was far-reaching. The project design model and corresponding terminology were used extensively throughout the project to characterize and define the process, as described in the Piping and Instrument Diagrams (P&ID) and process descriptions. The structure provided a common interface between the process engineers and the control engineers during the specification development stage of the project. The project design model provided a framework around which critical information could be collected and stored. This information included the number and type of units in the system, the number and type of procedural control items required for each unit, and the number and type of devices. Properties associated with units, procedural control items, and devices were also defined, along with their relationships to each other (Fig. 9.3).

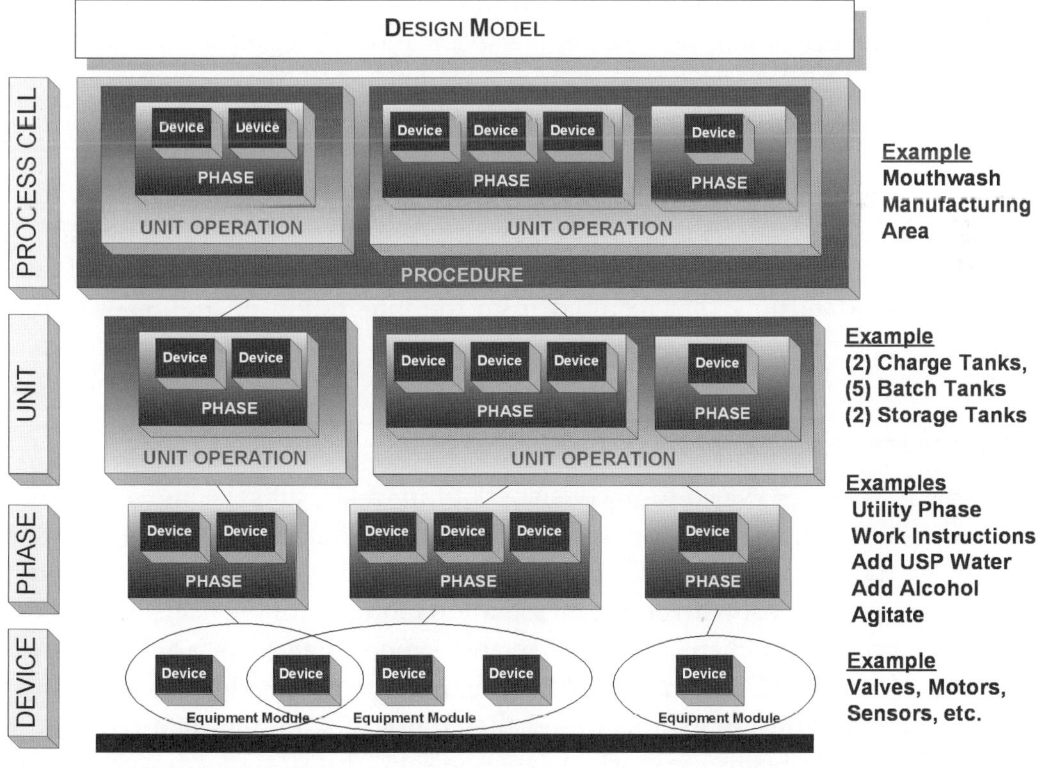

Figure 9.3. Design model.

Functional Requirements

In the first stage of the design and development process, functional requirements for the MBS were defined. The project design model was used as the framework for gathering the information. This information was collected and stored in a development database, using a series of forms provided with the toolset.

In the MBS, the process cell consisted of the Mouthwash Batch Manufacturing Area (MBMA) and was composed of the following units: two charge tanks, five batch tanks, and two storage tanks. After the units were defined, the project team needed to identify the procedural control items associated with each unit. Properly defining the procedural control items was critical to the flexibility of the system.

After appropriate procedural control items were determined, characteristics for procedural control items and devices needed definition. These characteristics included the type, required permissives, interlocks, alarms, and messages. Setpoint and report parameters were also defined. The relationships between each procedural control item and its corresponding units and between devices and their corresponding procedural control items were determined.

The information regarding the units, procedural control items, and Control Modules (CMs) was collected and stored in the development database and used to generate the functional requirements specifications. These functional requirements included report server requirements, batch server requirements, SCADA requirements, and real-time control requirements.

System Design

Much of the system software design was in place before the project started. Figure 9.4 shows the data flow that resulted from the integration of the major system components.

Report Server and Database

The MBMA Report Server design consisted of four main subsystems: report view and selection client, report generator, Web server, and database server. The design transferred data between the SCADAs, batch server, and database server to provide the database server with all the information required for the EBRs. Integration of a Web-based report client, Web server, and a report generator application was designed to meet the requirement of allowing users to request EBRs on demand.

The database design also included the ability to store the data required to implement flexible electronic work instruction sets. This design satisfied requirements to maintain flexibility of the system to deliver electronic work instructions

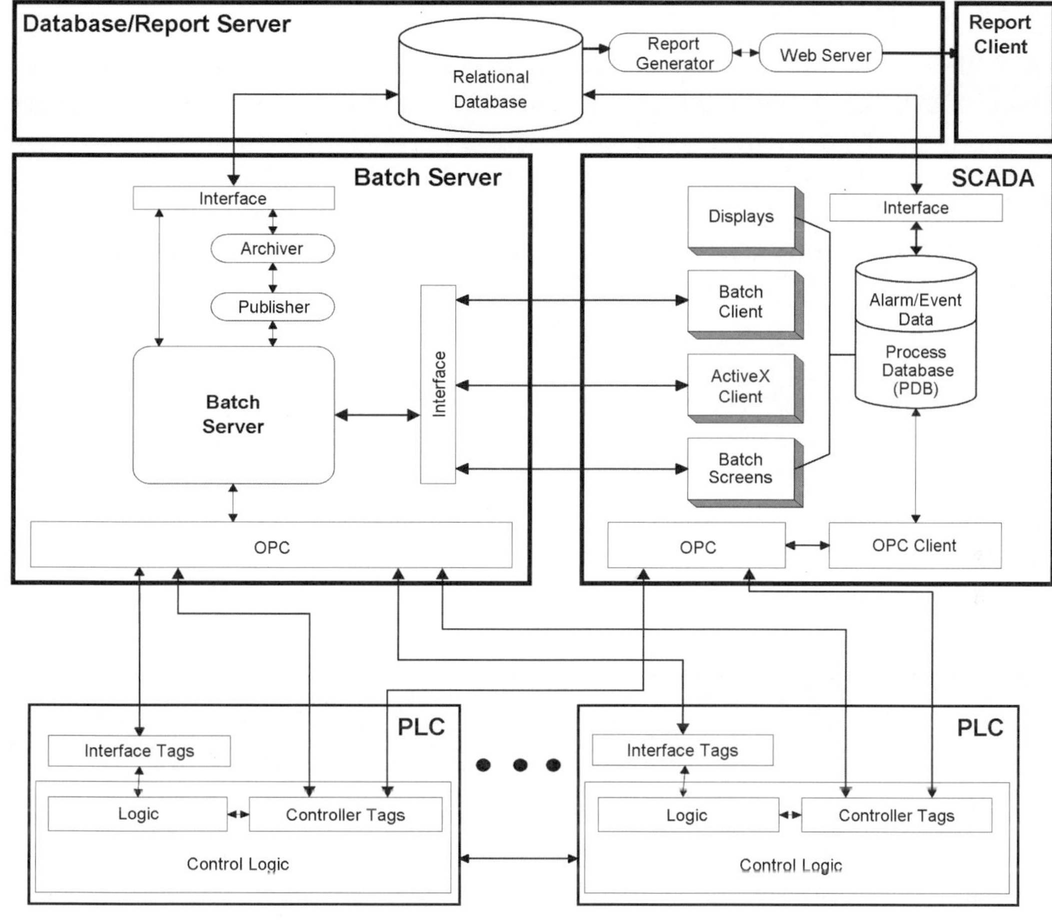

Figure 9.4. MBMA data flow.

for various formulas with minimal application impact and thus reduce testing and validation.

Batch Server

The batch server coordinates batch operations on the plant floor by communicating to the logic resident in the Programmable Logic Controllers (PLC). The server collects data from PLCs and writes it to a relational database as part of an EBR.

An operator display was used as the interface between the operator and the system to facilitate batch setup. From this display, an operator uses the bar code scanner to scan in product information, select a batch tank, confirm setup, and start the batch.

SCADA

The MBMA uses three SCADA terminals, located in Charge Suite 1, Charge Suite 2, and the Batch Tanks room. The SCADA workstation served as the primary user interface to the PLC. System parameters used by the PLC for control (e.g., permissives, interlocks, alarms, and device control) are settable via the SCADA workstation. The ability to set these parameters is secured via the SCADA software and compliant with the audit trail requirements of 21 CFR Part 11, where applicable. Figure 9.5 shows the main SCADA display screen.

Real-time Control

Each area is controlled by a PLC processor and associated Input/Output (I/O). The controllers provide real-time process control, data collection, and device level control of the batch.

There are multiple interfaces associated with the PLC. These include an interface to the batch server for recipe execution and data acquisition, an interface to

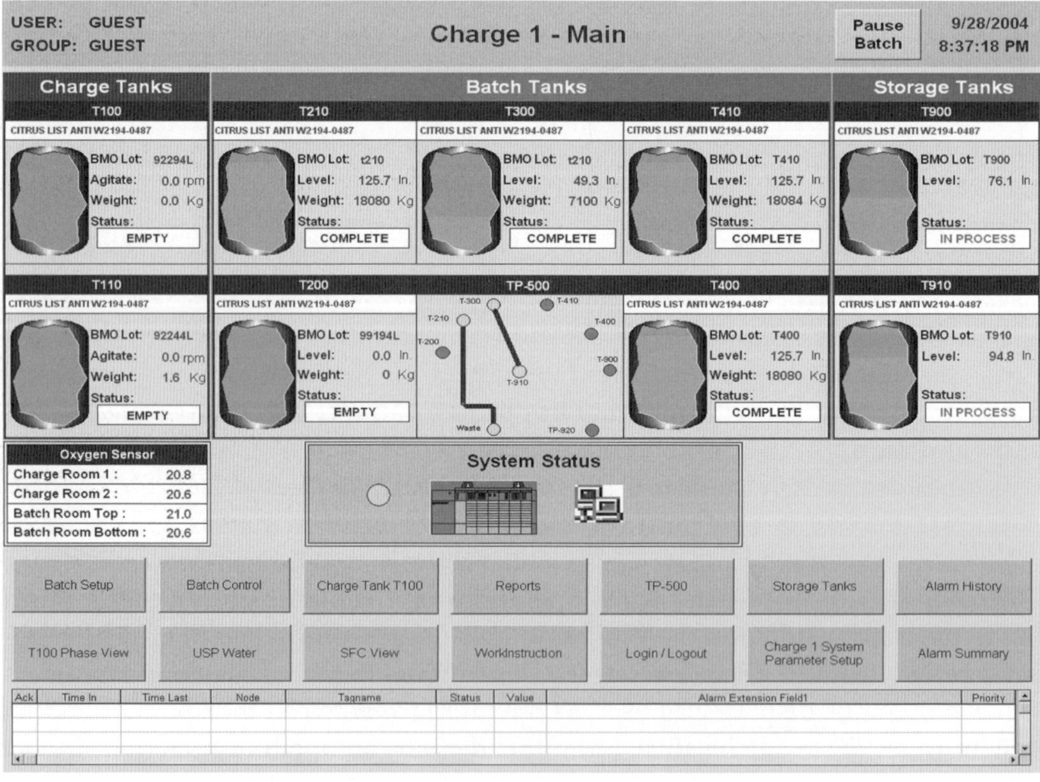

Figure 9.5. Main display.

the SCADA workstations for monitoring and control, and an interface to other PLCs on the automation network for bulk material transfer.

The control system was divided into four subsystems: Charge 1, Charge 2, Batch, and Storage. Here again, the general design specification provided by the toolset was leveraged by the PLC design specifications. Standard design elements used in all four PLC applications were specified, reviewed, and approved one time.

Data Structures: The Foundation

In order to build a sophisticated control system, the foundation must be solid and sound. The development toolset provides a set of time-tested data structures that serve as a foundation. The effective use of data structures is the cornerstone of any successful control system implementation. The data structures provided with the toolset are the result of careful object modeling and reasoning about the components of an automation system and the relationships among them. The benefits of the data structures used in this application include the following:

- *Consistency*. With many developers working on the project, the data structures served as an anchor that kept the developers from drifting apart.
- *Common interface*. The various objects in the system (e.g., devices, phases, and analog inputs) each present the same interface, respectively. This allowed SCADA developers, PLC developers, and batch developers to work in parallel, significantly reducing the overhead involved with keeping these resources synchronized.
- *Auto-generation*. The abstract concept of a data type was the central enabling factor for the auto-generation of software code performed by the toolset.
- *Scalability*. This application had a large number of devices, phases, and analog I/O but was no more complex than a system with only a small number of objects. This efficiency allowed the application developers to focus their energies on the few complicated engineering issues.

Implementation

With the functional requirements in place and the design approved, the toolset and corresponding database was used to auto-generate control code, tag databases, alarms, and configuration files for all levels, including the controllers, the

SCADA nodes, and the equipment model for the batch server. A considerable amount of time was saved during the implementation by utilizing the tools. Just as important was the consistency and quality achieved when auto-generating code directly from the database against a known set of design standards. This resulted in additional time savings during the integration and unit testing. The toolset was used to generate test protocols that were used to test the system prior to shipment. Test templates were created, reviewed, and approved for each class of object defined at the various levels of the object hierarchy. Once approved, the test templates, coupled with the object data stored in the toolset database, were used to generate test protocols. Again, a considerable amount of time was saved without sacrificing quality. The factory acceptance tests were reviewed and approved by the project team to ensure compliance with the functional requirements and design specifications for the system. Extensive testing was conducted prior to shipment. Change control procedures were followed to track deviations so that the time and effort invested in the Factory Acceptance Test (FAT) could be leveraged against the Site Acceptance Test (SAT) and subsequent validation testing.

Installation, Testing, and Qualification

An SAT plan was developed for each of the subsystems: Charge 1, Charge 2, Batch, and Storage. This allowed commissioning and testing to proceed in a staggered approach to accommodate the construction schedule.

The purpose of the SAT was to confirm that the required documentation was complete and that the control system was shipped, installed, and configured properly. Additionally, the SAT included specific operational tests that could not be completed during the FAT (e.g., testing the interface to existing plant systems, communication testing, and testing of system security when integrated with the plant domain). The SAT was then used to test any changes or modifications that occurred during commissioning.

Installation and Operational Qualification (IOQ) for the entire system proceeded immediately upon the completion of construction and commissioning activities. Qualification testing leveraged the SAT, where possible.

System Flexibility Demonstrated

The system flexibility was put to the test even before completion of all validation and testing activities The project team was already under pressure to get the system qualified and into production for the targeted product. However, various business drivers required the system to be tested and validated to manufacture a new formulation of mouthwash before site testing was completed.

Once again, significant advantages were realized because of compliance to the ISA-88 model during the design of the system. Changes were made to the recipe at the batch management level to accommodate the new recipe without changing the controller code or SCADA configurations. Because FAT, SAT, and site test-and-debug work all were performed on the module phase (i.e., recipe independent) design of the system, only recipe-specific retesting was required. The changes were implemented without major redesign and with minimal impact on qualification.

The system was able to go into manufacturing trials and Process Qualification (PQ) with the new product recipe. Qualification was completed, and the system was pressed into continuous operation to meet the demands of the new product.

System Highlights

Weighing System Interface

The site's existing weighing system electronically captures raw-material data and electronic signature information associated with the weighing of raw materials for use in various process cells at the site. The new mouthwash batch manufacturing system needed to provide an EBR that was inclusive of this preweighed material information. Additionally, the system was required to utilize information from the preweighed raw materials to confirm manual material additions to the batch. These requirements drove the need to design an interface between the weighing system and the MBS.

The interface consists of a combination of Off The Shelf (OTS) software and custom software. A query is assembled using data retrieved from a bar code scanner embedded in a handheld computer. The preweighed data included the material specification number, target and actual amount weighed, lot information, and expiration date of each raw material.

Once retrieved from the weigh system, the preweigh data resides in SCADA memory and can be displayed by the operator. Any differences or discrepancies with the batch production raw material list (retrieved programmatically from the batch server) are visually identified by the system at this stage, allowing the operator to rectify these differences by manual data entry with an electronic signature.

When the batch is started by the operator, the preweighing data is written to a set of tables in the production database, for later retrieval during work instruction steps of the batch. The bar code scan information performed for each ingredient, prior to adding it to the batch, is compared to the data in the production database to confirm that the correct material is being added. This system-based confirmation allows for a single, electronic signature requirement for manual additions, thus resulting in production cycle-time savings.

Electronic Work Instruction

The system design included the use of work instructions for the manual addition of raw materials. The data presented to operators in work instructions was the raw material data that was processed through the interface to the weighing system at the start of the batch.

The work instructions designed for use with the MBS can be used for any raw material, and for any product, therefore significantly reducing the cost and effort to change the formula of an existing product or to develop an entirely new product.

Although each product has a consistent set of raw materials, the weigh system may produce a different set of data from batch to batch. Therefore, the work instruction design had to be flexible enough to account for multiple containers of the same raw material. The raw material additions were ordered and grouped together for one electronic signature, based on batch recipe parameters. This feature not only reduced the amount of operator electronic signatures but also allowed any arbitrary ordering or grouping of the raw material additions without the need to change the work instruction, thus saving change control and validation efforts (Fig. 9.6).

Electronic Batch Report

The EBR design was based on requirements developed by a cross-functional group consisting of quality, production, and IT resources. The purpose of the report was to examine the various phases executed to provide a comprehensive history of the batch production. Time stamps of phase completion, electronic signature information, and electronic work instruction summaries were included in the report layout to provide production details.

An important requirement of the batch report was to simplify the quality review activities to reduce the overall cycle time of product release. To satisfy this requirement, obvious deviation indicators were incorporated into the body of the report to enable quality personnel to easily spot any out-of-specification occurrences. In addition, other high-level summary information about the batch was

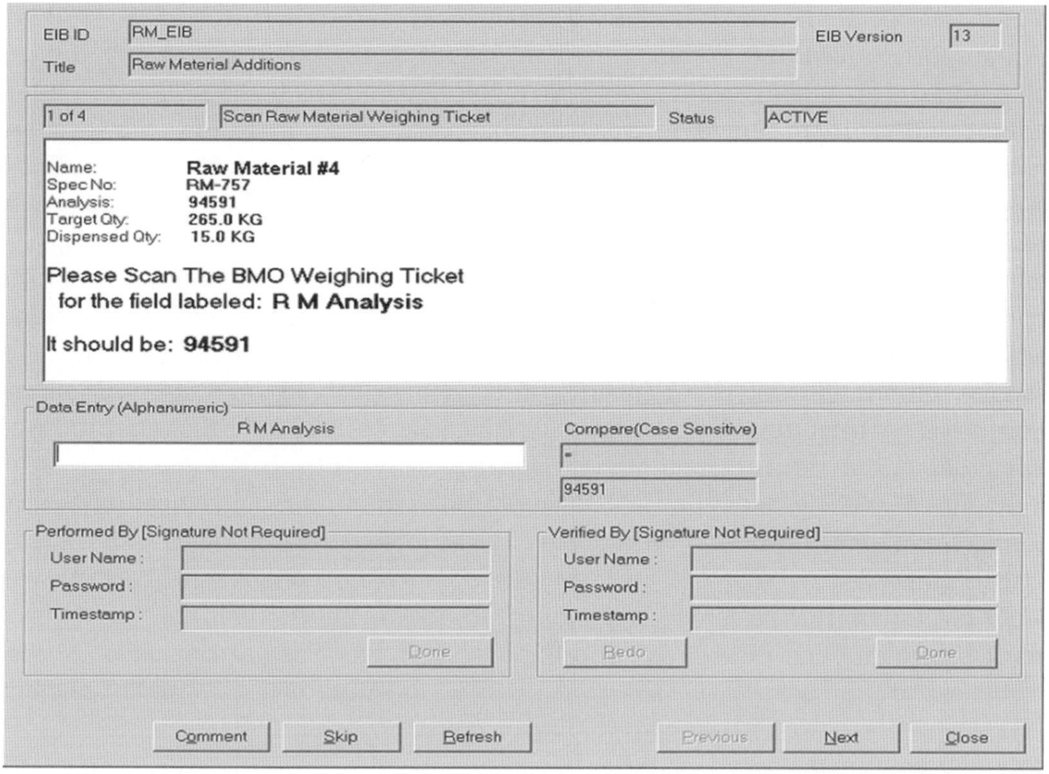

Figure 9.6. Work instruction.

rendered. This included a critical alarm summary, an ingredient and overall yield summary, and a raw material addition summary (Fig. 9.7).

The production reports were of interest to many different people in many different departments at the site. A Web-based reporting system was selected to satisfy this requirement. Selected reports were run against the database and rendered in the user's Web browser. Any user on the corporate-wide intranet can now access up-to-the-minute mouthwash production data.

Summary

By using innovative software development tools, well-established standards, and a sound development methodology, the MBS was implemented on time and within budget and immediately pressed into service to produce the new product. The development methodologies, tools, and system standards used were instrumental in maintaining the aggressive project schedule. This schedule accelerated

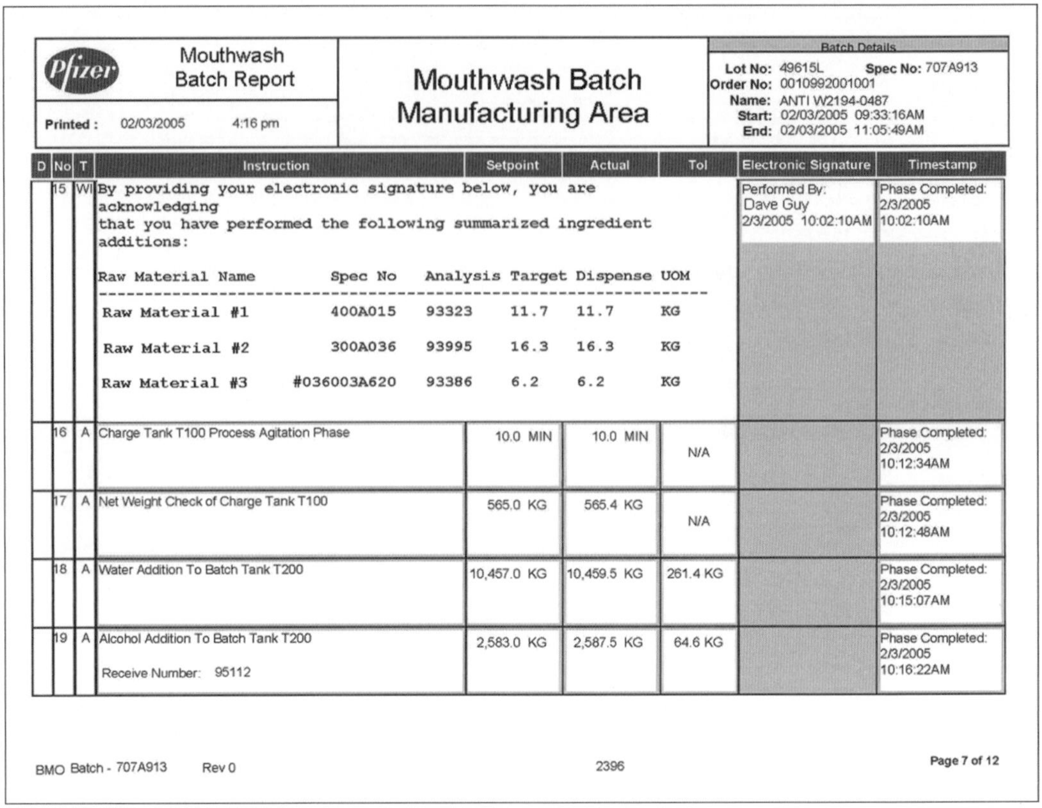

Mouthwash Batch Report	Mouthwash Batch Manufacturing Area	Batch Details
Printed : 02/03/2005 4:16 pm		Lot No: 49615L Spec No: 707A913 Order No: 0010992001001 Name: ANTI W2194-0487 Start: 02/03/2005 09:33:16AM End: 02/03/2005 11:05:49AM

D	No	T	Instruction	Setpoint	Actual	Tol	Electronic Signature	Timestamp
	15	WI	By providing your electronic signature below, you are acknowledging that you have performed the following summarized ingredient additions: Raw Material Name / Spec No / Analysis / Target / Dispense / UOM Raw Material #1 / 400A015 / 93323 / 11.7 / 11.7 / KG Raw Material #2 / 300A036 / 93995 / 16.3 / 16.3 / KG Raw Material #3 / #036003A620 / 93386 / 6.2 / 6.2 / KG				Performed By: Dave Guy 2/3/2005 10:02:10AM	Phase Completed: 2/3/2005 10:02:10AM
	16	A	Charge Tank T100 Process Agitation Phase	10.0 MIN	10.0 MIN	N/A		Phase Completed: 2/3/2005 10:12:34AM
	17	A	Net Weight Check of Charge Tank T100	565.0 KG	565.4 KG	N/A		Phase Completed: 2/3/2005 10:12:48AM
	18	A	Water Addition To Batch Tank T200	10,457.0 KG	10,459.5 KG	261.4 KG		Phase Completed: 2/3/2005 10:15:07AM
	19	A	Alcohol Addition To Batch Tank T200 Receive Number: 95112	2,583.0 KG	2,587.5 KG	64.6 KG		Phase Completed: 2/3/2005 10:16:22AM

BMO Batch - 707A913 Rev 0 2396 Page 7 of 12

Figure 9.7. Electronic batch record.

the time to market for the new mouthwash flavor. The accelerated time to market translated into real financial savings for the corporation. The new product launch date was not to be compromised. Otherwise, the corporation would have incurred added costs to utilize contract manufacturers if the site could not begin production to support the product launch.

The technologies used and system designs implemented provided a level of robustness required to support continuous production demands. Although minor operational enhancements have been implemented since validating the system, no significant control system modifications have been required. The new system used current, well-established technology that was known within the plant. As a result, minimum retraining was required, and a wide range of existing support options is now available.

The level of automation reduced labor requirements while decreasing the production cycle time, as well as the quality review and release cycle time. The use of specific technologies (e.g., bar code scanning) resulted in less operator intervention, thereby minimizing human error. Implementation of electronic signatures provided traceability and assurance of regulatory compliance. Additionally, electronic batch reporting eliminated time and effort required to create the required batch documentation. The design of the EBR aided in reducing the cycle time required to release the product for distribution and sale.

Finally, the controls were built on a solid foundation using a modular design that provided the flexibility needed to respond to changes with minimal impact on system design and validation. By building the system using ISA-88.01 principles, subsequent recipe additions and modifications have been made to the system with minimal reprogramming and revalidation efforts.

CHAPTER 10

Product Life-cycle Stages Linked Using ISA-88 and ISA-95

Presented at the WBF North American Conference, April 30–May 3, 2007, by

Baha Korkmaz
Regional Business Manager
baha.korkmaz@ips.invensys.com
Invensys Validation Technologies,
33 Commercial Street C41-2F,
Foxboro, MA 02035 USA

Arnold "Marty" Martin
Engineer
maretina@amgen.com
Amgen, 40 Technologies Way TB-1,
West Greenwich, RI 02817, USA

Abstract

Manufacturing industries are going through accelerated changes in their business models in order to be able to introduce products quicker and more effectively in the market. The typical manufacturing business model includes close integration of discovery, Research and Development (R&D), clinical process, manufacturing process, storage, and delivery processes. Integration of each stage of product life cycle can be harmonized by utilizing the ISA-88 standard. Once each process is internally controlled, the integration of each process to its surrounding processes is achieved via a combination of the ISA-88 and ISA-95 standards. This allows for a consistent set of models and terminology, optimizing and integrating the various product life-cycle stages.

What we see in the industry today is a change of priorities. Processing industries are realizing that product life-cycle management—including manufacturing automation—must have a high priority to achieve excellence both in business and operations. Technology is now available to achieve this excellence by using people to apply it to processes.

We started seeing global initiatives from all the major manufacturing corporations to integrate their systems and globally "close the loop." However, the focus of integration cannot be just Enterprise Resource Planning (ERP). We need to integrate the product life-cycle stages with each other, similar to the way ERP integrates business processes.

This chapter focuses on laying out a high-level functional model for Product Life-cycle Management (PLM) for ERP integration readiness by utilizing ISA-88 and ISA-95 standards.

Introduction

A new product progresses through a sequence of stages from introduction, growth, maturity, and decline. This sequence is known as the product life cycle. Knowing that the patent of a drug has a limited life span, it is desirable to maximize the maturity time of the product life cycle, compared to the introduction, growth, and the decline. Once the patent runs out and the generic manufacturing kicks in, the product may begin its decline. This is the time to introduce new drugs that can compensate for the loss during the decline phase.

PLM is the process of managing the entire life cycle of a product from its conception, through design and manufacture, to service and disposal. It is one of the four cornerstones of a corporation's information technology structure. All companies need to manage communications and information with their customers, using Customer Relationship Management (CRM), and with their suppliers, using Supply Chain Management (SCM), and the resources within the enterprise using ERP. In addition, manufacturing engineering companies must also develop, describe, manage, and communicate information about their products (via PLM). We can also consider the process of manufacturing the product and the process of project execution for manufacturing systems to be directly or indirectly integrated parts of the PLM concept.

Background

ISA-88 batch control standards have been applied successfully since the mid-nineties; they became an accepted set of models and terminology for all batch process control and automation projects. These applications are often limited to the process cell domain and mostly deployed at commercial manufacturing facilities and pilot plants.

Imagine that a large pharmaceutical corporation has hundreds of process cells that are mostly automated under the ISA-88 guidelines. Most likely, multiple vendor systems are being used to automate the process cells. In the meantime, the IT organizations deployed ERP, CRM, and PLM systems to automate most of the business activities. IT and automation organizations within the corporation have been aligning the models and terminology they use as their organizations evolve, thanks to ISA-88 and ISA-95 standardization initiatives. The gap between the IT world and the control world is closing. It is now time to close the loop between the layers of implementations and the stages of the product life cycle. It is also time to utilize ISA-88 beyond the process cell and introduce it to R&D, clinical trials, maintenance, upgrades, retirement, and PLM. When we begin using ISA-88 models and terminology beyond the process cell, ISA-95 standards become a partner to the ISA-88 standards.

Imagine that R&D and Manufacturing are so harmonized that the output of R&D becomes the input to the production process. ISA-88 is applied throughout the PLM phases to such an extent that even Process Analytical Technology (PAT) initiatives for product quality and release efficiency, Laboratory Information Management Systems (LIMS) functions, and all production stages are harmonized and performing their jobs successfully. This approach also minimizes the non-value-add time for the projects and production, aiming for true business and operational excellence. This chapter attempts to highlight the importance of ISA-88 standards for PLM.

What is Product Life-cycle Management?

PLM is the process of managing the entire life cycle of a product from its conception, through design and manufacture, to service and disposal. Documented benefits include the following:

- Reduced time to market
- Improved product quality
- Reduced transition costs between PLM stages (i.e., phases)
- Savings through the reuse of original data
- A framework for product optimization
- Reduced waste

- Savings through the complete integration of engineering workflows
- Improved revenue and profit cycle prior to patent expiration

PLM is not just about the creation and central management of all product data and technology used to access the information and knowledge, but also entails all processes that take place from the birth of the product to it's retirement. It has a large scope.

PLM should not be seen as a single software product but as a collection of software tools and working methods integrated together to address single stages of the life cycle, connect different tasks and stages, or manage the whole process. Figure 10.1 shows the high-level stages of a life science industry product life cycle.

Product life cycle may also be seen from a product-introduction and sales point of view. These stages are listed as follows:

1. Introduction
2. Growth
3. Maturity
4. Decline

For example, when a new drug is discovered in the life science industry, the corporation applies for a patent. While the patent approval process is active,

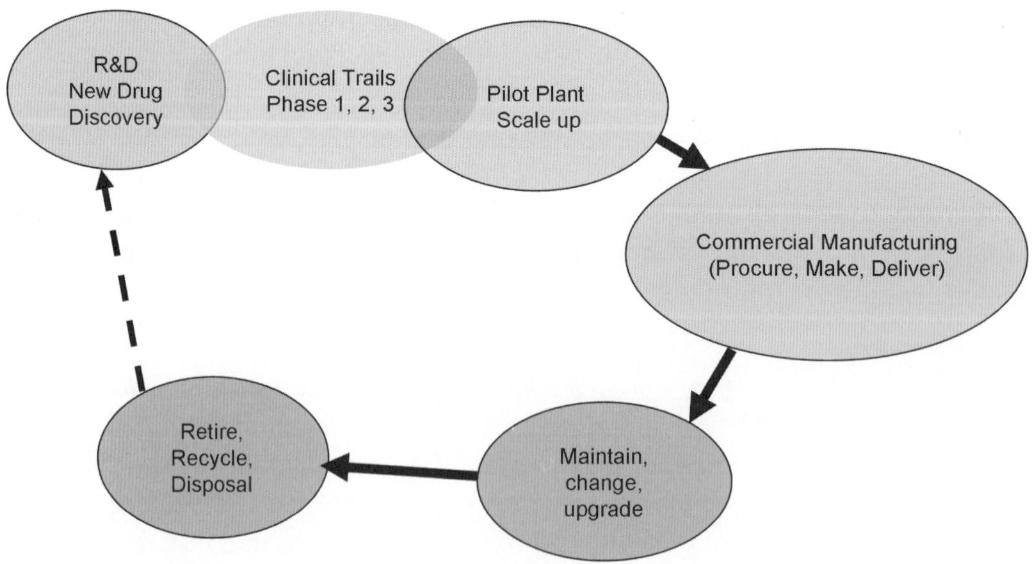

Figure 10.1. High-level product life cycle.

the introduction of the new drug is planned and executed. The Food and Drug Administration (FDA) approval cycle is activated and clinical trial phases are initiated. Pharmaceutical or biotechnology businesses have plenty of incentives to decrease the introduction time and maximize the growth and maturity time for the product. Once the patent is approved, the product must be in the market as soon as possible. Product sales are at their maximum level when the maturity stage is reached during the life cycle. The objective is to get to the maturity level as soon as possible and have the maturity level continue for as long as possible. When the patent expires, the product sales begin dropping. If it was a blockbuster drug, then it is very likely that other life science companies introduced similar drugs to the market in the meantime. These competitive pressures encourage the life science corporations to invest in PLM with ISA-88 guidelines, Manufacturing Execution System (MES) for electronic work instructions, site-specific automation above the process cell–specific automation, and other ERP initiatives.

All life science industry companies are under increased pressure to introduce new drugs to the market more frequently then ever. The companies that proactively approach the new technological trends and invest in automation, in PLM with ISA-88, and in MES and ERP integration with ISA-95 have increased their chances for survival and success.

PLM and Recipe Management

General and site recipe standard ISA-88.03 provides the best practice for the creation and maintenance of recipes. Process development and, consequently, the pilot plant process will determine the General Recipes (GRs) for products. GRs define the information required to manufacture a product independently from the equipment details of the manufacturing facility. Figure 10.2 shows the recipe life-cycle phases as part of the PLM.

Phase 1: Process Description

The process development stage creates the cell lines and purification process. After developing the process descriptions, a GR can be created with the information collected during the process development stage. It is important that R&D engineering personnel and scientists are trained in ISA-88.01 and ISA-88.03 standards. It is very desirable to create and maintain a common set of process models and terminology that is not only understood by automation engineering, but also by plant engineering, process engineering, and R&D. The knowledge transfer should occur from the process description to the GR for each new product.

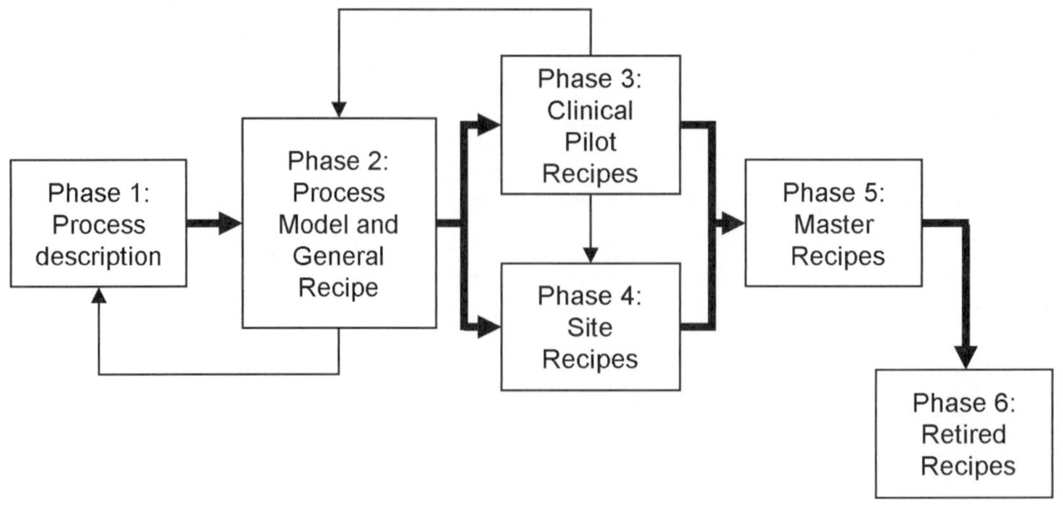

Figure 10.2. Recipe life-cycle phases.

Phase 2: Process Model and GR

A combination of graphical and textual process descriptions (e.g., procedural function chart) should be used to document all recipe hierarchy and procedural elements. Chemical, biochemical, and physical processing are described with the required bill of material and equipment requirements. Since there is no plant at this stage, the equipment specifications are generic and geared toward describing how the product is made without worrying about how the plant is laid out. At this stage, you don't care about the plant layout; you care more about the formulations, scale-up requirements, and so on.

A GR may be one large process description in which certain stages can be carried out in different locations and different buildings (i.e., process cells). This means that the transformation of the recipe from general to site and from site to master will potentially create many master recipes. For example, it is possible that fermentation and recovery take place at one site, while filling and packaging take place at another site.

This scenario will require at least four master recipes: Fermentation, Recovery, Filling, and Packaging. In reality, you may have many master recipes for one batch. PLM will drastically help to transfer knowledge that maintains the GRs from R&D to the pilot plant and manufacturing operations by utilizing the ISA-88.01 set of terminology and the process model.

Regardless of how small and how manual the R&D operations are, the ISA-88 concepts may be applied to several stages of R&D activities. The process modeling concept, equipment modeling concept, and reduced recipe concept (i.e., the separation of recipe and equipment procedures) would be useful. The ISA-88 model applies to any particularly sized process and lets you create any size recipe.

Phase 3: Clinical Phase and Pilot Recipes

There are multiple clinical trial phases while the new drug is being investigated by the FDA for approval. During the initial clinical phases, conversion of the GR to a pilot plant recipe would be required. This recipe will need some adjustments based on the tests. A pilot plant recipe is developed that is specific to the pilot plant equipment specifications. It is desirable to have the pilot plant automated similarly to the commercial production plants for the corporation. That makes it easier to transfer the recipe components from the pilot plant to the master recipes of the commercial facility. The GR and the findings from the pilot plant runs will provide sufficient input to develop the site recipes for each site. The pilot plant is the necessary platform for scaling up the laboratory process to commercial size.

Phase 4: Site Recipes

The site recipe is similar to the GR with the addition of site-specific information such as loading, storage, packaging, and distribution requirements; site material specifications; local language; local units of measure; and process cell details.

There should be one site recipe per site per product. It is also possible that one product requires multiple sites to be produced. For example, fermentation and recovery process cells could be in one site, and filling and packing and labeling could be in another site. For this example, there will be two site recipes. It is desirable to have an ERP system involved to manage the site recipes, since an MES is typically installed per site and will not be able to handle the total production life cycle unless it is integrated with the ERP system.

Figure 10.3 shows the recipe hierarchy for multiple sites for one product example. Site recipes may be maintained under the ERP system domain with a specification management system. A copy of the site recipe gets transferred to the site and its MES. The MES then creates and maintains the master recipes for each product.

Phase 5: Master Recipes

General and site recipes provide the equipment-independent version of the process as "product specifications" and "process descriptions." The specific equipment definitions for each product recipe are identified as part of the master recipe. Master recipes are not only product specific but also equipment specific. They are

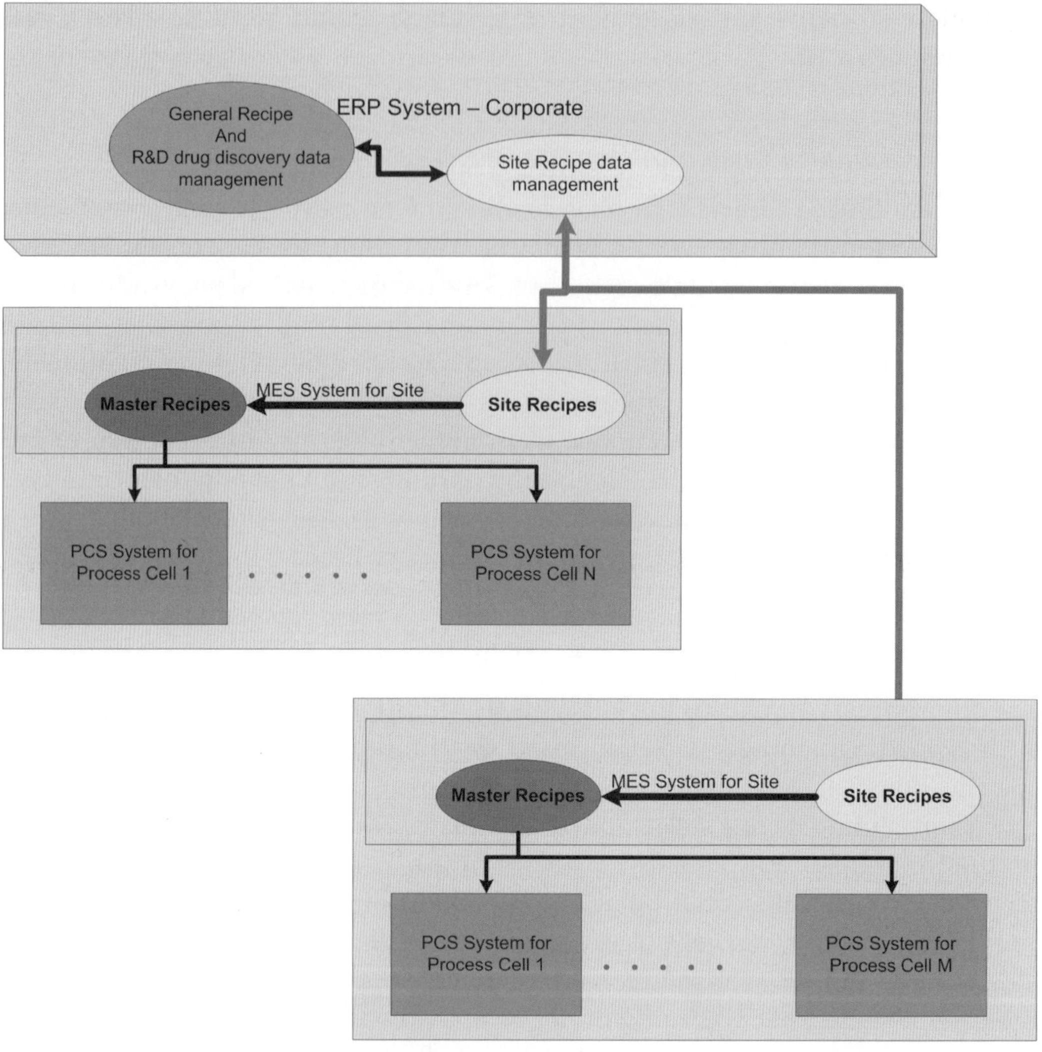

Figure 10.3. Recipe hierarchy for multiple sites.

reviewed and qualified for a specific process cell with specific paths (e.g., piping, Equipment Modules [EM], transfer panels).

It is normal that there are many master recipes within a site to produce the same product. Production scheduling and equipment resource management can be handled more flexibly if master recipes exist for certain units (e.g., a Clean-In-Place [CIP] skid and its lines, buffer prep units, media prep units).

A MES system can then manage the ISA-88 level process and provide the unit supervision to optimize the batch process. Master recipes may also reside in a Process Control System (PCS) or a combination of MES and PCS. MES and PCS integration with PAT, LIMS, and a deviation management system (e.g., Corrective And Preventive Action [CAPA]) will support the batch review and release process. The objective is to release the batch shortly after the batch is complete.

Phase 6: Retired Recipes

When the product has passed its maturity and decline phases, corporate will eventually decide to retire the product. Retiring the product may require the following actions as a minimum:

- If the manufacturing plant was designed for a single product, then it may be redesigned to become a multiproduct facility to increase efficiency and flexibility to meet the market demands for products.
- Store the retired general, site, and master recipes for the retired product. It is important to be able to reuse some parts of the recipes for a future product. Keep the reusability and libraries in mind.

It is also possible to retire some master recipes, even if the product is not retired. This option will happen if there is a significant change to the process, and the master recipes then retire the old versions and begin using the new versions.

Manufacturing System Automation as Part of the PLM

The life science industries begin losing market share and revenue when the FDA approvals process begins. The faster the new drug gets introduced into the market, the faster the patient care will begin, and the faster the revenue stream will flow. There are plenty of improvement areas for operational excellence to optimize PLM for a drug product. Several improvement tips are as follows:

- Optimize the knowledge transfer from R&D to Manufacturing by using ISA-88 based data and knowledge transfer systems.
- Minimize non-value-add processes (e.g., manual transfers, intermediate inventory storage areas, human delays, warehouse and shipping).
- Automate the manufacturing process wherever it makes sense. Completely paperless manufacturing should be a stretch goal.
- Align process, technology, and people for every discipline within the enterprise.

- Implement PAT technology along with ISA-88 modeling to improve quality and batch release cycles.
- Integrate LIMS into PCS and MES systems to minimize the inefficiencies of multiple entries and transportation times. Optimize the work instructions, including the LIMS activities.
- Implement 21 CFR Part 11 for electronic signatures.
- Implement a total Electronic Batch Record (EBR) handling system.
- Integrate ERP, MES, and PCS systems by utilizing ISA-95 standards:
    - PLM
    - Specification management (e.g., general, site, and master recipes)
    - Resource management (e.g., equipment, material, people)
    - Production planning and scheduling
    - Quality Assurance (QA), EBR, and batch release processes
    - CAPA
    - SCM
- Implement predictive maintenance to the equipment.
- Implement production scheduling and equipment resource management.

Manufacturing automation is one of the most critical areas. Corporations started to invest more in manufacturing automation to close the gaps between ERP and other business systems and manufacturing systems. ERP, MES, and PCS must be integrated to get the most return possible during the growth and maturity phases of the product life cycle. It is even important during the decline and retirement to optimize the return on investment. The following section will address project life-cycle optimization by utilizing ISA-88 standards.

ISA-88 Based Automation Project Planning and Execution

How can ISA-88 help minimize the project duration so that the company can start helping the patients and having return on investments as soon as possible? At the same time, it is essential to maintain high quality during the project life cycle.

Pre-project activities play a key role in setting up the right specifications for the design and construction including optimum automation of the asset. Utilizing industry-wide standards and standard methodology helps to shorten this pre-project cycle. While pre-project and project activities utilize Good Automation Manufacturing Practice (GAMP) guidelines for quality and proper deliverables of each life-cycle milestone, ISA-88 models and terminology should be utilized for the optimum automation components. For the automation components, this cycle contains the following activities:

- Develop a project plan and milestones:
    - Document the project plan, including the project phases and deliverables for each phase.
    - Document the Validation Master Plan (VMP), including all validation deliverables.
- Identify the schedule:
    - Develop the schedule based on the project plan and scope definition.
    - Develop the validation schedule based on the VMP.
- Develop the right scope:
    - Document the User Requirements Specifications (URS) based on GAMP.
    - Document Functional Requirements Specifications (FRS) based on GAMP.
- Develop the site recipes:
    - Develop current state process maps (if it is an automation project for an existing facility).
    - Develop future state process maps.
    - If the GR is documented, use it with the site specific information.
- Evaluate and select a vendor:
    - Arrange prototyping activities to evaluate the vendor's capabilities (i.e., products and services).
    - Document each vendor company's capabilities and financial health.
    - Create a vendor assessment matrix for comparison.
    - Perform vendor audits.

The team is expected to understand and apply the ISA-88 based modular batch control standards and modular manufacturing principles. ISA-88 models also help organize the project team so that the project activities enforce the utilization of ISA-88 throughout the project. Following these principles will ensure efficiency, quality, and subsequently, the success of the project.

Once the project is kicked off, it can be executed in parallel and phased fashion. The ISA-88 modularity must be closely followed. In addition to the project cost, the total automation life-cycle cost must be taken into account. Figure 10.4 shows that system purchase, order entry, manufacturing, testing, and shipment take place in parallel with the application engineering effort. There must be multiple teams formed for the project, with clear deliverables and milestones. An MES project team and a PCS project team may be separate. It is important to separate each team into a process-specific team and an equipment-specific team. The process-specific team is responsible for developing process maps that show recipe components, including recipe phases and master recipes, for each product. The equipment-specific team will be responsible for developing all control-system-related components, such as the following:

- Input/Output (I/O) logic
- Interlocking
- Control Module (CM) classes and CMs
- EM classes and EMs
- Equipment phase classes and equipment phases
- Unit classes and units with unit supervision

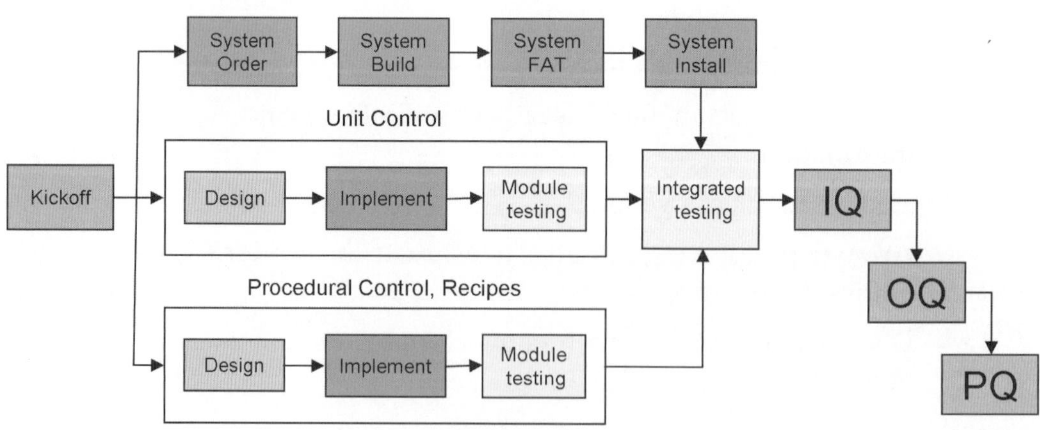

Figure 10.4. Testing.

In addition, the equipment-specific team is responsible for other displays, databases, tags, and so on that are related to the equipment-specific applications.

Each team will be responsible for testing their modules and developing test documents. Once each module class is tested and approved, the library elements are created and may be used to populate the rest of the implementation. The testing of each instance will be simpler, since the class-based application is already tested and approved. There is no need to test the functionality each time if it is a copy of a class, but the inputs and outputs to the module still need to be tested and verified in each instance.

The integration testing should take place at the installation site. Based on a GAMP risk-based approach, it is possible to combine the Installation Qualification (IQ) and Operation Qualification (OQ) protocols and execution in order to speed up the validation process. Process Qualification (PQ) protocols and execution will then verify the functions that are documented in the URS. It is very important that both GAMP and ISA-88 are utilized throughout the life cycle of the project.

Table 10.1 shows the suggested team structure for large-scale automation projects that may or may not include MES components. This is a matrix representation of the project team. It is possible that the same person will be assigned to more than one team. Team matrices are created to maximize the project efficiency and utilize the best Subject Matter Experts (SMEs) and process engineers available for the project. This example shows that control system SMEs are leading the control design, implementation, testing, and approval for each process cell, while process-knowledgeable project engineers are leading the process cell integrated solutions such as Seed Train, Inoculum, Production, and Harvest. This structure ensures the clear definition of accountabilities and responsibilities. This also follows the phased approach to minimize wait times. For example, as soon as fermentation unit class work is finished, it can be tested and approved while the harvest unit class work is still in progress. The columns on the table show the various tasks. This can be expanded to include items such as EMs, unit control, equipment phases, interlocking, CMs, display development, master recipes, and so on. The content of the columns and rows depends on the project plan and scope definition.

This type of project team structure also allows project management to identify the cost versus revenue data on a per process cell, per unit class, or per task basis. It also helps to manage the project schedule. Overall, there is a great chance to finish the project on time, within budget, and with satisfactory quality.

Table 10.1. Project teams				
	Control	*Interlock*	*Human Interface*	*Recipe*
Process cell 1 Seed Train	Team a	Team e	Team i	Team m
Process cell 2 Inoculum	Team b	Team f	Team j	Team n
Process cell 3 Production	Team c	Team g	Team k	Team o
Process cell 4 Harvest	Team d	Team h	Team l	Team p

PAT and ISA-88

PAT is promoted as a risk-based framework for innovative pharmaceutical development, manufacturing, and QA, while maintaining and improving the current level of product QA.

The technology today allows the life science industry to incorporate PAT into the drug discovery and scale-up process, manufacturing process, and QA. Applying ISA-88 to this new path improves the efficiency and quality within the product life cycle. The new drug discovery process and the type of drugs that are being discovered today require cutting-edge scientific and engineering knowledge. There are at least two ways of applying ISA-88 to PAT. One is that the PAT process can use the ISA-88 modularity, models, and terminology to start, execute, end, and report the PAT results. The other way is the integration of PAT with the master recipes, control recipes, and advanced process control that is supervised by ISA-88 based unit supervision.

The FDA guidance report on this subject states, "Effective use of the most current pharmaceutical science and engineering principles and knowledge—throughout the life cycle of a product—can improve the efficiencies of both the manufacturing and regulatory processes." PAT is considered to be a process for designing, analyzing, and controlling manufacturing through real-time measurements of critical quality and performance attributes of materials and processes, with the objective of product quality. Since PAT is an integrated (even embedded) part of process control, it can also be handled and managed by using the ISA-88 standard. An ISA-88 recipe can contain PAT elements, such as formula, algorithms, and so on.

Other control activity model elements are utilized for PAT (e.g., production information management, process management [executing PAT-specific recipes], unit supervision, and process control). The ISA-88 architecture and modular approach allows building the quality into products via PAT. This approach supports the understanding and use of the relevant multifactorial relationships among material, manufacturing process, and environmental variables and their effects on quality. The data and information to help understand these relationships are obtained through preformulation programs, development, scale-up studies, and data collected over the life cycle of the product. Again, utilization of ISA-88 is essential to organize and execute all these activities, including the data acquisition, calculation, and reporting.

Embedding PAT into ISA-88 components will help increase quality, safety, and efficiency. This is likely to come from the following:

- Reducing production cycle times by using on-, in-, and at-line measurements and controls, where possible
- Preventing rejects and reprocessing
- Considering the possibility of real-time release
- Increasing automation
- Facilitating continuous processing to improve efficiency and manage variability

Conclusion

Applying ISA-88 to the entire product life cycle is not a new concept. There have been multiple papers written on this subject since the late nineties. This concept is being implemented by a few large life science corporations. The success of this approach will depend on the following:

- Proper training on ISA-88 and ISA-95 is absolute necessity for the success of this approach.
- Strong commitment to the standards from users and vendors is required.
- It is a business process that requires both people and technology to be committed to the process for success.

Further Reading

Instrumentation, Systems, and Automation Society. 1995. *ANSI/ISA-ISA-88.01-1995: Batch control part 1: Models and terminology*. Research Triangle Park, NC: ISA.

Korkmaz, Baha, and Velumani A. Pillai. 2000. Automation's evolving role in the pharmaceutical industry. *Pharmaceutical Engineering* (January / February).

U.S. Food and Drug Administration. 2004. *Guidance for industry: PAT—A framework for innovative pharmaceutical manufacturing and quality assurance*. Washington, DC: Government Printing Office.

Wikipedia. 2010. Product life cycle management. http: / / en.wikipedia.org / wiki / Product_lifecycle_management.

CHAPTER 11

Jazz Up Your Batch Projects

Presented at the WBF North American Conference, March 5–8, 2006, by

Frede Vinther
Senior Consultant
frha@nne.biz
Automation Solution Architects, NNE A/S,
Gladsaxevej 372, DK-2860 Soeborg, Denmark

Abstract

Enabling pharmaceutical companies to make investment decisions as closely as possible to the "product on market date" is essential for their business. This is not merely achieved by having a set of standard software modules available; the whole engineering and project execution process has to support this acceleration, from concept to operation.

One of the large time consumers in any project is communication. There are basically three stakeholders in any given project: the pharmaceutical company (i.e., the client), the Architect and Engineering (A&E) company, and the System Integrator (SI) company. Here, it is very important to focus on the contents of the communicated information and, especially, what information on which level is important to each stakeholder.

Another time consumer is lack of a functional breakdown of the actual process to be engineered and put in operation. This problem can be solved by applying standards such as ISA-88 and ISA-95 and by applying modular design and reuse. However, this does not solve all functional issues.

This chapter specifically addresses these two issues by demonstrating how essential well-executed, front-end engineering and planning are to the overall project efficiency. It describes the efficiency that can be achieved by using tools that support object orientation, combined with a set of standard functions.

Introduction

This chapter describes an approach that will really jazz up your batch projects, especially if your industry is pharmaceutical and your concerns are time to market, standardization, and validation.

The approach has a two-fold focus—humans and technology. First, we address the human factor by describing the main players involved in execution of a project, identifying their objectives, and exploring how to manage the differences between these. Secondly, we focus on the technology that will support the management of the human factor.

Background

What similarities do you see when you compare the plants of a large pharmaceutical company? What similarities do you see when you compare the plants of a large pharmaceutical company that are used to produce the same product?

More often than not, the answer is merely "the same company logo."

The thing about pharmaceutical plants is that they all have their own local history that also quite often involves a list of previous owners. This is the main reason why one specific pharmaceutical company may have a number of plants producing the same or very similar products, yet each plant is built according to its own local concepts and principles.

These concepts and principles not only cover the steel structures, machinery, and automation systems, but also all the Standard Operating Procedures (SOPs), Quality Assurance (QA) management, personnel training programs, and so on. If we were to compare the results of all these different processes aimed at the same purpose to, say, a car manufacturer, it would probably look like Figure 11.1.

Both cars depicted in Figure 11.1 fulfill the basic functionality required by a car—transportation, shield from weather, steering wheel, brakes, head lights, and so on. Yet they are obviously very different. The need for better coordination and consistency is crucial to the process.

The same applies to the pharmaceutical plants. They are also "handmade" according to specifications, and each and every plant is almost "one of a kind."

OR

Figure 11.1. It is obvious that these two cars are different, yet they are both "handmade" according to specifications.

This method of engineering and building pharmaceutical plants is neither cost and time effective with respect to producing the plant nor effective to maintain a number of such plants.

Planning a Project

When preparing and planning for a new project (whether it is a plant rebuild, plant addition, or new plant), it is very important to approach the task with the previously described history of "handmade" plants in mind. All involved players in the project will have their own "history" of building plants, concepts, and so on. Each player has specific areas of interest and focus. It is important to use this knowledge from "history." In order to achieve efficient and fast project execution, it is essential to focus on the project activities to be carried out to complete the project (including the scope and content of each of these activities and deliverables) and design the plant accordingly. Figures 11.2 and 11.3 illustrate the importance of Conceptual Design (CD) and Basic Design (BD) to the cost of a project.

Work done with pharmaceutical clients yields the points shown in Figure 11.4, which need to be addressed during the "front-end definition." The planning focus for a project must address the front-end definition engineering to ensure that the investments committed are the correct ones and that they address the overall project objectives, as shown in Figure 11.4.

Planning in Praxis

Based on a Project Activity Model (PAM) that covers all the phases of the project execution, the following is a highly productive method for planning and defining the document register for the entire project, as well as the planned completion of each of these documents (Fig. 11.5):

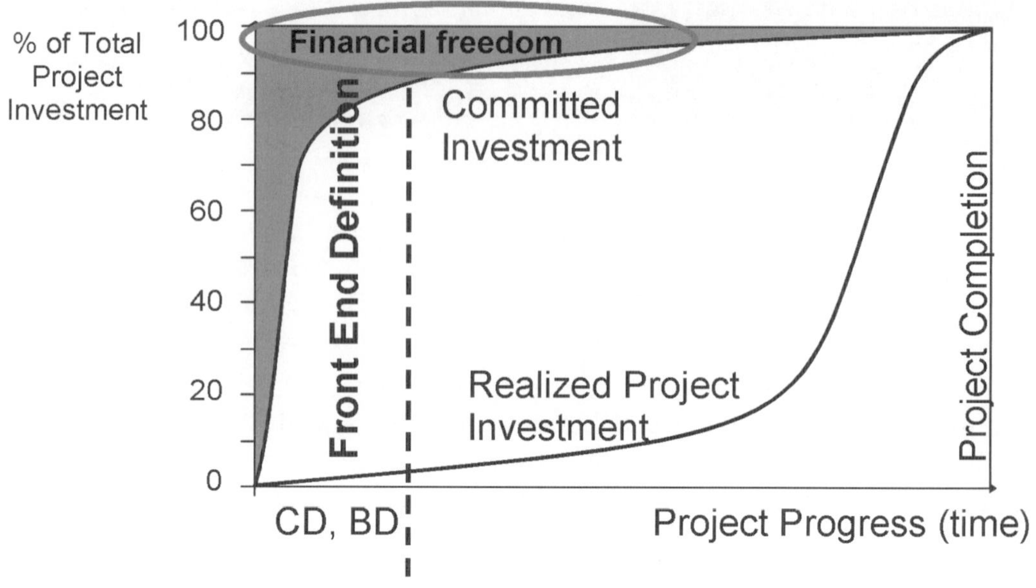

Figure 11.2. The early decisions have the most impact on the investment.

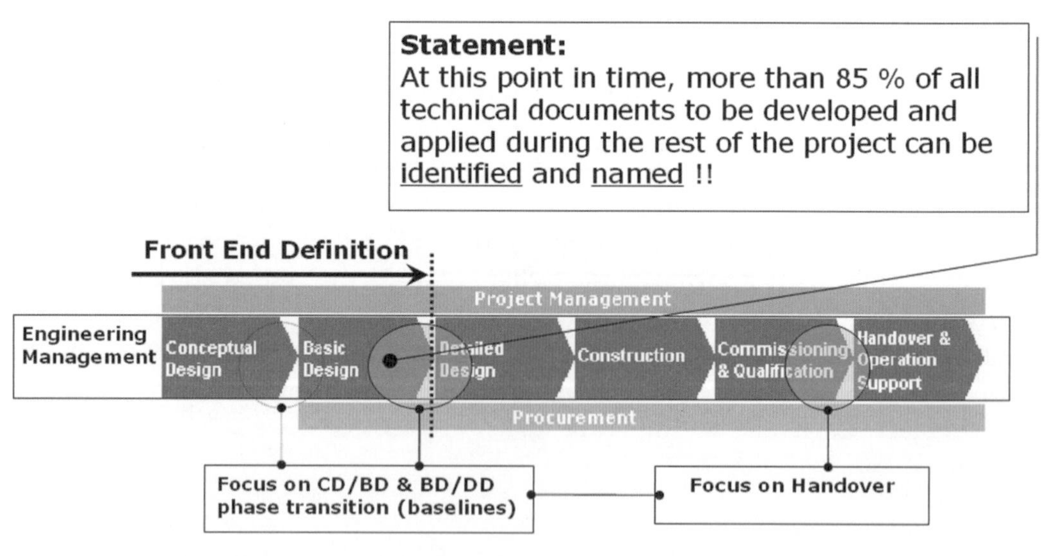

Figure 11.3. Front-end definition locks down 85% of the design.

Not defining clear objectives is the most frequent cause of project failures.

Thus, defining relevant objectives is fundamental for establishing a project.

High quality objectives are defined by "SMART":

- **S**pecific
- **M**easurable
- **A**ccepted
- **R**ealistic, but ambitious
- **T**ime framed

Clear objectives from the beginning of the project create an unambiguous understanding in the project team about the goals of the project, and just as important it becomes easier to inform stakeholders about what the project is all about.

Figure 11.4. Objectives for defining objectives.

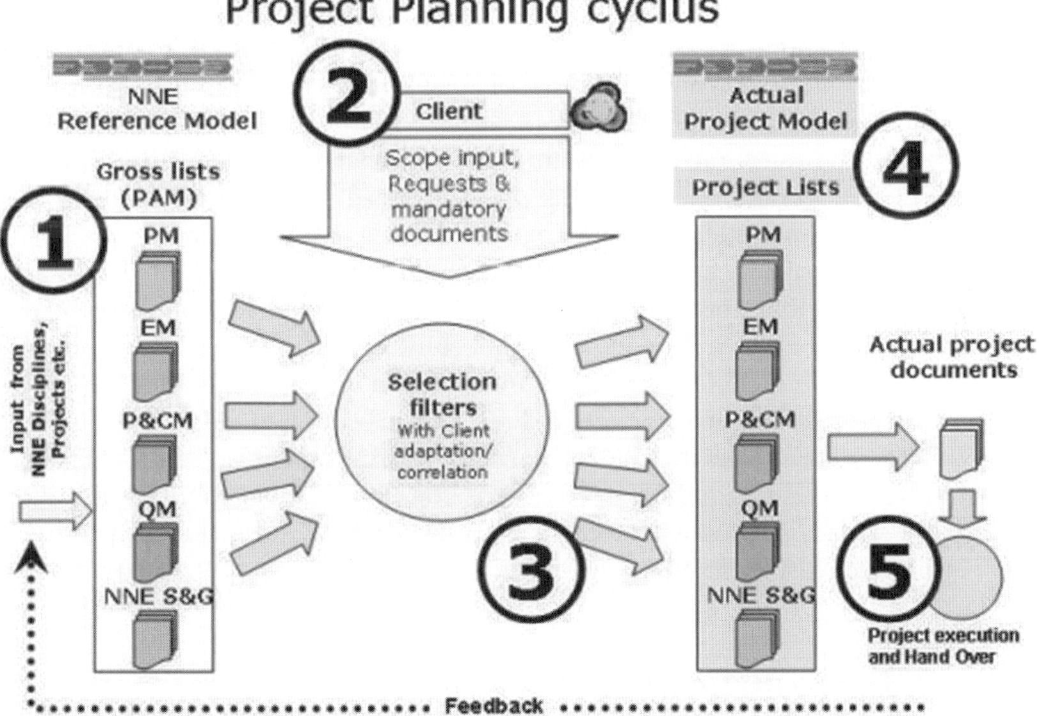

Figure 11.5. Map of product planning, with a feedback loop.

1. Start with the A&E reference model (PAM including planning tools with gross lists, etc.).
2. Perform an A&E scope of supply selection activities (selection data and required technology).
3. Adapt and correlate client input and technology requirements to A&E reference model items.
4. Define and document the actual project model, which includes identifying activities that differ from the A&E reference model.
5. Execute the activities and produce actual project documents.

In the workflow, the starting point is the reference model (see the upper left in Fig. 11.5), which contains a listing of possible documents and includes a link to their content (e.g., standardized documents and document types, based on feedback from executed projects). Combined with the client scope inputs and requests, a subset of documents is selected for the actual project.

The actual selection of documents for the specific project is made from a gross list of documents in a Project Life-Cycle Document (PLCD), maintained in a spreadsheet. Figure 11.6 is an example of such a list, where a number of project selections have been made already.

When all the selections have been made together with the client, a project-specific PLCD is generated. From this point forward, the project-specific PLCD will have its own life cycle, decoupled from the gross list document.

The PLCD is not only for selecting the document register, it also contains rows for software—both standard packages and the application software to be implemented. It is also used to plan for when the individual documents and items must be approved (shown as "App" in the PAM phases columns). It is a very good tool for the project manager to use to follow up on progress during the execution of the project.

With the project-specific PLCD in place during the front-end definition, almost all technical documents are defined. The documents describing how the QA is to take place and which protocols and test plans will have to be produced and executed are also defined. Now a plan is in place.

Scope of the Functional Specification

With the PLCD in place for the project, the number of documents to produce is in place but not the content of each type of document. This section focuses on the Functional Specification (FS) scope and content. Figure 11.7 illustrates the

Document Categories	Deliverables	Client deliverables	Project Selection	Owner	Format	Unique/ Multiple	Discipline	PAM Phases CD	BD	DD	Con.	C&Q	HO
	Basis of Project Decision												
	Manufacturing Objectives & Owner Philosophies												
	Project Objectives			Client	Report	U	PM	App					
	Reliability Philosophy			Client	Report	U	PM	App					
	Maintenance Strategy			Client	Report	U	PM	App					
	Operating Strategy			Client	Report	U	Process	App					
	Block Layout			A&E	Drawing	U	PM	App					
	Project Strategy			Client	Report	M	PM	App					
	Logistic Strategy			Client	Report	M	PM	App					
	Automation and IT Concept			A&E	Report	U	Automation	App					
	Production-/Process Concept			Client	Report	U	Process	App					
	Technology Concept			Client	Report	U	PM	App					
	Front End Definition, FED (CD,BD)												
	Process- / Building Interdisciplinary												
General													
	Specifications & Guidelines, see PAM-01485		yes	A&E	Specification	M	EM	App	App				
	Interface specification		yes	A&E	Specification	M	EM		App				
	Site Design Data			A&E	Report	U	EM	App	App				
Overview Documents													
	One page strategy, Project			A&E	Drawing	U	EM	App					
	Process Module Diagram (PMD)		yes	A&E	Drawing	U	Process	App	App				
	Building Module Diagram (BMD)			A&E	Drawing	U	Architect	App	App				
Requirement Specifications													
	Overall User Requirement Specification (OURS)			Client	Specification	U	Client	App					
	User Requirement Specification (URS)		yes	Client	Specification	M	Client		App				
	Installation Requirement Specification (IRS)		yes	Client	Specification	U	EM		App				
Design Specifications													
	Main Document list		yes	A&E	List	U	EM	App	App				
	Interface Description			A&E	Specification	M	EM		App				
	Overall Operational Flow Diagram (OOFD)			A&E	Drawing	U	Process	App					
	Operational Flow Diagram (OFD)			A&E	Drawing	M	Process	App					
	Equipment Layout		yes	A&E	Drawing	M	Mechanical		App	App			
	Plotplan		yes	A&E	Drawing	M	Mechanical		App	App			
	Overall Distribution Plan			A&E	Specification	U	Mechanical		App	App			
	Overall Pipe Routing Plan		yes	A&E	Diagram	M	Mechanical		App				

Figure 11.6. Sample of a PLCD.

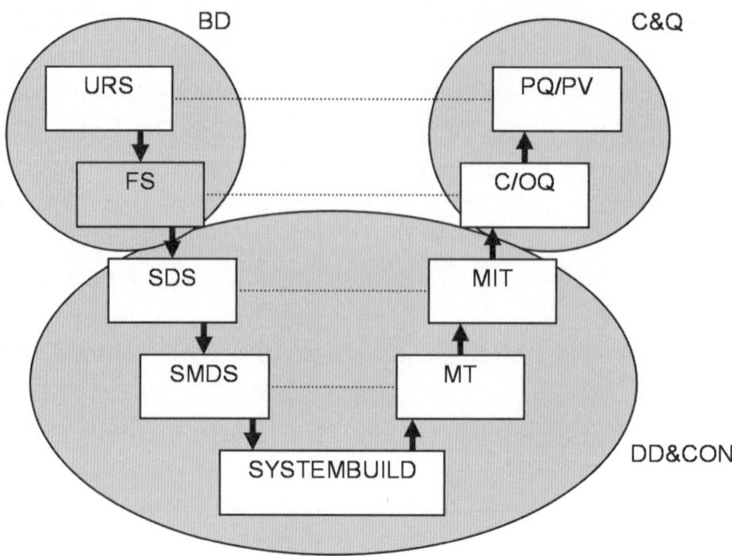

Figure 11.7.

position of the FS when put into the GAMP 4 V-Model frames with PAM activities identified.

The defining characteristic of the V-Model is that it is based on the physical plant to be built. The specification starts from the User Requirement Specification (URS) and is detailed more and more, down to the Software Module Design Specification (SMDS) and lower levels. The testing starts from the detailed level by a module test, followed by the module integration test, up to commissioning and qualification.

For automation, the FS is the main specification document among the client, A&E, and the SI companies. Therefore, special attention must be taken when defining the scope of this document. Normally, the FS is written and owned by the A&E company.

Before going into details about the content of the FS, it is important to realize what must be covered by the FS. Figure 11.8 illustrates the breakdown of the computer-controlled production process.

The FS must be grouped according to this breakdown. In other words, there must be a specific FS for the "automation system," covering items that are generic and plant-wide in nature, such as network, computers, operator panels, backups, and other overall system functions (e.g., electronic signatures, batch control capabilities, standard system software, standardized application software).

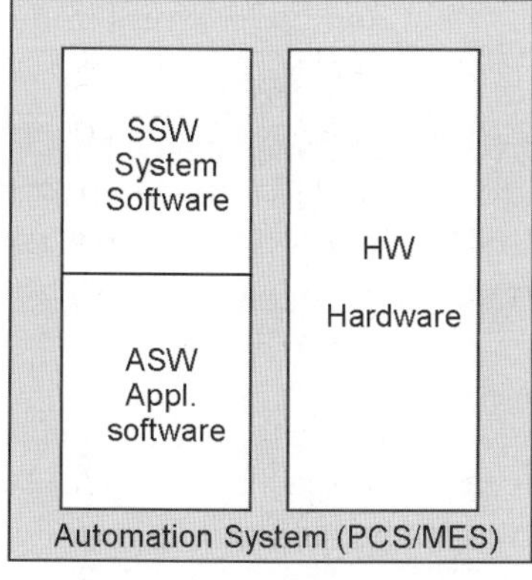

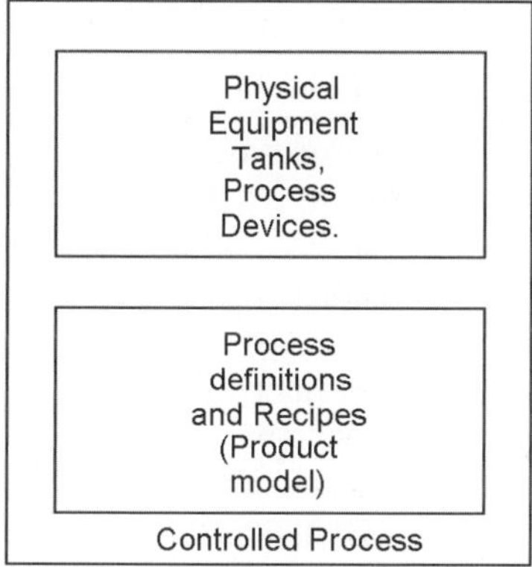

Figure 11.8. Total computer-controlled production process.

There must also be specific FS for the "controlled process," covering the equipment structuring and breakdown into units and EMs, the process functions of the units and EMs, and the functions to run the process (e.g., the recipe procedures).

In the URS, it is also important to identify the requirements that are not related to the production process itself but are instead related solely to the scope of delivery of equipment (e.g., count of free Input/Output [I/O] points for expansion, specific required field instruments). These are not to be validated at all, but only to be verified by the client to secure the actual delivery. Normally, these requirements are covered by the FS for the "automation system."

FS Content and Communication

Table 11.1 contains a number of statements about what the FS must cover and be used for, as seen from each stakeholder's viewpoint. It is obvious that the stakeholders have very different viewpoints on the content and use of the FS and that the FS must address a wide range of issues that it must tie together in the SMART way.

It is important for the FS to be written with process terminology in a clear, simple, and structured way, in order to tie the actual production process together with the functionality of the plant and the humans needed to operate the plant. This way, the FS will only address the functionality of the process defined by a terminology that is clear to the client. As such, it is a "generic FS," without the details of the automation system, that is eventually used for implementation. This allows automation system vendor selection to be postponed until after the approval of the FS.

Table 11.1. FS specifications			
Stakeholder	*Content*	*Uses*	*Concerns*
Client	Define process, equipment, functions, and process and system capabilities in such detail that the product will have good quality in a consistent and reproducible manner with minimal cost and resources.	Assure that the URS are met. (Client must review and approve.) The client will use it as the basis for the Operational Qualification (OQ).	It may be too high level to secure the product, thus the client will not be able to produce the quantity and quality that they require. It may not contain enough specific and measurable items for client to validate the plant.
A&E	Define the process, equipment, functions, and capabilities in such a high level that they just meet the client URS. Thus, the A&E can keep the engineering cost down to a minimum. Define the structure of the plant functions and breakdown.	The FS contains the client-approved scope of delivery. Structure the automation system and software to be configured.	If it gets too detailed, then it will never be finished. And once the software is implemented, it could be too cumbersome and time-consuming for the A&E to test and get the tests accepted by the client.
SI	Define the background, process, and the functions to implement and the structure to use for implementation.	The basis for writing the detailed design specifications that are used for implementation and test of code.	It may be too high level for the SI to find the real requirements for the design. It may be too detailed for the SI to be able to use their own tools efficiently.

An FS on this level will define the structure of the plant or process breakdown with clear and measurable objectives. In terms of ISA-88, it must define the physical breakdown into the equipment hierarchy. It must define the procedural breakdown of the process in terms of recipe procedures, unit procedures, and operations, to the level of identifying the equipment phases.

This way, the language used for communication between the client and the A&E is kept in process terminology (e.g., Agitation EM, Tank Jacket EM, Dosing EM, start agitation with specified speed, maintain temperature in tank at specified setpoint for the specified time, etc.).

With respect to communication to the SI, again, an FS at this level contains the clear objectives for the detailed design to be developed, where the requirements are clear but still at a level of detail that leaves room for efficient design writing and implementation.

By developing the FS on this generic level, written in clear process terminology, it becomes very easy to communicate with the client, as the level of communication is focused solely on process functions, not on software bits and bytes.

Once the FS has been approved by the client, the client can perform the risk assessment and start working on the protocols for commissioning and validation (i.e., OQ). Since the FS is written with SMART principles, the client can use the FS itself as the actual test schema for the validation, and the A&E and SI can start the detailed design that is now completely decoupled from the client.

Since all the functions of the process have been approved in the FS, there is no need to continue the client's involvement in the more detailed specification work, thus relieving the client resources to focus on their own tasks, in place of following the SI to the very detailed level of specifications.

At this stage, it is very important that the FS is placed under strict change management control, as any changes in this will influence both the SI's detailed design and the client's commissioning and OQ activities. As such, the FS is one of the main tools for managing the scope of the automation project.

It is essential that the contents of the FS are kept on this process functional level and not blurred with automation-specific details that will end up in the client commissioning and OQ protocols. These protocols will need to be verified at too high a quality level—involving the FS in this process could bring additional costs and project delays.

The A&E and SI will continue the detailed design by making the Software Development Specification (SDS). If the FS is used as the textual starting point, then a copy of the FS needs to be used for the SDS. This is to make sure that the FS is maintained on the process-functional level only.

During the execution of the detailed design, there may be a need to update the FS with changes, new functions, or a reduction of functions. In these cases, it is very important that the update of the FS is made according to the strict change management of the FS, as shown in Figure 11.9.

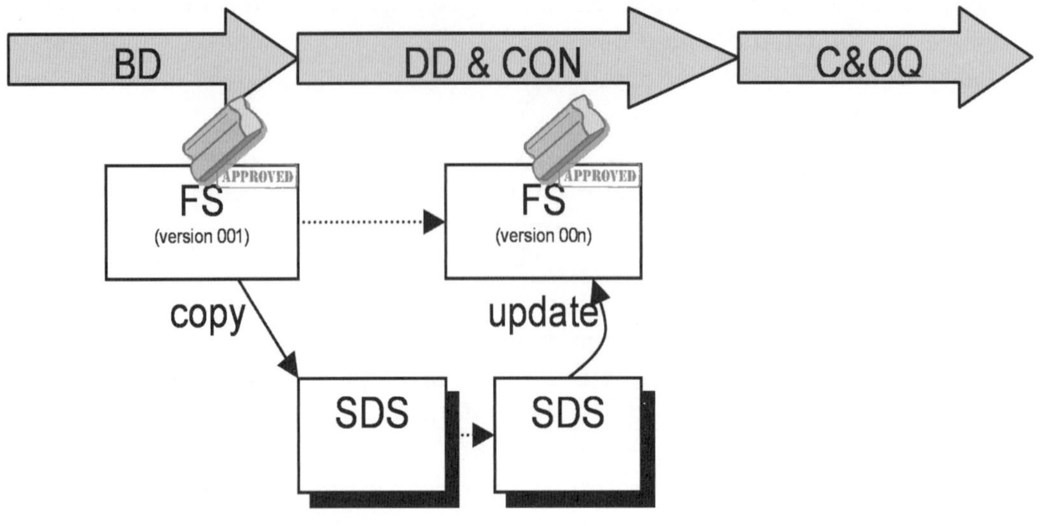

Figure 11.9. Now the FS is in place.

Using Tools

Traditionally, the specifications are sectioned according to the process segmentation and written in documents or spreadsheets. This leads to the fact that a number of automation functions are specified in more than one place (e.g., agitation function), as the function is used in a number of processes. On the testing side, this also leads to the same function being tested multiple times, whereby a lot of time is used to verify the same function multiple times without bringing more quality into the engineering.

If, instead, the FS is written with a tool that utilizes a class-based approach that also is structured according to ISA-88, then a real reduction in engineering man hours can be achieved. This class-based approach can be exemplified by a tank temperature control, as illustrated in Figure 11.10.

The temperature in the tank is controlled by means of a jacket on the tank, whereby the temperature of the jacket itself is controlled by manipulating the supply of either steam for heating or chilled water for cooling. This type of temperature control of tanks can be used on a number of tanks in the project, or it may even have been used on tanks in other projects for the same client or even for a different client. If the functionality of the temperature control is identical, then there would be no problem for the A&E and SI to actually reuse both the specifications and tests performed once to document the functionality of the tank for the specific client project as well.

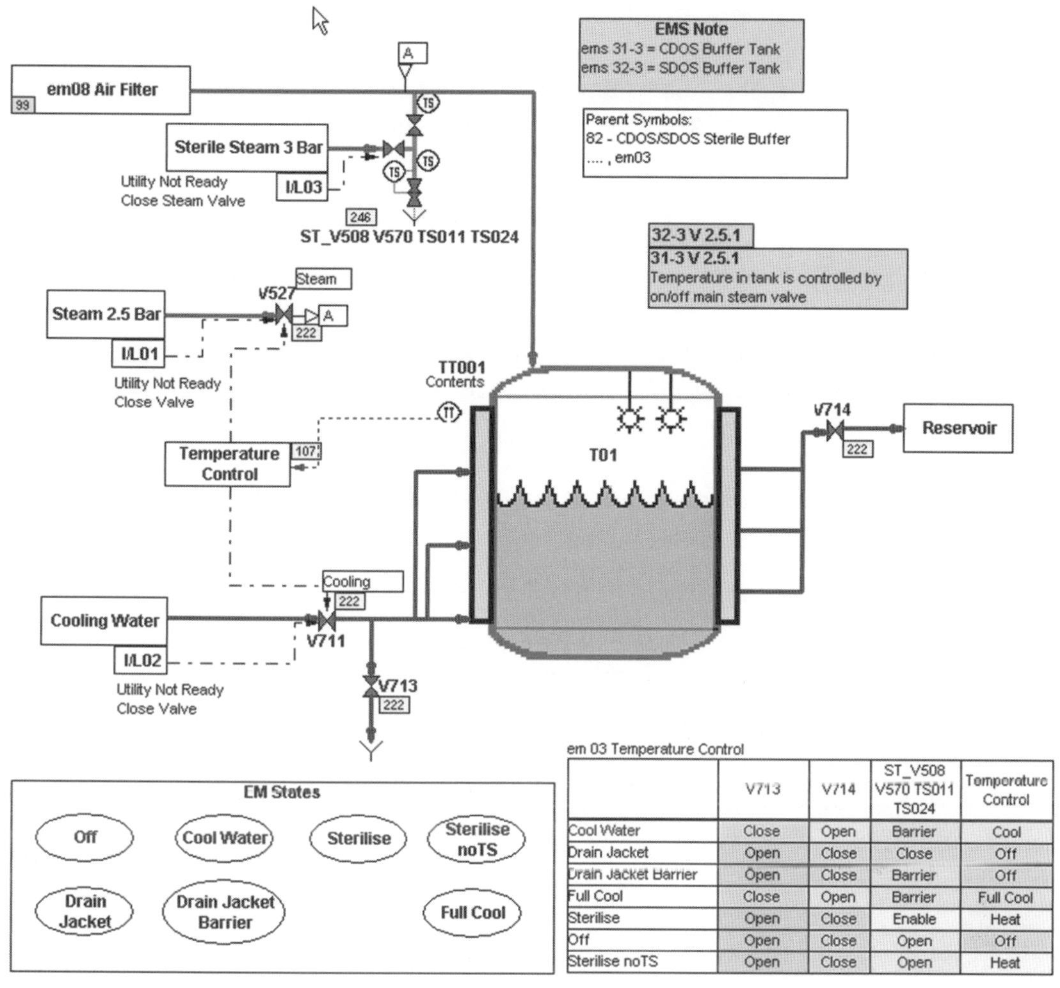

	V713	V714	ST_V508 V570 TS011 TS024	Temperature Control
Cool Water	Close	Open	Barrier	Cool
Drain Jacket	Open	Close	Close	Off
Drain Jacket Barrier	Open	Close	Barrier	Off
Full Cool	Close	Open	Barrier	Full Cool
Sterilise	Open	Close	Enable	Heat
Off	Open	Close	Open	Off
Sterilise noTS	Open	Close	Open	Heat

Figure 11.10. Tank temperature control.

The whole idea is that if there is a function that is used more than one time then it should only be designed once in one specification; it should only be coded once and functionally tested once.

By applying this class-based approach covering specification, software, and testing, the supplier can perform the engineering for the specific client project. This approach is exactly like the approach for basic Control Modules (CMs) for valves, motors, control loops, and so on. No additional specification writing or module testing is necessary when the SI uses the CMs supplied from the automation system vendor, as this has already been performed by the automation system vendor. In theory, there is no upper limit to the extent of such a CM, but in practice, the highest level of class in this context will be the ISA-88 unit.

The tool described previously and the benefits achieved by using it in fast-track projects were previously described and presented at the 2002 European World Batch Forum in Mechelen, Belgium, by Andrew Knott (Flour) and Francis Lovering (ControlDraw), "Analyze, Design, Develop and Validate—Fast." Now the SMART FS is in place.

Conclusion

By employing well-structured front-end definition planning, making the FS "generic" and "SMART," and combining them with a class-based specification tool, the following benefits can be achieved:

- Early identification and naming of >85% of all technical documentation
- Well-defined scope and resource planning
- Easy communication with the stakeholders
- Clear, SMART objectives
- Ability to reuse functions within the project and with future projects
- Functions that are only validated once
- Reduced cost, time scale, and risk

CHAPTER 12

The Challenge of Integrating Multiple Batch Systems to Global Business Systems

Presented at the WBF North American Conference, March 24–26, 2008, by

Frede Vinther
Sr. Automation Specialist
frha@nnepharmaplan.com
NNE Pharmaplan A/S, Vandtaarnsvej 108-110, DK-2860 Soeborg, Denmark

Abstract

Integration of batch control systems to business systems is essential to enable the business to manage and respond to the demand for products. This integration has been assisted by the ISA standards—particularly ISA-95.00.05, which describes the business-to-manufacturing transactions.

This chapter is based on a number of integration strategies and concept consultancies performed for various clients over the last year. The focus is on the challenges that need to be addressed and covered to enable a successful integration in the world of biotech and pharmaceutical manufacturing, where new manufacturing methods and technologies are introduced on a daily basis. One major challenge to overcome is to look at manufacturing and automation as a whole and not just as computer system functions. Overcoming this is not only a technical issue but also a cultural and personal preference issue. These and other issues are described in this chapter

Background

Over the last decade, the pharmaceutical and biotech industries (pharma for short) have been changing the ways in which they run their businesses, especially with respect to regulatory compliance. Regulatory compliance has been impacted dramatically by changes in regulations applied by organizations such as the Food and Drug Administration (FDA) and the European Medicines Agency (EMEA). This has not only impacted the cost of building and running production facilities but also impacted the cost for Research and Development (R&D) and the whole development chain, from R&D to approved product. These changes have taken place without any real increase in approvals of New Molecular Entities (NMEs), as illustrated in Figure 12.1. The result is that drug prices for consumers have increased tremendously.

PricewaterhouseCoopers (PwC) has forecast the growth in the world economies' Gross Domestic Product (GDP) for the next decade, as shown in Figure 12.2. The G7 countries will grow just 40%; whereas the E7 countries will triple their GDP. If all these countries spend the same proportion of their GDP in 2020 as today, then the global pharma market for medicines will be worth about $800 billion in 2020. (The global market in 2004 was about $518 billion.)

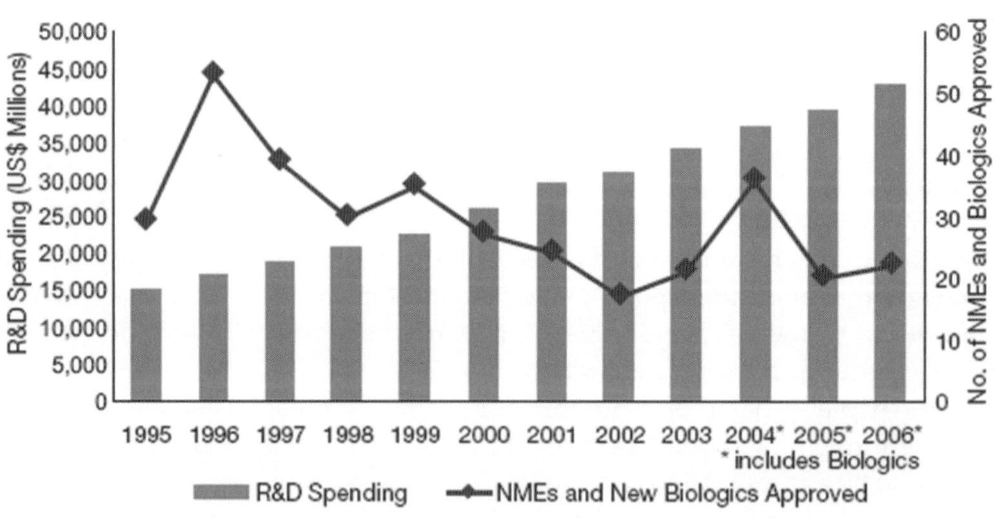

Figure 12.1. R&D spending and number of approved NMEs.

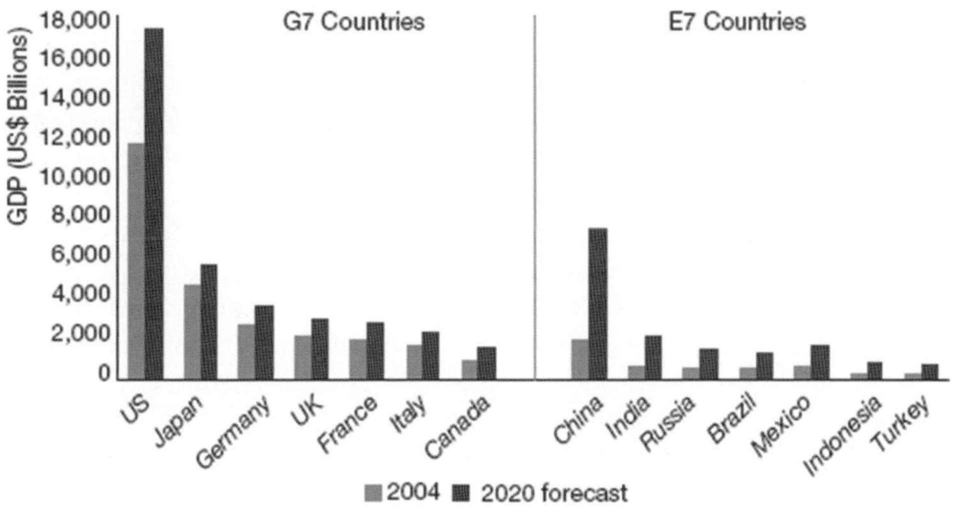

Figure 12.2. Forecast GDP for 2020.

Another major obstacle for fast development and fast approval of NMEs is the requirements made by legislation that make the whole process very slow and inflexible, as illustrated in Figure 12.3. Dr. Scott Gottlieb, the FDA's deputy commissioner for medical and scientific affairs, recently noted the following about NME development by big pharma: "The highly empirical, statistical method that currently predominates is inflexible; it restricts innovation and results in 'overly large' trials that yield information about how large populations with the same or similar conditions are likely to respond to a treatment. But doctors don't treat populations, they treat individual patients."[1]

Governments have already started questioning the price for medicines, and they will push much harder for lower prices in the years to come. But pushing

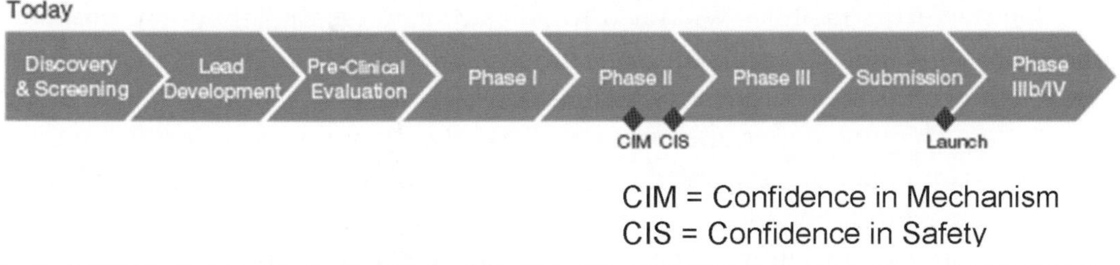

Figure 12.3. The regulatory process today.

pharma prices does not do it alone; governments themselves have played a major role in pushing the medicine prices to the current level. Governments will need to update legislation to adhere to and enable a much more globalized thinking with respect to medicine and fighting diseases. This will enable faster, more reliable, cheaper drugs and well-functioning and very dynamic drug development and production.

One example of legislation that obstructs bringing new medicine to the market is the Japanese legislation that today requires clinical trials for NMEs to be conducted on native Japanese people. This adds another 5 to 7 years of clinical trials on top of any U.S. or European approval before the drug can hit the Japanese market.

In order to lower the consumer prices for medicine, the entire way that public and private health care is organized and the way of treating of diseases, understanding diseases, and developing and producing medicine will need to be changed.

Alignment and Discrepancy: The Human Factor

Turning the focus toward the manufacturing of drugs and the whole supply chain and its logistics, the globalizing world and its demands for medicine will foster a dramatic change. The era of big pharma searching for blockbusters and building facilities solely to manufacture these will vanish.

Instead, medicines will need to be developed for disease subtypes and targeted toward different patient subpopulations. These medicines will only be tested on patients who suffer from those conditions, thereby reducing both the number of trials and the size of the trials required to prove efficacy. With proper legislation, the development process of the future will be much more refined, as illustrated in Figure 12.4.

This mechanism of targeted treatment will drastically reduce the time to market for a specific treatment but, at the same time, drastically increase the number of closely related but still distinguished treatments and medicines. The targeted treatment will have a "breathtaking" impact on all we currently know, do, and use as best practice in relation to manufacturing and supply of drugs.

Manufacturing facilities will need to be extremely flexible; they will probably need to be geographically distributed to secure a safe supply chain to the individual patients and, at the same time, govern the same high security for the medicine as we know today. The whole manufacturing and supply chain will be much more complex. To secure patient safety and fast delivery, it will need to be automated, fully integrated, and interoperable. The pace of establishing manufacturing facilities and adjusting established facilities to enable production of related and nonrelated medicines requires the individual process stages to be Commercial Off The Shelf (COTS)

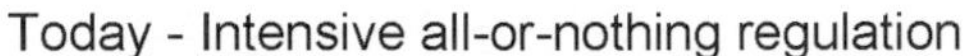

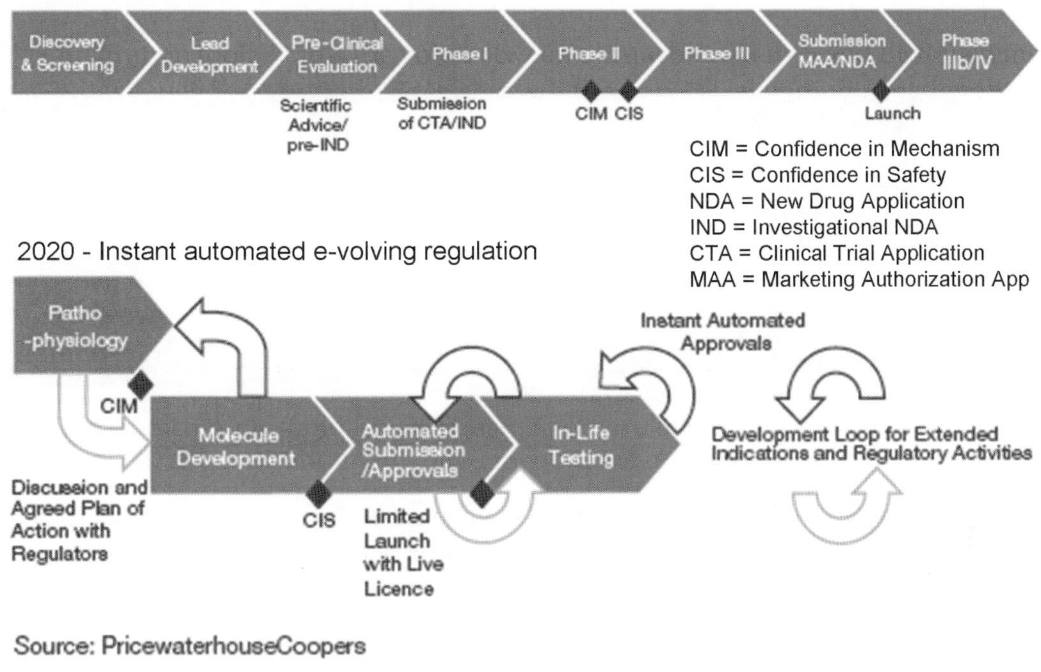

Figure 12.4. The possible regulatory process in 2020.

from individual subsuppliers of processing equipment and then be integrated at the pharma manufacturer's site with virtually no effort.

This use of COTS really "hits the nail on the head" with respect to major obstacles for the pharma industry and its subsuppliers. To achieve this goal, the companies and the people working there need to overcome the "Not Invented Here (NIH) syndrome."[2] This is a very big task to deal with, as NIH certainly is not solely between companies but is very well adopted inside companies where departments and individuals nurse and use it on a daily basis. Some would call NIH "resistance to change."

Manufacturing Scheme

To enable the previously described flexibility, manufacturing facilities need to undergo some substantial changes. Pharma needs to transition from the traditional build-to-order manufacturing to build-completely-modular manufacturing, as illustrated in Figure 12.5.

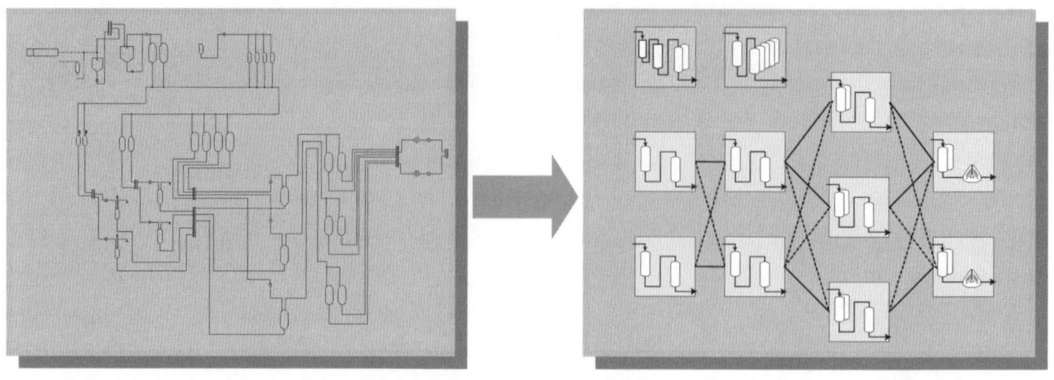

Figure 12.5. The transition to modular process construction.

Each process stage is a complete module—Vendor-Packaged Equipment (VPE) that includes all mechanical, instrumentation, and specific controls needed to perform the process function of the module. Each and every one of the VPEs or process modules should be "self-contained" with respect to processing, and the process interface between these process modules should be very simple or, basically, from a mechanical point of view, nonexistant. This means they cannot be connected via pipes; instead, they should have a very loose coupling, where the output of one process module is transferred to the input of the next process module in bags, Intermediate Bulk Carriers (IBCs), or other transportable containers.

Besides process modules that perform the actual chemical or microbial processes, there must exist a form of product storage modules; these storage modules will, for now, be called *farm modules*. They exist in two flavors: One type is dedicated to a specific process module or set of identical process module types that are located near the same storage station and primarily have product buffering or holding functionality (i.e., the farm module). The other type is much more flexible and can support multiple types of process modules that can be moved between storage stations (i.e., the smart farm module).

Both types of farm modules should have the capability of controlling and maintaining products, as well as environmental conditions, for the bags, IBCs, and other types of product containers that are placed in the farm module, as illustrated in Figure 12.6.

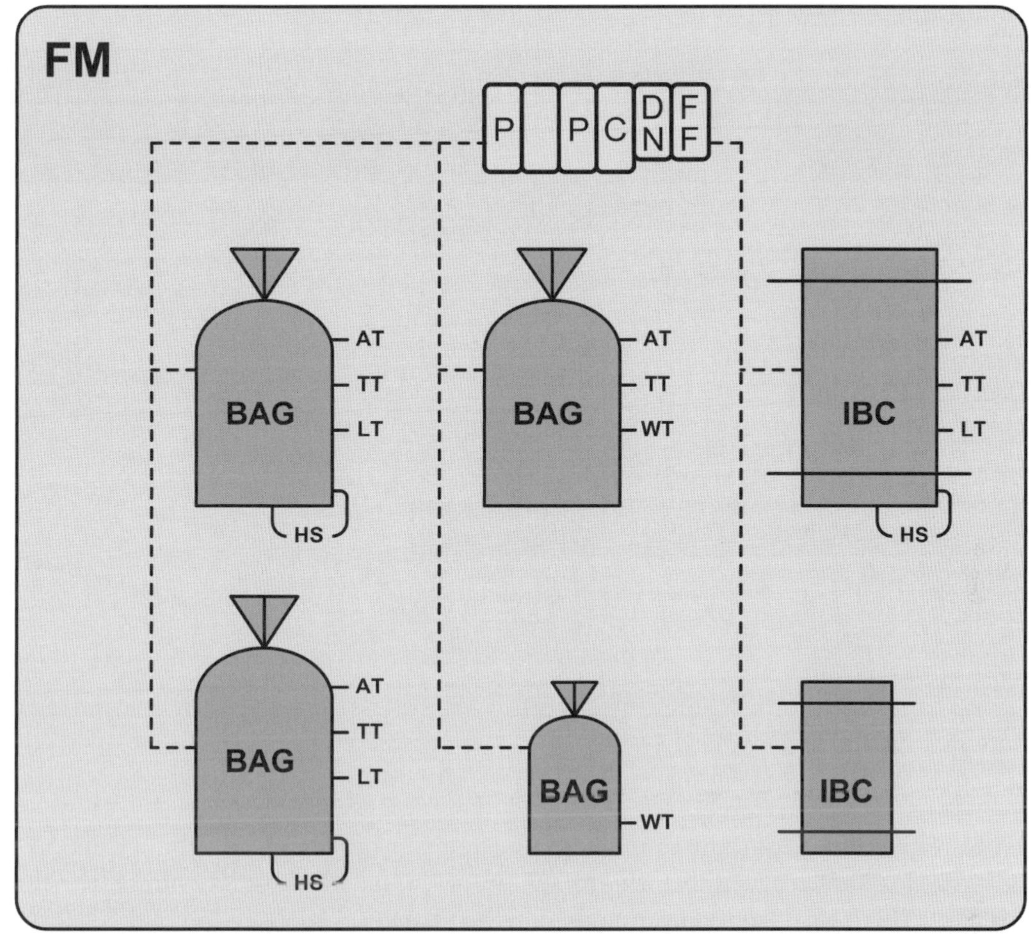

Figure 12.6. A farm module.

Manufacturing Operations Management

From an operation management point of view, the manufacturing scheme explained previously clearly adds higher demands for flexibility, as we use in most big pharma companies today. I executed a study for a pharma client, aiming to incorporate flexibility from a schedule-oriented and cost-avoidance point of view. We investigated the change of the production setup of a manufacturing facility from one product to another similar but distinct product, where we focused on the time to market for a product. Focusing on the time from the start of the changeover from the previous product to the time when the next product was produced and released, we found the following information (Fig. 12.7).

It is clear that such a reduction in downtime really lowers the cost of manufacturing, as the utilization of a manufacturing facility will annually be increased from

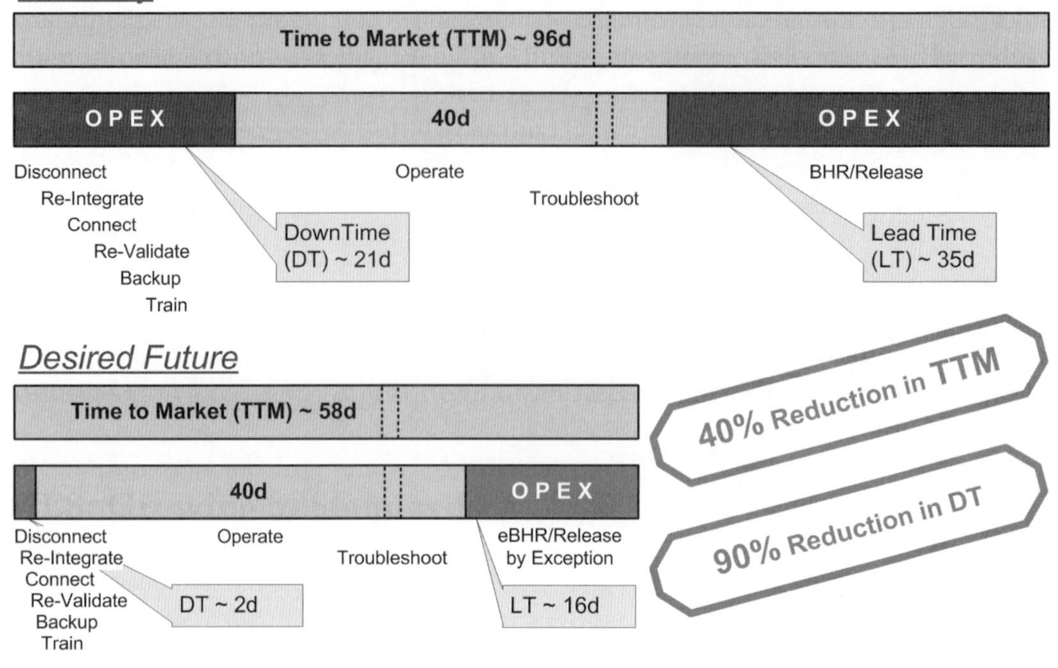

Figure 12.7. Analysis of downtime.

65% to 95% if the downtime and operation times are propagated. The reduction in lead time does not really affect the utilization of the facility; it solely affects the time to market. This was described previously in the "Background" section of this chapter.

Reducing the downtime for product changeover to 2 days is foreseen as possible, partly due to a sliding changeover of equipment.[3] This estimate also depends on if the ISA-95.03 Manufacturing Operation Management (MOM) and its associated controls and quality-related activities that are required to perform the product changeover are based on prevalidated process modules, including their controls (i.e., VPEs). It also depends on if the manufacturing facility MOM has plug and play for interfacing and integrating with the control for the VPEs.

During the client study, a number of architecture scenarios were identified, ranging from existing low-automation levels to the limitations of today's computer system technology, VPE control and interface capabilities, and people's NIH mind-set (Fig. 12.8).

When going all the way to procedural coordination only (see the right side of Figure 12.8), the coordination of the process modules, including any intermodule coordination, should be done by the MOM. This puts very high demands on the capabilities of the MOM.

Minimal Automation | Paper-based Execution | Paper On Glass | Full Automation with Phase level VPE Control | Full Automation with Unit level VPE Control | Procedural coordination Only

BS	=	Business Systems e.g. ERP, LIMS, DMS, etc
MES	=	Manufacturing Execution System
MOM	=	Manufacturing Operation Management
PCS	=	Process Control System (Cell State Engine + Distributed Control)
VPE-B	=	Flexible Batch Control (Unit State Engine) packaged with Equipment
VPE	=	Vendor Packaged Equipment (for argument, include BAS and small utils in this object)
HIST	=	Continuous and Event Historians as appropriate
MANUAL	=	Manual interface
IF	=	Custom-built application interface, or manual interface

Figure 12.8. Degrees of automation.

It is questionable if the MOM defined previously can be accurately described as a "system." In my opinion, the MOM will most likely be a combination of COTS applications running on a common infrastructure, and the MOM will use this infrastructure for information exposure and gathering. It is clear to me that the concepts and technology for such an MOM must have its roots in the IT world and not in the traditional automation or batch world.

Standards

Over the last decade or so, the International Society for Automation (ISA) as a community has developed ISA-88 for batch control and ISA-95 for enterprise and control system integration. These standards have been well adopted in the pharma industry.

The good part is that numerous applications have been put into operation, covering functionality for integrating the business systems with the automation systems, with respect to scheduling of production orders, discrete and batch execution, and monitoring the production performance for business use. The functions have been divided according to the ISA-95 functional hierarchy model using standard definitions and terminology (Fig. 12.9).

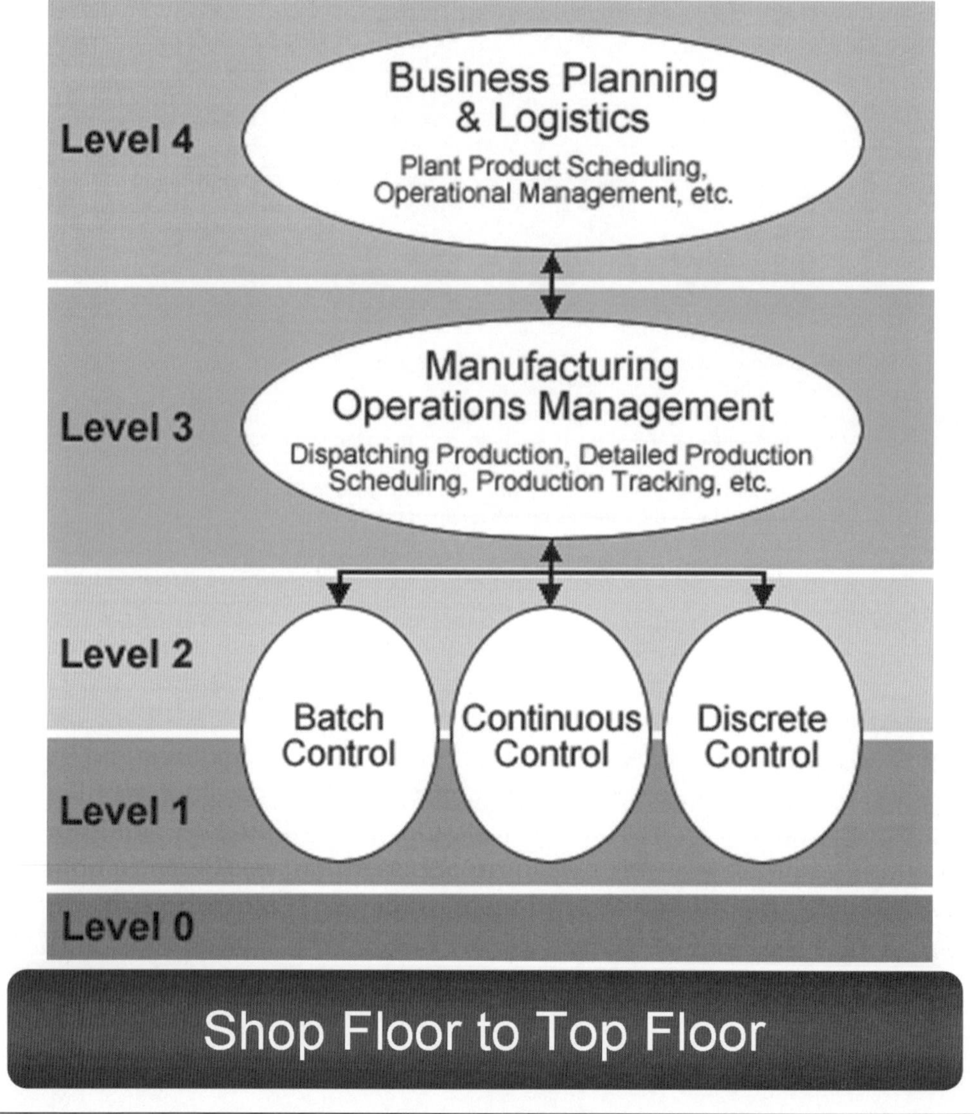

Figure 12.9. The ISA-95 functional hierarchy and its common description.

The bad part is that most of these installations and applications have been built to order. Another problem with these standards is that they are not aligned to complement each other; instead, there is quite some overlap, as described by the ISA JWG 88-95 in "ISA TR88-95.01: Using ISA-88 and ISA-95 Together." This technical report explains that many of the architectural scenarios described previously are actually impossible to define or to apply with consistency to either of the two standards.

On the enterprise-control system integration side, initiatives have recently been made to extend the standard to include methodology for the actual computer-based integration of business-to-manufacturing transactions (ANSI/ ISA-95.00.05-2007). Still, this standard is merely a framework for integration and does not define an actual plug-and-play functionality. As such, it will still foster applications that are built to order and not COTS.

In the same range of standards, the ongoing Make2Pack (http://www.make2pack.org) addition to ISA-88 (available as "ISA-TR88.00.02-2008 Machine and Unit States: An Implementation Example of ISA-88") is perhaps more interesting, as it strives to address and standardize the interface for machines-to-control systems (i.e., a subset of the VPEs described earlier in this chapter). However, with the challenges for pharma that are described here, the ambition level for this standard is way too low.

Path Forward

In light of the challenges that will come to all of us within a short time frame if the forecasts described and referenced in this chapter are even partially true, the way of integrating systems vertically (i.e., shop floor to top floor) and horizontally (i.e., Globalized Supply Chain) needs to change. This integration must function in near real time and have a full-fledged plug-and-play nature, interface, and communication capabilities. It must also provide security and robustness for execution, as well as the ability to retrieve and document the specific medicine supplied to the individual patient (in the same way that the WBF community performs automation today).

From the automation viewpoint, a number of initiatives will need to be started to create a path forward that incrementally will enable us to fulfill the visions stated in this chapter. Some of these initiatives could include the following:

- Shift into a much higher gear on standardization work.
- Analyze and establish a business model for the new way of doing medical care in forums other than pharma companies.
- Create a common standard for ISA-88 and ISA-95 areas, or redefine the ISA-95 standard to define the complete MOM as a generic business and software model that is not specifically targeted toward batch but will still accomplish the task of batch control.

- Investigate and explore technologies and standards used in other IT areas to create a basis for understanding how to enable integration across systems, software platforms, and technologies.
- Extend the current work done by Make2Pack to create a real compliance standard for integration of VPEs to the MOM (e.g., the plug-and-play standard).
- Secure the "buy-in" from automation, IT-system, and application vendors, since without them, the process will fail.

These technological initiatives can only be successful if all participating stakeholders will truly support the overall vision and, thereby, also make initiatives to break the current NIH culture that keeps us all from using already-existing products, research, or knowledge because of their different origins. This may quite well be the most difficult task of all to overcome. Remember that forecasting is very difficult:

> To sail a steam-boat from New York to Liverpool is completely unrealistic. One could just as well plan a tour to the Moon.
>
> —Professor Dionysius Lardner, one of England's leading experts in steam machines, 1835

References

1. See the FDA's speech at the 2006 Conference on Adaptive Trial Design, Washington, DC.
2. According to Wikipedia, this can be defined as a persistent sociological, corporate, or institutional culture that avoids using already existing products, research, or knowledge because of their different origins.
3. While the previous product is being finished on one process module, that process module can be reconfigured, or if required, it can be replaced by another process module while the production finishes in the following process modules. Even the startup of production for the next product can be initialized while the previous product is still being produced in some process modules.

CHAPTER 13

The Road to Full MES Integration: Practical Experience from the Pharmaceutical Industry

Presented at the WBF North American Conference, March 5–8, 2006, by

Niels Andersen, M Sc (Physics)
Automation Specialist
nca@nne.dk
NNE A/S, Gladsaxevej 363, DK-2860 Soeborg, Denmark

Abstract

This chapter describes the successful implementation of a Manufacturing Execution System (MES) on top of a Process Control System (PCS) in the world's largest insulin Active Pharmaceutical Ingredient (API) plant, operating with more than 300 units with more than 500 recipes and 600,000 parameters. The functionality of the actual MES database application is described, along with its evolution from early versions in finished-goods production to a modular and flexible large-scale MES application within API production.

The solution posed by this chapter relies on ISA-88.01 models and terminology. The architecture introduces kernel modules, encircled by libraries of software extensions. The general database model has proved very robust. The simple principle of recursive operations, combined with a simple execution principle of an operations state machine, forms the basis for a very versatile MES integration platform and a powerful batch modeling system. Furthermore, this chapter includes effective tools for master recipe and report definition, facilitating combined development.

For integration, the importance of a well-defined interface to distribute different aspects of control between database MES and real-time PCS applications is described. Finally, this chapter describes the impact of the modular design and reuse on execution in fast-track projects, which recently contributed to substantially reduced project execution times for both pharmaceutical new and upgrade projects.

Project Background

Two years ago, NNE A/S handed over the Insulin Bulk Plant (IBP) project to the customer Novo Nordisk A/S, a world leader in diabetes care. The plant is the world's largest insulin API production plant, with more than three hundred units, tanks, centrifuges, columns, and so on. The insulin factory is a so-called two-lane, full-length super highway, producing two products from fermentation through recovery and purification to crystallization (Fig. 13.1).

Figure 13.1. Aerial view of the IBP.

During the conceptual design, it was decided to install a PCS with an MES on top for combined batch execution. The decision was based on experience from similar projects where it had been difficult to engineer the recipe control and unit synchronization at high levels within a PCS system alone. Furthermore, additional requirements to provide data as the basis for continuous process optimization and automatic, paperless batch and quality documentation would be difficult to implement in the PCS. The system should also prevent errors and defects by keeping track of all operations and by automatic identification of materials, equipment and personnel. Finally, it should ensure and demonstrate regulatory compliance, including compliance to CFR 21 Part 11.

After screening the market for suitable standard MES solutions, it was decided to build on experiences drawn from an in-house database application implemented in a formulation area a few years earlier. It was further decided to split the execution responsibility between the real-time PCS system and the database application at the ISA-88 Process Operation level. The design and implementation of the PCS system took its starting point in a similar application, which was broken up in operations, leaving both procedural and recipe control to the MES system, along with the handling of manual operations.

System Architecture

The building blocks of the MES application consisted of subsystems for handling master recipes, material definitions, material control, equipment modeling, and batch execution. A mapping proved that the existing toolbox covered some two-thirds of the functional requirements, exposing gaps at the reporting, material handling, and execution functionality, regarding arbitration and queuing. Also, a proper OPC-based interface (http://www.opcfoundation.org) between PCS and MES had to be developed.

A subsystem was added to handle report definition and generation. Much of the added functionality was developed for general-purpose tools and organized in a number of library modules. The configuration of the system defines master recipes, reports, units, materials, and so on. The entire application may be described as a number of concentric circles (Fig. 13.2).

The inner circle depicts the six kernel subsystems for the following:

- Master recipe handling
- Equipment modeling
- Material definition

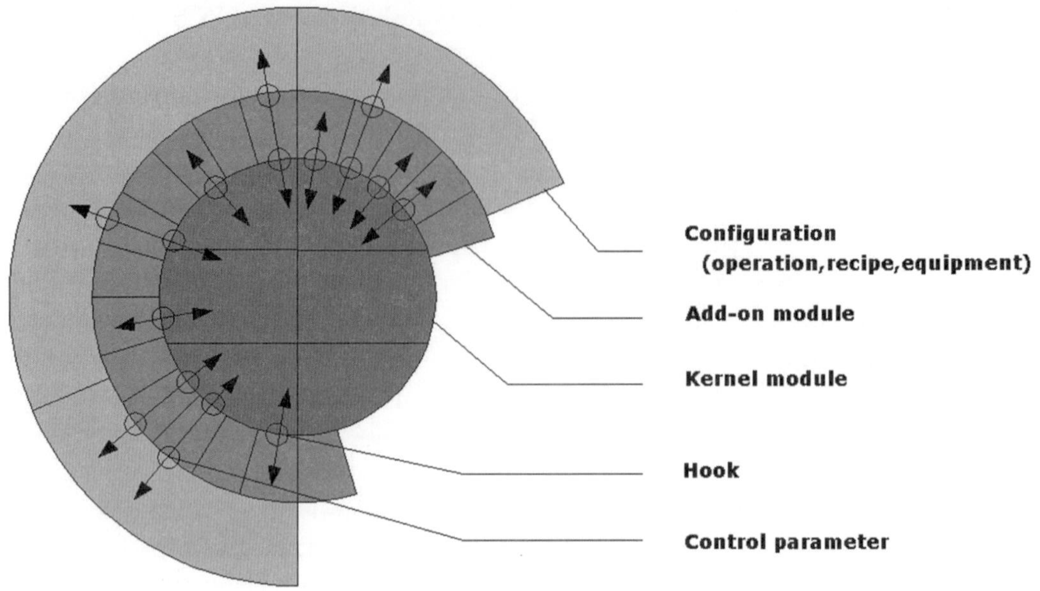

Figure 13.2. NNE MES architecture.

- Material control
- Batch execution
- Report definition and generation

The kernel provides the basis for an MES system within any kind of batch production. The middle circle illustrates the libraries of add-ons necessary to operate a bulk factory. It provides manual operations and material handling with barcode identification for quality sampling, weighing, charging, filter testing, and so on. This level—the library of add-ons—defines a bulk factory MES application. Finally, the outermost circle shows the configuration of materials, equipment, locations, master recipes, procedures, and reports. This layer makes the system a specific insulin API production MES application.

The interactions between the layers are typically controlled by hooks and control parameters. The kernel contains a number of so-called hooks (i.e., calls at predefined events to library procedures that have been registered through a standard hook interface table).

Since the database application is generally not suitable for collection of time series data, a data historian was added. In a coarse overview, ignoring interfaces to barcode readers, scales, and any other peripheral devices, the complete architecture can be outlined as depicted in Figure 13.3.

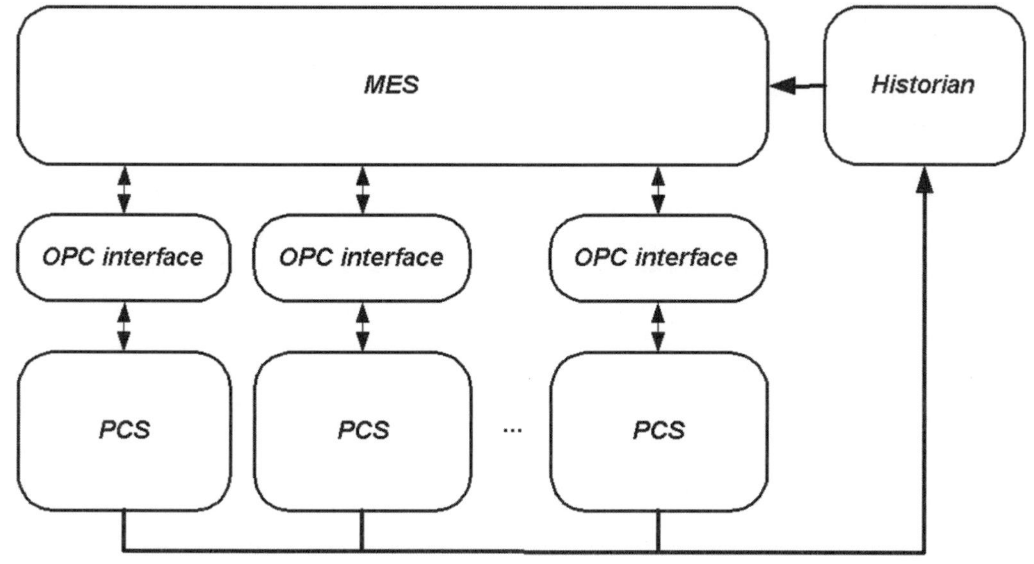

Figure 13.3. Parallel PCS architecture.

Figure 13.3 shows the division of the PCS into a number of independent systems that allow for individual, parallel implementation, test, and validation during project execution; this increased the maintainability during the operation.

Data Model and General Execution Principle

The data model of the MES solution was inspired by ISA-88. The interface between MES and PCS is at the operation level, meaning that MES controls the sequence of operations within procedures and PCS executes these operations one at a time. However, the MES data model relies on the ISA-88 principles of collapsibility and expandability of the procedural model, leaving only the recipe level and a recursive operation level. To distinguish between basic automatic or manual operations and their containing operations, the term "compound operation" is used for the containing operation. An example is shown in the next section of this chapter. This implies that the master recipe programmer freely can model both the traditional recipe and unit procedures into a number of compound operation levels.

Every operation has a status associated with it. The individual states are specified as part of the operation definition, along with the possible transitions between them. During execution, an operation will pass through the relevant states, driven by a state machine. The state machine is the core of the batch execution. In principle, it just waits for an external basic operation status change to occur, which can only be caused by one of the following completion events:

1. The start of the control recipe
2. The execution of a manual operation
3. An automatic operation as a response from the PCS system

A few mandatory states exist with any optional state in between. Operations are aligned with sequences by status relations. Typically, the next operation will start when the previous operation is completed, but branches can be defined at any intermediate state. Operations are combined into a hierarchical treelike structure through relations. Relations may have conditions associated with them, and operations may have start conditions.

At execution, the state machine responds to an operation status change in a database transaction, which evaluates the entire operation tree of resulting operation status updates through their relations. Having evaluated all transitions to be legal and all conditions to be fulfilled, it commits the new state of the recipe. This, in turn, results in activation of either a manual operation invoked through a hook or an automated operation invoked through the interface to PCS.

Master Recipe Handling

The master recipe definition is created using a graphical tool. Figures 13.4 and 13.5 show the general Clean-In-Place (CIP) recipe procedure and part of the CIP unit procedure, displaying only the first three rinses. The CIP concept is that the consumer unit starts its CIP procedure, followed by start of the valve matrix (x-unit) operation, and finally the procedure is executed at a CIP substation that contains the sequence of rinse operations with water, base, and acid in proper amounts and temperatures.

The CIP procedure (i.e., sequence) consists of a number of instances of the same rinse unit operation. The operations are instances of the same basic operation. They have identical alias tags (CIP_RINSE) but different instance tags (RINSE_1,_2). The rinse operation source has a number of parameters defining media, amount, temperature, and so on. The value of the media is overwritten in each operation instance (e.g., the first rinse with water, the second with base). Combining rinse operation instances to procedures and procedure instances to recipes, you may choose to perform overwriting at various appropriate levels. The sequence of medias remain unchanged irrespective of which unit is to be washed. Hence you can overwrite the media parameter at the instance level. This resolves the media values at the unit-procedure compound-operation level; the amount of media depends on the type or characteristics of a unit, so amount parameters are overwritten at the recipe level.

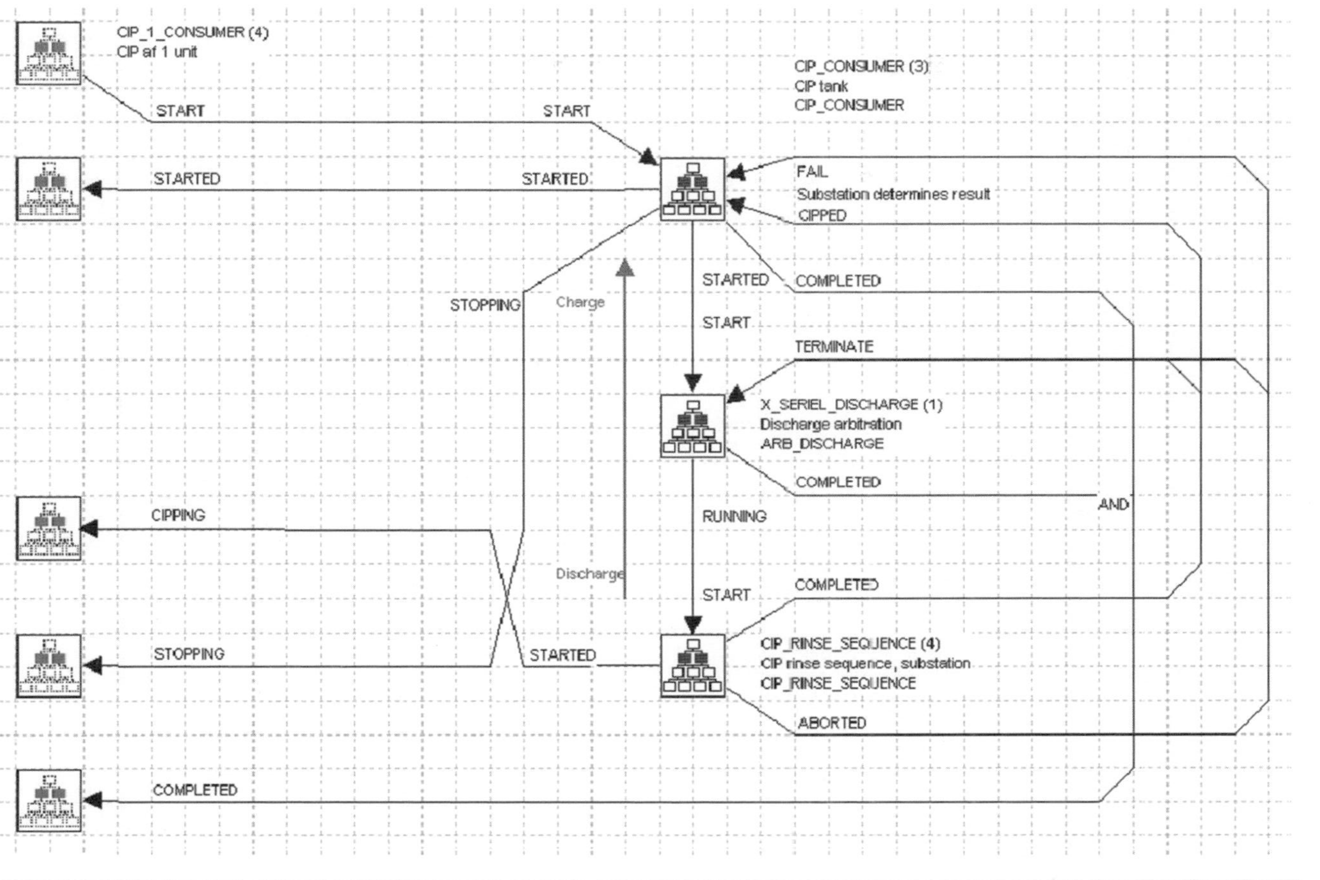

Figure 13.4. CIP recipe procedure, CIP_1_CONSUMER. (Icons on the far left refer to the state of the parent compound operation.)

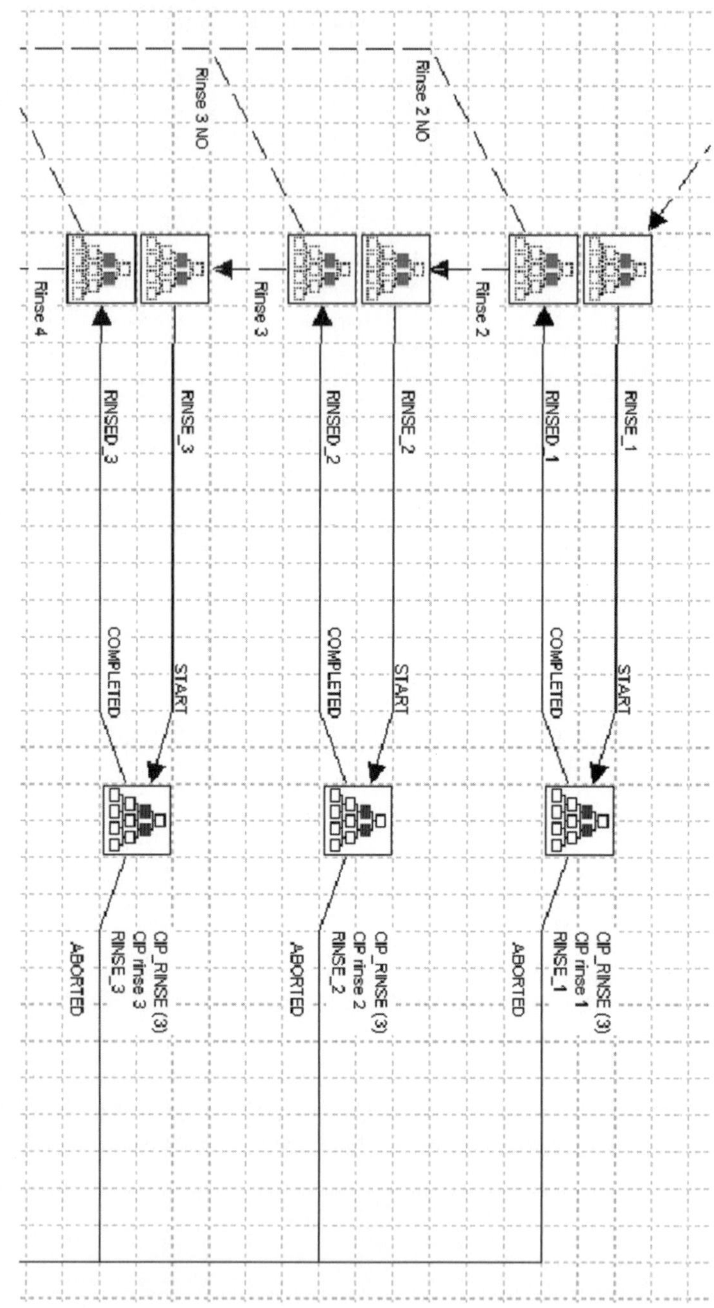

Figure 13.5. Part of CIP unit procedure, CIP_RINSE_SEQUENCE. (Icons on the far left refer to the state of the parent compound operation.)

The overwriting principle also applies to equipment types, which are assigned to master recipes and their operations. As shown in Figure 13.4, the principle allows one compound operation to control the CIP of all units, irrespective of their type, tanks, centrifuges, x-units, and so on. This compound operation controls the cleaning of more than three hundred units. The compound operation is instantiated in approximately one hundred master recipes, since one master recipe can clean any unit of a distinct type. The actual unit is specified at control recipe generation, where equipment is also allocated to the CIP substation and the x-unit.

Typically, operation start conditions concern locking and cleaning. The condition names refer to functions stored in the database, which will return "true" or "false," depending on the value of the equipment status in question. Cleaning operations will omit the condition on cleaning status, while it is mandatory in production operations. Actually, the start conditions apply only to the first unit operation, which, if started, modifies the current status parameters and, for example, sets the locking status and the current batch number. Subsequent operations may then depend on the start conditions, specifying the same batch to prevent batch mix-up. Finally, when releasing the unit, the last operation resets the locking when completed. The equipment status modifications are defined in the master recipe and prevent batch mix-up.

Operations, their parameters, and journal parameters are specified in the master recipe editor using alias tags. The translation of these to actual tags in a PCS is done using interface configuration files. The modular building blocks, master recipes, compound operations, and basic operations are all subject to extensive version control. An operation version is either in draft, approved, or withdrawn status while master recipe versions run through a life cycle, as shown in Figure 13.6.

Note that it takes two electronic signatures to approve a master recipe—one from production support and one from Quality Assurance (QA). A master recipe cannot be set for approval until all the operations within it have been approved. Each step is accompanied by a revision note, normally referring to the change request that initiated the change. Using the master recipe handling tool, it is possible to print these revision notes and a change log listing all differences between any two versions of a master recipe. When approved, an item is locked for further updates.

To modify a master recipe, you have to create a new version. If only recipe parameters need to be changed, this will suffice because of the overwriting principles. If any control aspects ruled by relations at an operation level need to be changed, then new versions of compound operations must be created and inserted to substitute the approved ones. For configuration control, you may generate so-called baseline reports that explain which master recipes were approved at any point in time or during any time interval.

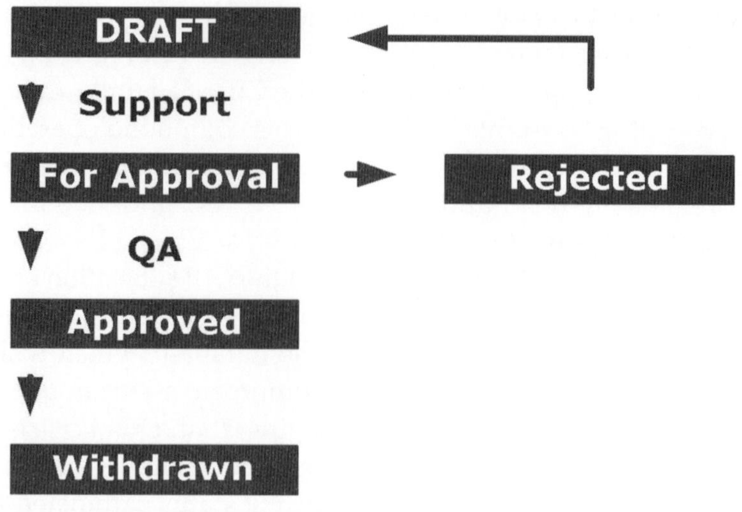

Figure 13.6. Document life cycle.

Equipment Modeling

The equipment modeling system represents the equipment hierarchy, using both types and instances. Both have status and parameters associated with them. In the IBP project, however, equipment parameters were kept in the PCS system. Again, equipment is a recursive data entity, allowing for equipment entities within equipment entities.

Equipment types can be modified by characteristics. The types are used to model equipment at the master recipe level. A master recipe may be specified to run in an area. When generating a control recipe, the merge package will translate equipment types into actual instances of equipment belonging to the actual area chosen to execute the control recipe. At the time of the IBP project, minor modifications in the equipment modeling system introduced connections (the model equivalent to pipes), which could be used for the arbitration of equipment.

Material Definition and Control

This subsystem handles the definition of article numbers, bills of materials, reservation, picking, goods reception, and dispatch. Certain materials will, upon reception, have their concentration and similar critical parameters recorded together with standard information on expiry dates, quality, and amount.

The subsystem has a complete storage control keeping track of the materials at locations in load carriers, again in a recursive manner allowing for bottles in boxes or sacks on pallets. During setup, you may specify articles to locations, articles to load carriers, and carriers to locations, thus modeling, for example, cold storage for certain articles.

Materials are received, stored, picked, consumed and produced, stored, released, and dispatched using bar-code scanners. Materials must have adequate quality status to be released for production. For production in the bulk environment, no reservation or picking will be performed during weighing and charging in the raw-material and solvents dispensing area. These functionalities are automatically performed by the specific add-on operation in order to comply with the general principles of the material control subsystem.

Also, in this bulk environment, most materials flow in pipes. The inlet of batches from one tank to another is controlled by the PCS, which in turn will journalize the batch number and amount as a journal entry on the active unit operation in the MES. These journals are then registered in a common journal type in the batch tracking system of the MES, allowing for unified tracking of automated and manual consumption of materials. The batch tracking can display both downstream and upstream relations between batches.

Batch Execution

Basically, the batch engine consists of a state machine. It controls and coordinates the execution of all control recipes, which are generated on the basis of information regarding master recipes, equipment, bill of materials, and batch size. The execution is responsible for comprehensive and reliable recording of process and quality data, allowing for analysis and optimization of the manufacturing processes.

The HMI supports monitoring and operator intervention. In general, all recipe and operation statuses, parameters, journals, and equipment status information are available to the operator, generally as lists that are typical of database applications (usually conveying more data than the operator can comprehend).

For large recipes with prolonged execution, the immediate allocation of equipment performed at control recipe generation is not optimal, since equipment may not be available at the time of execution. Recipes, therefore, contain operations to bind equipment dynamically during execution based on equipment connections and start conditions regarding availability and cleaning. If no suitable equipment is available, then the operation is queued. The queuing mechanism is typically used for queuing the alternate execution of production and cleaning batches.

The batch execution subsystem is responsible for invoking any manual operation hooked into the state machine (e.g., filter handling, sampling, weighing, or dispensing). The exception handling includes facilities for disabling relations between operations at individual levels and for setting an operation in a special error status from which it can freely be set to any appropriate status. A suspend mechanism will allow operation changes from the PCS system, despite problems with execution of subsequent operations within the entire transaction.

Report Definition and Generation

Experiences from earlier projects had shown that batch report definition often fell behind, and reports were commonly defined using standard spreadsheet or database tools. Now the idea was to combine the recipe definition with the report definition and to have both recipes and reports defined at approximately the same time during project execution. Most importantly, the recipes and reports were to be defined in a tool suitable for end-user operation. This led to the introduction of a report definition and generation subsystem.

The report structure is built by referencing operations from the master recipe and defining how the operations must be reported in relation to each other. For each operation, a report component is defined. The content builds on parameters and journals, as defined in the recipe editor. A *report component* is most easily described as the method to report a compound operation. As compound operations build treelike recipes, components will also build treelike report definitions, which become chapters, sections, and lines during reporting. In the same way that a single top-level compound operation defines all three hundred or more CIP recipes, its report definition can also specify all associated batch reports.

Batch reports are subject to the same version control and life cycle as the master recipes. The system checks for consistency and warns against approving reports associated with unapproved master recipes. The reporting system has an interface to the historian for curve data, which is extracted at time intervals that are defined by the started and completed time stamps.

Operator Access Control

User access control was established through a plant-wide shared system to enforce only authorized access to critical operations. It supports electronic signatures, as required by FDA CFR 21 Part 11. The system is shared by MES and PCS, so users have their access rights defined in one place only.

Integration

At the interface to the process control system, OPC was introduced as the communication standard. The MES and PCS are connected through an interface program. Actually, there are over twenty instances of the program serving each a cell in the plant.

At startup, the interface program reads a configuration file that defines the units and their operations with parameters and journals, each mapped from MES alias tags to PCS OPC names. The interface creates and enrolls OPC groups within a PCS OPC server.

On the MES side, the interface program will regularly poll a database view that provides information about the actual operations set to start in the PCS system. When matching unit and operation alias tags in its configuration, the interface will activate the corresponding unit operation in the PCS system.

MES covers the non-time-critical aspects in the database. The PCS and historian systems are responsible for the real-time aspects, such as continuous data collection, event logging, equipment monitoring, alarm handling, trend curving, and phase sequence control.

The protocol for executing an operation is simple: The interface selects the operation from the list of operations to start, changes its status to "starting," invokes the PCS system (which, in turn, reads all the production parameters through the interface), and updates the operation status to "started." While running, a PCS will journalize a number of entries before finally setting the operation status to "completed."

Since the recipes mix manual and automatic operations, special operations are inserted at the beginning and end of a unit procedure for allocation and release of the unit. The start operation is kept active and may serve as a repository for journals regarding critical alarms and events during execution of the unit procedure.

The cleaning management is a good example of how to make the two systems cooperate, using their inherited characteristics of database and real-time systems. Both systems have registered a unit-cleaning status. It is the responsibility of the PCS system to run timers on clean states in order to degrade these when holding times expire. The MES system must record the actual cleaning status to prevent production on unclean equipment. Consequently, PCS informs MES through the interface of changes in equipment cleaning status, even if no batch is running. Note that the actual status is always returned from PCS to MES.

On the other hand, PCS is not aware of the resulting cleaning status after a CIP operation on the unit, since it is controlled by operations and parameters running on the CIP substation. Therefore, the special start and stop operations have hygienic parameters added to their list of parameters, indicating to PCS which status to set after "start" or "end." For example, in a production procedure, the hygienic parameter at start sets the "production" cleaning status, and at the end, it sets the "used" cleaning status. In both cases, PCS updates the current cleaning equipment status accordingly. Also, at start, a check is performed to ensure the synchronization of the two systems. If not synchronized, then PCS will abort the procedure.

For CIP procedures, the cleaning results are calculated in MES, which looks up the resulting cleaning status in a table, depending on the cleaning status at start, the action, and its result. The resulting cleaning status is transferred to PCS as a parameter value during the stop operation.

This completes the example of how a transfer of responsibility from a PCS to an MES reduces the complexity of the PCS code.

Highlights and Benefits

The IBP project produced some 600 master recipes and reports, covering more than 17,000 operations with a total of about 600,000 associated parameter values. During the past 2½ years, many recipes have migrated to higher versions, mainly due to parameter changes. The largest production master recipes now hold version numbers in the mid-twenties.

A production control recipe in the purification area typically spans some 30 units, while CIP and solvent mixing only spans 3 units. The MES database contains data from approximately 125,000 executed control recipes, of which some 6000 are production recipes. This corresponds to 1.83 million automatic unit operations. The journal data table contains 275 million entries, of which some 20 million are entries directly linked to production data, while the rest are related to tracking of execution. The current size of the database has exceeded 250 GB, and it grows at a rate of approximately 80 GB per year.

So what are the benefits? First of all, it works. It is flexible, modular, and coordinates the execution and collection of data in this huge factory, providing insight to the process and generating consistent documentation. The separation of control responsibilities between PCS and MES at the operation level not only made parallel development possible during the project, but also resulted in a robust architecture that has avoided most of the potential problems with coordination, timing, performance, and so on, which could have become a major concern for such a large system. The reason for this seems to be that the splitting was done at a level that combines the real-time action of PCS and the database transaction scheme of MES in a synergy that exceeds the performance of the individual components.

The MES system includes batch-tracking capabilities—both upstream and downstream—which help during batch release. The reporting system has been extended with a deviation-reporting facility, which is able to extract records of journals deviating from predefined acceptance intervals.

The system is robust and users have gained so much confidence with it that they consider abandoning all CIP reporting, keeping only deviation reporting and, otherwise, relying fully on the MES system for the cleaning of units. Ongoing projects also seek to dispose of log books and replace them with manual operations during batch execution.

The system has been easily expandable over time. Additional equipment has been integrated into the system without problems. The system modularity also proved its flexibility in a recent upgrade project, regarding one of the purification lines. As the piping changed, so did the recipe procedures, while the compound operations reflecting process modules and units remained unchanged to a large extent. Recipes were easily cut at relations between compound operations, rearranged, and reassembled into new production recipes and overwritten with new production values.

An interface to corporate Enterprise Resource Planning (ERP) or other material control systems has not been developed—nor has an electronic documentation management system. Instead, a new generation of corporate projects for Lean manufacturing is coming up. The database provides data for optimizing projects. The data are simple to extract, either for reporting or for further aggregation at higher-level IT systems. For all these extensions, the database is ready and available.

History

It should be obvious that you cannot create a system like this just by looking at requirements. You must have the necessary experience to build on. IBP was the peak of a long haul within the MES domain. Looking back at the evolution of the MES integration toolbox, some decisive milestones can be pointed out.

The first experiences came from an intensive automation project that was aimed at establishing the first fully automated "paperless" factory. The factory was designed to manufacture disposable insulin devices. The production comprised the full range of finished pharmaceutical processes, from formulation through filling, inspection, assembly, and packaging.

Originally, the database was introduced to fill the gap between a central ERP system and Distributed Control Systems (DCSs). The database contained just the batch engine. It was introduced to assembly and packaging cells in 1998, but it was soon realized that the combination was insufficient to model the more complex recipes used in the formulation cell. The logical formulation model at that time included no less than six layers of operations within operations, which led to the construction of the graphical recipe editor.

The "paperless" factory, of course, also included an interface between the original MES and a document management system for both batch reports and deviation handling. The aseptic area started production in 2000.

The solution included an interface to an independent, third-party material control system. The experiences resulted in the design of a material control subsystem within the database itself for the next generation of projects over the years 2001 to 2003. Although the automation projects were not quite as ambitious as the original, the MES toolbox proved to be a good, scalable solution for semi-automatic manufacturing, providing easy-to-operate interfaces for the operator.

Impact on Project Execution

The fast and safe execution of this large project led to more contracts. Currently, MES solutions have been introduced in a number of purification plants for which the project execution time is continuously decreasing, reducing the time to market. The first successor was completed in just 14 months from contract to handover.

The main reason for this is the modular engineering concept, which allows reuse of compound operations and report components. The combined modular development of recipes and reports has proved to be a very productive way of engineering an application.

The master recipe tool has demonstrated its abilities as an intuitive specification and design tool, as well as a powerful engineering tool that, in combination with the execution engine, has the flexibility to meet and execute complex recipe structures. It provides a good starting point for maintenance of the solution after it is handed over to the customer.

Project execution in parallel modules allows for fast-track engineering. *Fast track* means shorter time to market and therefore faster return on investment. Fast-track engineering makes it possible to initiate capacity expansion projects later and thereby reduce the risk. A library of validated compound operations to control any modular building block is a reassuring starting point.

The validation principle is bottom-up. Compound operations are validated off-site, progressing upward toward the recipes. Recipes are commonly validated on-site. For families of recipes that use the same top-level compound operation, a so-called golden-master recipe is selected for test, while the other members of that family are validated automatically against the golden one.

For a pair of identical twin factories, we even succeeded by installing a copy of the entire MES database from one to the other, emptying it for production data and otherwise modifying just one single installation parameter—the PCS prefix identifying the plant coordinates. The entire MES, including configuration, was consequently validated by an installation qualification only.

With the configurable simulator at hand, even batch reports can be extensively tested off-site by using the system abilities to simulate PCS feedback and journalization of process data.

The Future

The integration toolbox fills the void between old-fashioned PCS and ERP systems. These days, vendors on both sides squeeze the gap by extending functionalities normally belonging to the MES domain, and vendors of standard MES packages provide more and more features to bridge the gap.

The system-integration toolbox may, however, still find a niche in a world of upgrades and stepwise implementation of MES functionality, due to its unique flexibility, especially regarding recursive operation levels.

However, from an engineering consultant's point of view, solutions based on commercial MES systems are preferable. Our experience within automation integration calls for broad flexibility in the MES domain, which we hope to be able to find in the upcoming standard systems.

CHAPTER 14

MES Roll-out in a Regulated Environment: Reducing the Costs of Validation Based on Risk Assessment

Presented at the WBF European Conference, November 13–15, 2006, by

Wim De Bruyn
Lecturer IT
wim.debruyn@hogent.be
Hogeschool Gent, Schoonmeersstraat 52, 9000 Gent, Belgium

Bert Van Vreckem
Lecturer IT
bert.vanvreckem@hogent.be
Hogeschool Gent, Schoonmeersstraat 52, 9000 Gent, Belgium

Abstract

Manufacturing Execution Systems (MES) are used more and more in regulated environments (e.g., pharmaceutical, food, chemical production) to control the production processes. When implementing an MES solution in a regulated environment, rigorous and well-documented validation is a mandatory part of the roll-out.

It is important to perform risk management using a systematic approach. Possible unexpected events need to be identified and organized. Actions need to be taken in order to prevent or at least minimize negative consequences. To reduce costs in this process, it is vital to prioritize the most risk-sensitive parts, not only in the manufacturing process, but also in the IT development process.

This chapter seeks to help you in defining or evaluating a strategy to assess risks during the implementation of an MES solution and, making use of the evolving standards, using the method of risk-based validation. Guidelines to reduce risks in an MES roll-out and to take advantage of standardized approaches will be explained and illustrated with real implementation examples. A multi-tier validation approach is proposed, which results in lower costs for validation and a continued lower cost in supporting this validated state.

MES: Growing to Standards

A Manufacturing Execution System (MES) is a dynamic information system that drives effective execution of manufacturing operations. Within the IT framework, it can be treated as a layer of abstraction between low-level Manufacturing Control Systems (MCS) and high-level Enterprise Resource Planning (ERP) systems. MESs close the information gap between the management processes at the ERP level and the production processes and plants at the automation level. The growing popularity of MES has resulted in a number of industry standards such as the ISA-88 standard for batch control and the ISA-95 standard for the integration of enterprise and control software. Both standards define models and terminology that can be used for building MES software and interfaces with the ERP and MCS layers. Examples are the OPC-XML (based on ISA-88) from the OPC Foundation and Business To Manufacturing Markup Language (B2MML; based on ISA-95) from the WBF, which form a standard alternative to the proprietary interfaces such as SAP's IDOC. These new standards have made it possible for MES software to evolve from custom-made systems toward standardized, interoperable packages, as shown in Figure 14.1.

MES in regulated environments offers many advantages. The advanced data acquisition, analysis, and reporting facilities in an MES allow for full traceability of produced goods, and this is essential in complying with increasingly strict regulations. MES can help optimize production, resulting in lower costs and higher profit margins. When implementing an MES solution in a regulated environment, validation is vital. Does the MES comply with all relevant regulations (enforced by governments, certification, etc.)? What risks are associated with the MES? How will you react if things go awry? Can you prove during an audit that you thought very hard about all this?

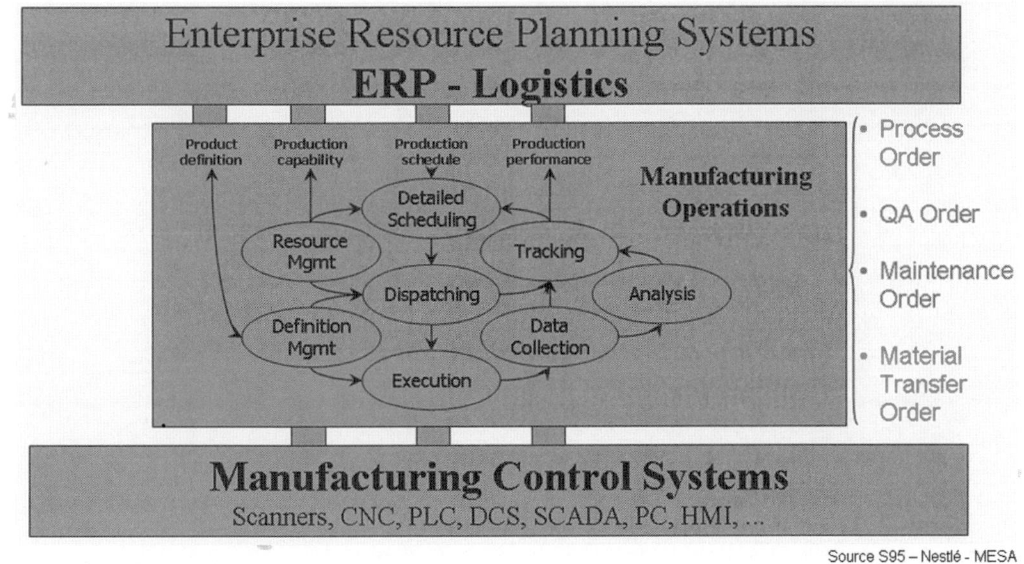

Figure 14.1. The ISA-95 manufacturing operations management model as a basis for MES.

Good Automated Manufacturing Practices

The standard model for validation in the pharmaceutical industry is laid out in the *Good Automated Manufacturing Practice (GAMP) Guide for Validation of Automated Systems*. GAMP is effectively a methodology for specifying what a process should do and then ensuring that the delivered solution exactly meets the initial design specifications. To define the life cycle of an automated system, the GAMP guide uses the more than 50-year-old "V-Model," a widely used model for commissioning and qualification of engineering installations, published as a standard by the Federal Republic of Germany, as shown in Figure 14.2.

In 2004, an up-to-date version of the model—the "V-Model XT"—was published, which takes new methods and technologies like component-based development or the test-first approach into account. The GAMP guide states that, during each phase, the user needs to formulate the qualifications that the product (in our case an MES) should comply with. However, not all responsibility lies with the user; during the coding phase of the project, the supplier is expected to rigorously test the written code. The cost of validating applications is usually estimated at 10% to a staggering 50% of the total cost of implementation. Late introduction of test and validation activities leads to an enormous backlog of validation tasks, with the danger of falling back into retrospective validation. Validation after the fact is less and less accepted and requires even bigger costs and time to redo the work that should have been done before.

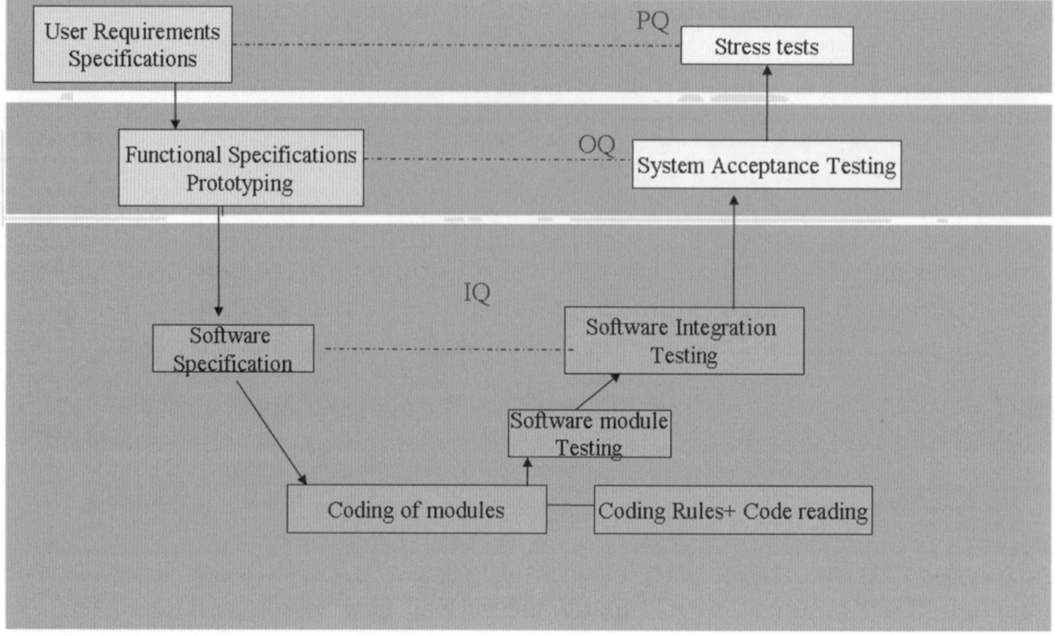

Figure 14.2. The GAMP V-Model.

If we can significantly reduce the amount of validation work, but keep the same quality and validation level, imagine the costs we can save! To achieve this goal, prioritizing is essential. You must first choose the validation items that are most important and focus on these. We propose a risk-based validation strategy that follows this principle.

Risk Management to Structure, Focus, and Reduce the Validation Effort

Risk management is a challenge for any organization and can be defined as the continuous systematic approach to identify possible, unexpected events and the organization of measures and actions to prevent or at least minimize negative consequences. In addition, risk management also seeks to make use of or introduce positive consequences. In order to manage risks effectively, it is important to classify them according to some relevant characteristics. For example, the probability that a certain risk occurs in a project greatly influences its importance. Highly probable risks should be tackled before very improbable ones. The frequency of appearance is another thing to consider. Some risks have greater consequences than others, and this has to be taken into account when taking countermeasures. A final consideration is the detectability of a risk and whether countermeasures can even be taken (i.e., avoidance).

We need a strategy to address these risks. What are the risks? Which ones are "show stoppers"? Which ones are acceptable? Which actions are taken in case of certain events (e.g., stopping the project)? All possible risks need to be identified, assessed, and measured in order to control them.

Identifying risks is easier said than done. As Goethe once said, "The biggest difficulties are where we don't look." The events on September 11, 2001, are a dramatic example; people were not conscious that a terrorist act of that scale was possible. Too often, the risks are assumed, and one doesn't look for them creatively. To compile a list of all possible risks in a project, one should organize brainstorming sessions or interviews, invite experts, or perform simulations. The risks should be classified according to their origin (e.g., internal, external, juridical, regulatory), risk objects (e.g., supplier, project engineer), consequences, and other characteristics. It is important to write these down and document the risks formally.

Once you have a list of risks, it is time to put some numbers on them. First of all, determine the likelihood (i.e., low, medium, or high) of an adverse event occurring within a certain time period or number of transactions. After that, it is important to identify the impact of a risk (i.e., low, medium, or high). Not only the immediate effects, but also the impacts on financial and business management, regulatory compliance, and so on, need to be taken into account (Fig. 14.3). The GAMP guide uses three numeric classes of risks, as shown in Figure 14.4.

The probability of detection is important (Fig. 14.5). If the risk is identified and detected, appropriate measures to avoid the consequences of the risk can be taken (i.e., mitigation), and as a consequence, the risk can be managed with a Corrective And Preventative Action (CAPA) plan and placed under control. Of course, real cases are more complex and are solved using elaborated spreadsheets or specific risk-management software, such as the one shown in Figure 14.6.

Of course, the weight factors, estimation methodology, consensus-forming process, sequences of actions, escalation procedures, prioritizing, maximum high-priority risk items, and so on should be documented well and be treated as part of the general risk approach. See Figure 14.7 for a review of the assessment steps.

By measuring risk parameters, risks can be brought under control and managed. Contingencies are added; special attention is given to the transfer of ownership; risks are monitored; prevention is utilized (as much as possible); risks with low severity are accepted; corrections to the processes are be made (i.e., avoidance); and some risks can be transformed into opportunities.

Appendix M4 of GAMP, "Guidelines for Categories of Software and Hardware," provides an overview to categorize risks for the following software products:

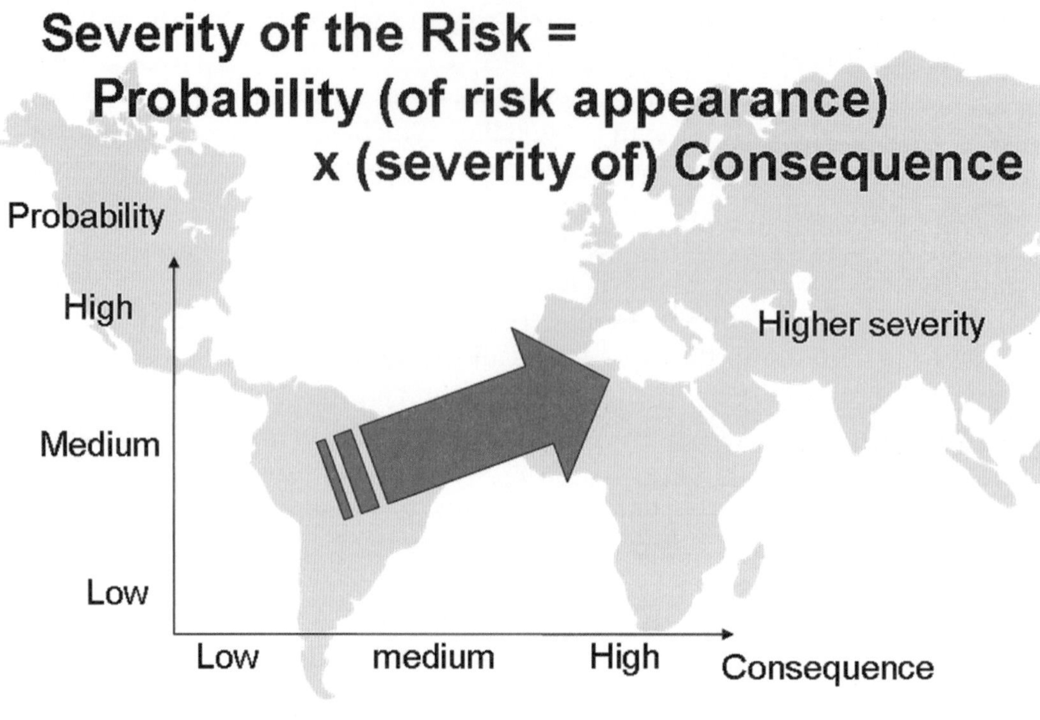

Figure 14.3. These metrics can be used to classify the risk according to severity.

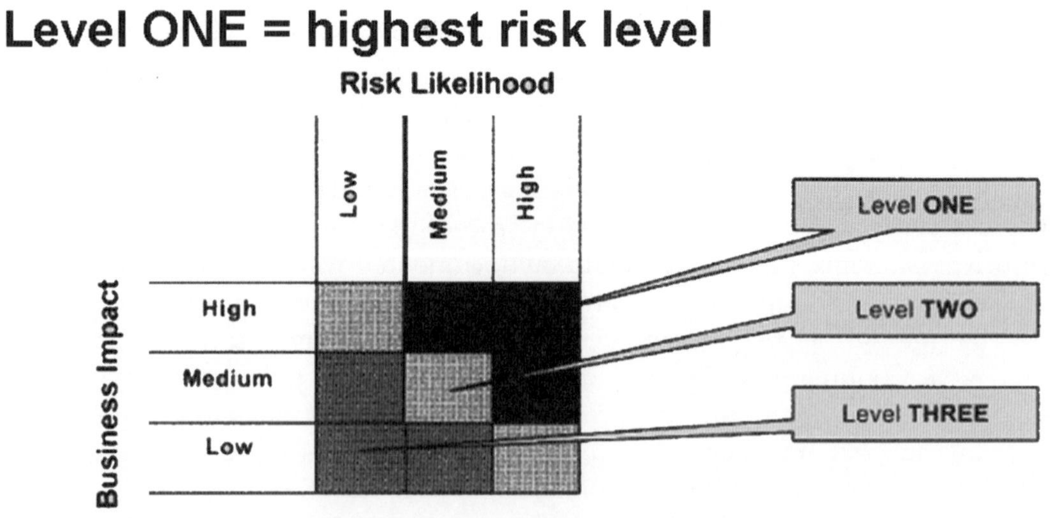

Figure 14.4. Risks should be addressed in order of decreasing severity.

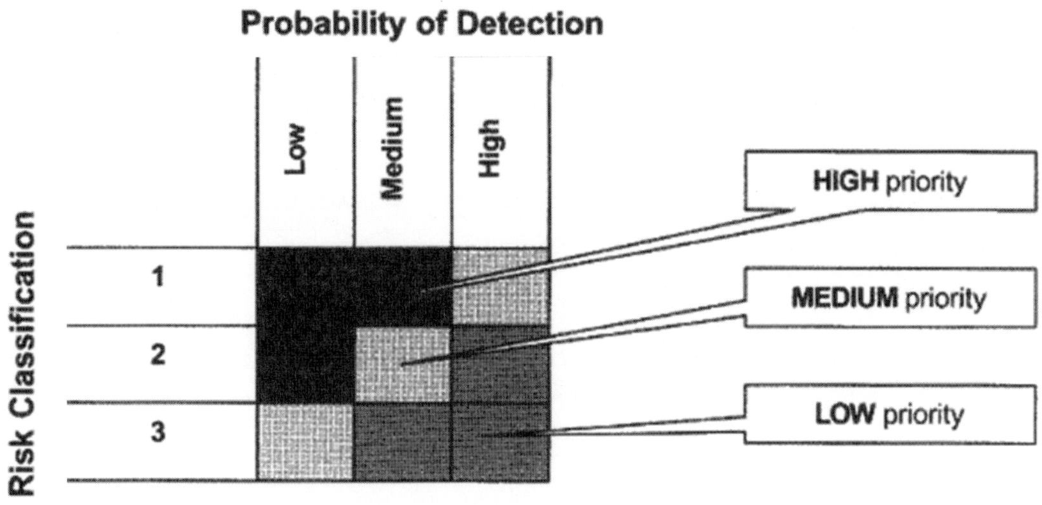

Figure 14.5. Classifying risk by probability of detection.

Project Title / Risk Assessment Overview						Project Number:			
Assessment Scope / Assumptions Made									
Function	Sub-Function	Assessment of Risk							Measures
		Relevance	Scenarios	Likelihood	Impact	Class	Detect	Priority	
Risk Assessment Approved by:									

Figure 14.6. A risk assessment form.

- Operating systems
- Controllers and intelligent instruments
- Standard software packages
- Configurable software packages
- Custom-built systems

Although it's a lot of help to start the risk assessment of a certain product, it's clear that a complete MES system (e.g., hardware, software, network, Input/Output

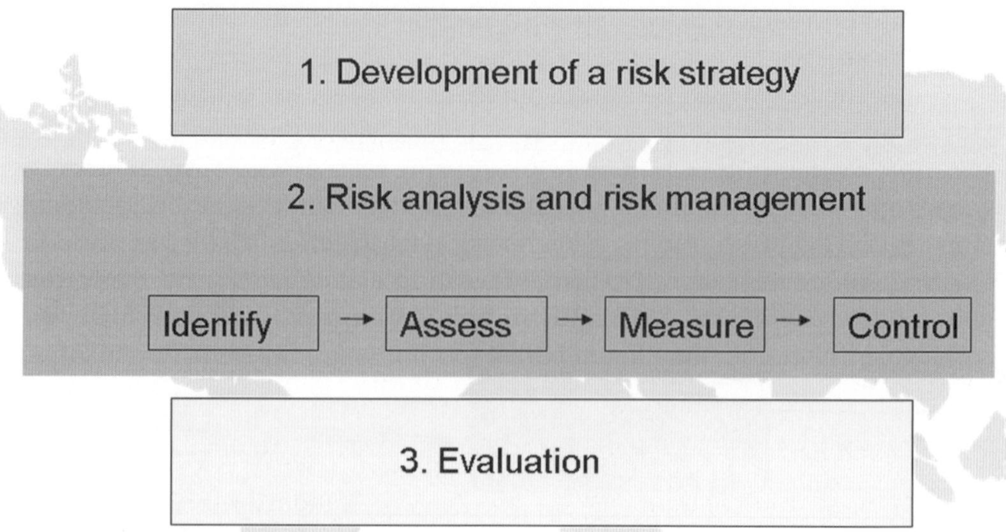

Figure 14.7. Steps in risk assessment.

[I/O] points) is assembled from modules, libraries, routines, components, dependencies, and operating system components that fall into different categories and require different validation effort levels. No specific validation is done for some general operating system parts, although their features are functionally tested and challenged indirectly while the application is being tested. Of course, change control should be applied to manage different upgrades and patches and should be formally documented.

Another example is Process Analytical Technology (PAT), with instruments incorporating firmware that is configurable and, even in some cases, programmable. Network links to databases (e.g., calibration data), network links to applications (e.g., monitoring), backups, connection monitoring, and implementation in feedback loops will require different levels of validation. Custom software requires a full validation of the software development life cycle. Evidence of every step and test in the software development should be documented.

It's clear that with MES we come to a multi-tier validation plan. The functional risk and the developmental risk of the software modules need to be evaluated.

Multi-tier Validation Plan

After having a risk and benefit analysis, we need to rank the appropriate system validation effort and depth:

1. All high-risk systems, all systems in which risk outweighs benefits, and unique (i.e., custom) software
2. All remaining systems in which medium risk is associated with an equal level of benefits and all software consists of a prepackaged, widely used core that is customized or set for unique use
3. All remaining systems in which benefits exceed risks and all standardized or prepackaged software

If we translate the ranking of validation efforts to a typical, standardized MES software structure, then we obtain the following:

1. Developed add-ons for special functionality
2. Configuration of data and specifications in standard MES
3. The standard core MES software

The following list describes the validation effort needed for each level:

1. Complete validation—trace matrix tracking, design through testing, normal and border case testing, code review, and planned revalidation
2. Normal performance and user acceptance testing—vendor audit and periodic status (level) review
3. User acceptance testing—examination of published reviews and periodic status (level) review

Ideally, a prior approval or acceptance of a validation strategy should be obtained by asking for a review by experienced consultants or a review by someone from the Food and Drug Administration (FDA) or its local equivalent. A MES or Laboratory Information Management System (LIMS) solution in pharmaceutical production can be seen as a medical device that incorporates software as well as hardware.

In Sandy Weinberg's *Good Laboratory Practice Regulations* (3rd edition), an example is given from a major pharmaceutical LIMS user (in the United States) who saved more than $2 million (81%) by applying a three-tier validation. Comparable savings in maintenance of the validated state are common.

In the Roche Penzberg biopharmaceutical factory (Germany), savings of more than 50% have been made in the validation of a green field production plant by using standard MES components. Also, the modeling phase standardization resulted in lower setup costs.

We remain strongly interested in your feedback, questions, or interactions concerning this subject. The Department of Business Information and ICT of the Hogeschool Gent (in association with University Gent), Belgium, is starting up an MES center of expertise that will provide academic research (e.g., PhD or master's degree) and studies in the industry for practice training of ICT students and continuous education. You can reach us at wim.debruyn@hogent.be and bert.vanvreckem@hogent.be.

Further Reading

Instrumentation, Systems, and Automation Society. 1995. *ANSI/ISA-ISA-88.01-1995: Batch control part 1: Models and terminology*. Research Triangle Park, NC: ISA.

International Society for Pharmaceutical Engineering. 2001. *GAMP4: Good automated manufacturing practice guide for validation of automated systems*. Ed. Sion Wyn. Tampa, FL: ISPE.

International Validation Forum. 1995. *Validation compliance annual: 1995*. New York: Marcel Dekker.

Manufacturing Enterprise Solutions Association. MESA International resource library. http://www.mesa.org/resourcelibrary.

SAP. SAP Web site. http://www.sap.com.

Weinberg, Sandy, ed. 2002. *Good laboratory practice regulations*. Exp. 3rd ed. New York: Marcel Dekker.

Wingate, Guy, ed. 2004. *Computer systems validation, quality assurance, risk management, and regulatory compliance for pharmaceutical and healthcare companies*. Boca Raton, FL: CRC Press.

WBF. WBF Web site. http://www.wbf.org.

CHAPTER 15

Fast and Efficient Configuration and Integration of Automation Solutions from a Global Perspective: A Practical Approach

Presented at the WBF North American Conference, March 5–8, 2006, by

Leif Poulsen
Senior Specialist
lpou@nne.dk
NNE A/S, Gladsaxevej 363, DK-2860
Soeborg, Denmark

Abstract

Fast and efficient configuration and integration of automation solutions is a prime requirement in the pharmaceutical industry. The key to success is to use standardized modules that can be preconfigured and prequalified before implementation on-site, possibly by suppliers in low-cost regions. This will save time and money for both implementation and validation and ensures flexibility for future adaptations. In order to ensure full exploitation of this principle, it is necessary to set up an efficient execution framework that supports all project phases with the necessary standards and guidelines.

This chapter describes how you can adopt modular engineering and set up the overall framework for fast and efficient configuration and integration of automation solutions worldwide. It comprises tools for analysis and specification of requirements based on ISA-88 and ISA-95 models, functional specifications using computer-based modeling tools, configurable library modules for various platforms, and computer-based tools and methods for testing according to the Good Automation Manufacturing Practice (GAMP) guidelines. These are all tied together in an overall Automation-Project Activity Model (A-PAM) for handling of the necessary documents. An illustrative example covering the implementation of a new Active Pharmaceutical Ingredient (API) plant shows how it works in practice.

The Market Challenge

Fast launch of products is very important in the pharmaceutical industry. After many years of heavy investments in Research and Development (R&D), it is vital to get to the market fast when the approval of a new drug is in place. As outlined in Figure 15.1, it is normal to start up the design and construction of new production facilities for a new drug upon successful completion of Clinical Trial Phase II. However, due to uncertainty about Clinical Trial Phase III and uncertainty about the real market potentials for the new drug, it would be preferable to postpone the investment in new production facilities as much as possible. Most pharmaceutical companies, therefore, ask for late start and very fast execution of new plant projects. This requirement has become the prime challenge for engineering companies.

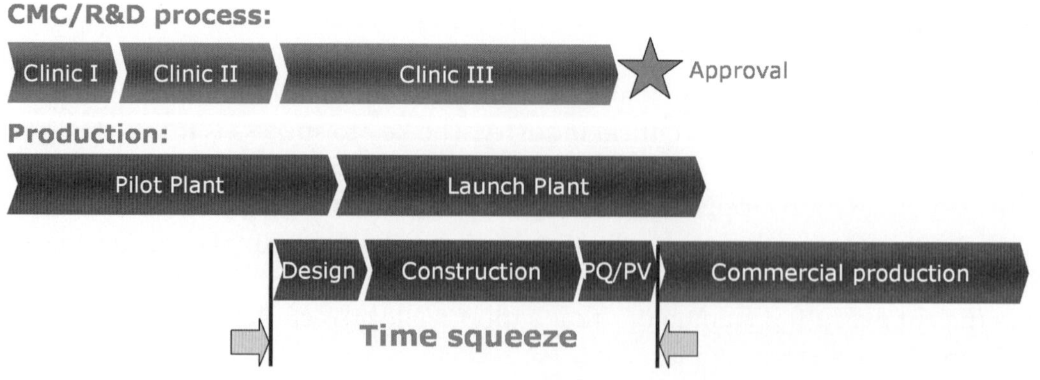

Figure 15.1. Fast launching of products is vital in the pharmaceutical industry.

The Modular Approach

The modular approach is the most obvious response to the requirement for fast implementation. If the process equipment can be designed as large, standard LEGO bricks, it is possible to make a customized solution very quickly. Figure 15.2 illustrates how this approach can be applied in all phases, from conceptual design to handover and support of the solutions. Each step in the overall project implementation model is briefly described throughout this section.

Conceptual design involves the definition of the modular structure by use of simple and clear module diagrams. It is important that the project scope is unambiguous yet still defined at a sufficiently high level to allow fast changes and evaluations of alternatives. Changes can quickly be propagated, since space allocations, building structure, utility requirements, and investment costs are all derived from the modular structure.

Basic design includes more detailed specification of the requirements for each module. In this phase, it is possible to apply systematic reuse of module specifications, both within and across projects. Reference or standard modules can be modified to meet the specific demands of the client. Besides being cost-efficient, the reuse provides a more exact cost estimate and excellent benchmarking possibilities.

Detailed design includes detailed specifications and models for implementation of each module. The modular structure confirmed during basic design must be

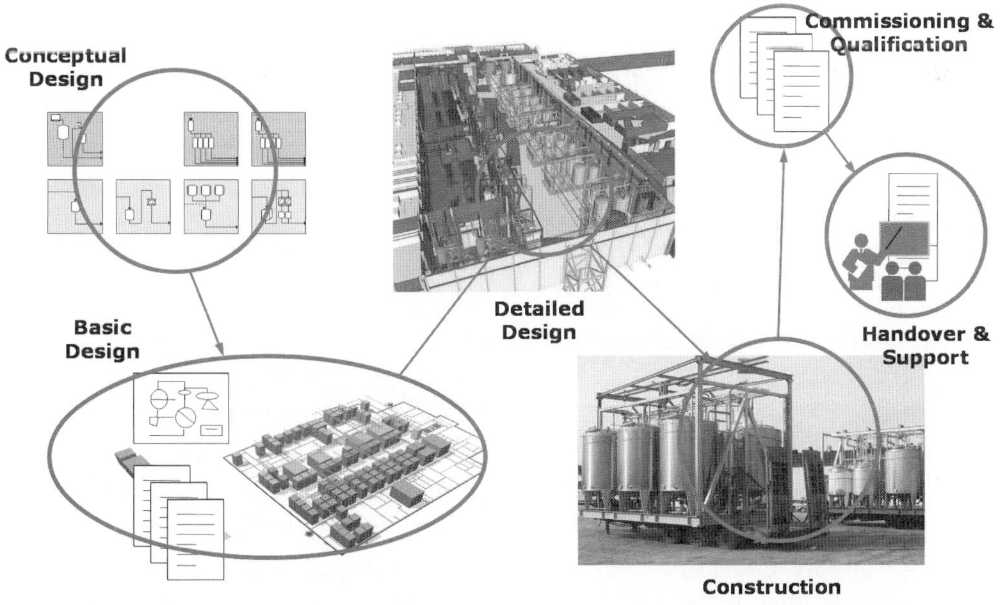

Figure 15.2. Modular engineering from conceptual design to handover and support.

retained throughout the entire project—in the cost structure, contracting, and project organization—thus giving everyone clear objectives that match a resulting process or building function. All aspects of the facility must be designed according to the modular structure. Due to the strict interface management, it is still feasible to add or substitute modules—as long as the physical production or delivery time allows it. The modules can be designed by different vendors (as their capabilities allow).

Construction based on modular engineering allows for the simultaneous off-site construction of each single process module and building part on different locations, depending on competencies and cost. The units are then transported to the site and lowered into the plant, where the final Installation Qualification (IQ), Operational Qualification (OQ), and Process Qualification (PQ) are performed.

Qualification can be performed quickly and efficiently, since a major part of the equipment is commissioned and qualified in parallel and potentially off-site, without dependencies to adjacent upstream and downstream equipment. Modular engineering even allows for qualification activities, such as IQ, to take place off-site. A plant based on modular engineering has many advantages for the end user:

- It is inherently flexible and expandable.
- The structure of the process equipment is easy and repetitive.
- The automation system follows the same structure.
- The interdependencies in operation (e.g., Clean In Place [CIP]) are minimized.
- The validation documentation is clearly structured.

Because modular engineering is founded on the reuse of well-known solutions, it increases the odds for a fast approval process and ensures reliable production from the day the facility is handed over.

The Definition of Standard Modules

The definition of standard modules can be based on recognized engineering standards like ISA-88 and ISA-95. As outlined in Figure 15.3, a complex plant can be decomposed into a number of predefined process modules (e.g., fermentation lines). Each process module can then be broken down into a number of predefined units (e.g., tanks). Each unit can be broken down into a number of predefined Equipment Modules (EMs; e.g., tank-filling systems), and each EM can be broken down into a number of predefined Control Modules (CM; e.g., valves). Each level

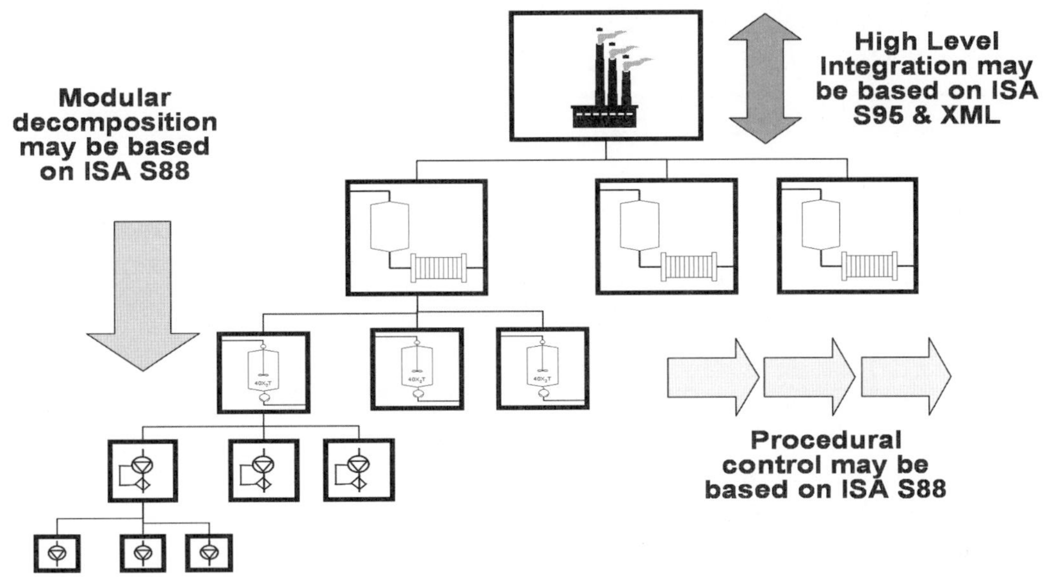

Figure 15.3. Decomposing a complex plant into simple modules based on recognized engineering standards.

in the decomposition of the equipment part has its equivalent for the automation part, and in this area, it is possible to reuse predefined configurable software modules, kept in a library for each of the most commonly used platforms.

Among the modular building blocks, the process module is especially interesting for fast configuration of new, customized solutions (Fig. 15.4). A process module may be defined as an independent module that can handle the execution of the batch among several units, including the CIP of units in between process batches and in parallel with production. Several batches may be executed simultaneously in a process module. (This definition corresponds to a process cell in ISA-88.01 terms.) A process module must be defined by a cross-disciplinary effort in order to cover building, mechanical, and automation aspects. The definition of interfaces covering the necessary flow of material, energy, and information is very important. The success of modular engineering is fully dependent on well-defined interfaces. For the automation part of a process module, this covers batch control functions able to execute and document each batch defined by a recipe. Normally, the recipe is downloaded from a higher-level system (e.g., a Manufacturing Execution System [MES]) that receives the necessary data for the reporting upon batch completion.

As outlined in Figure 15.5, the great advantage of using standardized process modules is that very different total solutions can be established by repeated use of a limited number of standardized process modules.

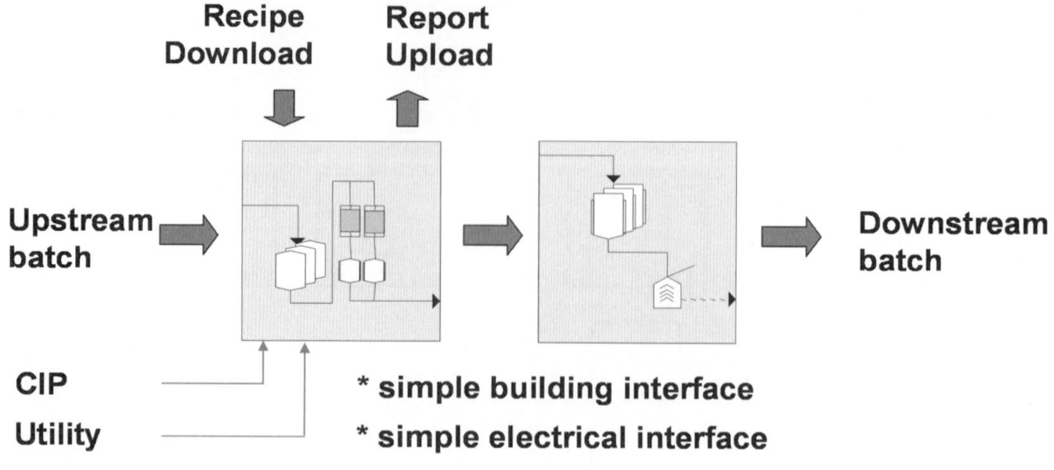

Figure 15.4. The main building block is the process module.

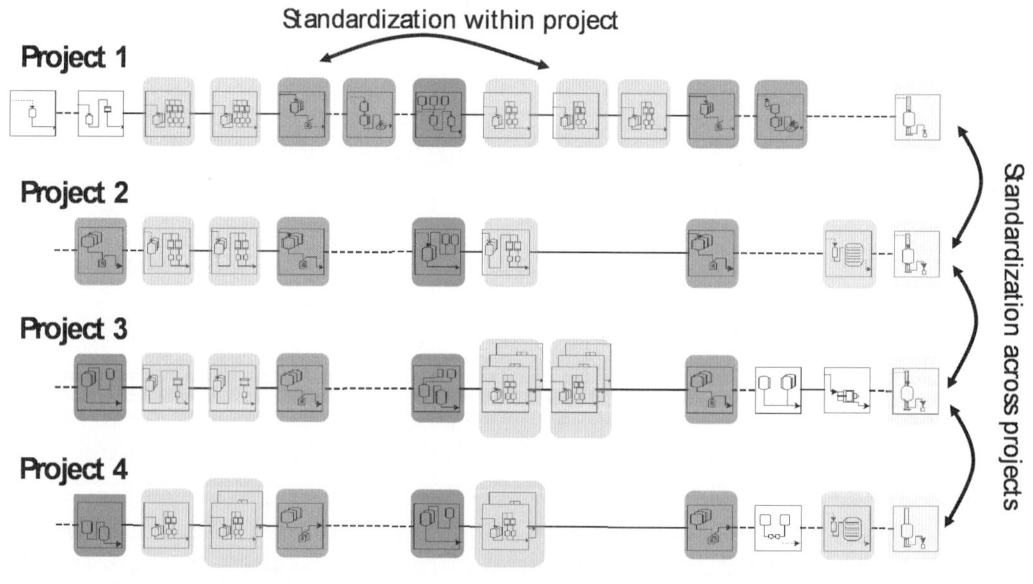

Figure 15.5. Standardization of process modules within the project and across projects.

The Automation Execution Framework

In order to implement projects quickly and efficiently, it is necessary to operate with predefined project activities, each supported by document templates, specifications, guidelines, and support tools. An example of a general A-PAM is outlined in Figure 15.6.

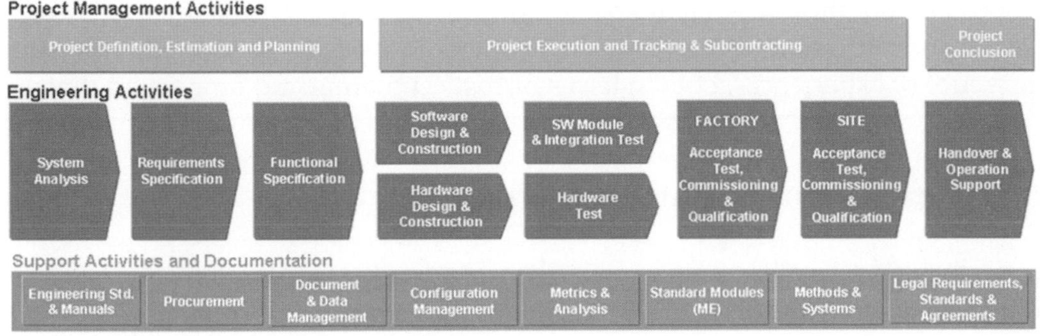

Figure 15.6. A-PAM, based on GAMP 4 terms.

The purpose of the A-PAM is to ensure fast, efficient, and predictable implementation of automation solutions and to continuously improve performance by the collection of experiences and good practices from executed projects, thus making this information available for upcoming projects in a usable and effective manner. The A-PAM model, which is based on GAMP 4, comprises three parallel activities:

1. The project management activities describe how to specify, execute, and hand over the project in close collaboration with the customer and subcontractors.
2. The engineering activities describe how to specify, design, implement, test, and document the solution in compliance with predefined standards and procedures.
3. The support activities describe how to support the project organization (e.g., by provision of predefined standard automation modules) and how to collaborate with subcontractors and partners about the implementation.

All activities defined in the A-PAM are supported by predefined document templates, specifications, and guidelines based on previous experiences, including illustrative examples of how to do things.

It is very important to consider the predefined transition requirement between the project activities that states which documents need to be completed or available before you are allowed to proceed to the next activity. This is especially important if you want to hand over some of the project to a subcontractor.

The full benefit of the A-PAM model is gained by using it from the very beginning to the very end of each project and also during the preparation of project proposals and project evaluations. Using the predefined structures of project phases and process modules provides a splendid basis for systematic recording and reuse of experiences, both in terms of cost estimating and time scheduling.

The Automation Engineering Tools

The fast project implementation must be supported by efficient project management and engineering tools. Figure 15.7 outlines a number of tools that can help manage complex automation engineering projects.

Project planning is supported by tools that help estimate the necessary amounts of resources and necessary time for each project phase, based on past experience. Projects are planned around a database comprising key information about each standard module making up the complete solution.

Project monitoring is performed by a Web-based tool that allows project managers and coordinators to update the current project status for each part of the project; this tool provides project managers with key information about scope, price, time, quality, customer satisfaction, and team performance—all of which are illustrated by colors indicating if intervention is necessary (e.g., green is good, yellow is warning, red is alert).

Analysis and specification are supported by a computer-based modeling and specification tool based on ISA-88, comprising a library of predefined EMs and procedural elements for fast and easy set up of customized models and specifications.

Design and implementation is supported by predefined design documents for process modules associated with precoded and prequalified code modules for the most important implementation platforms.

The testing and prequalification of standard software modules is supported by an automated test environment with predefined test procedures and automatic logging of any deviations. This makes it very fast to test and qualify any changes to the software modules and efficiently keeps track of the complete set of software versions.

Documentation is handled by a document management system with a predefined setup of document folders that correspond to the breakdown in process modules and to each document with predefined attributes, for fast retrieval and updates.

The Global Supply Network

The process modules for a specific plant may be supplied from a global network of collaborating suppliers, as outlined in Figure 15.8. For example, a local customer in United States needs a new plant, and they analyze and specify the requirements

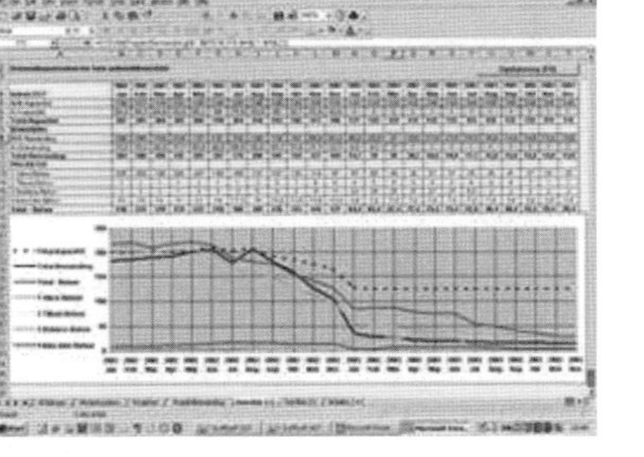

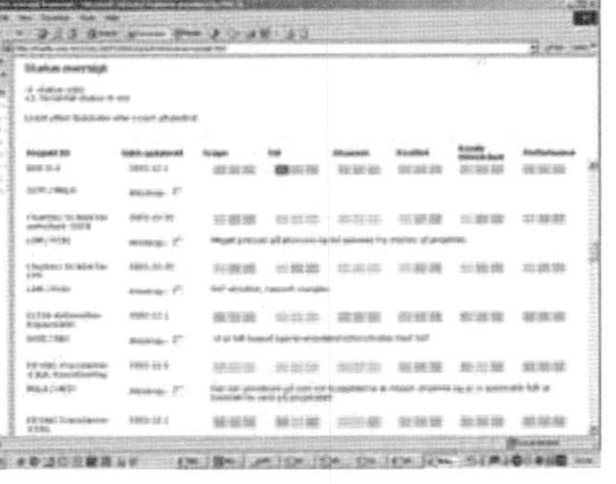

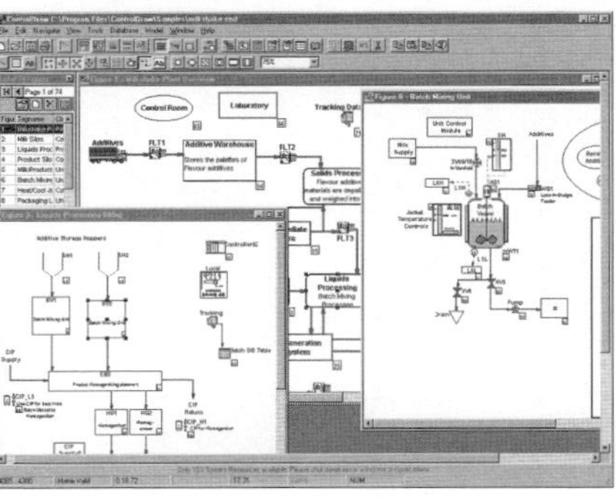

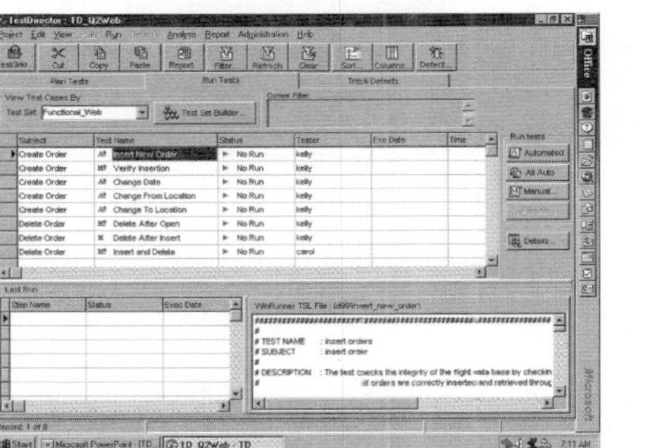

Figure 15.7. Examples of computer-based project management and engineering tools.

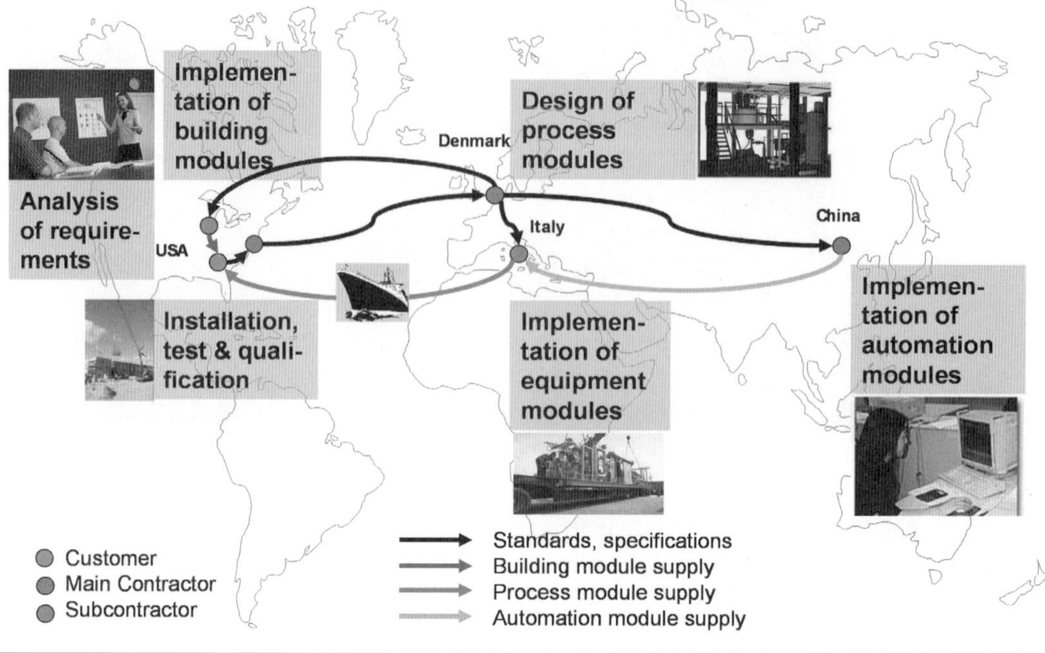

Figure 15.8. Example of global network for supply of configurable modules.

for this in collaboration with the main engineering contractor, who has an office close to the customer. The requirement specifications are handed over to a central back office of the main contractor in Europe, where a customized design based on predefined equipment, process, and automation modules is defined. The design specifications are then handed over to subcontractors that take care of the implementation. For the automation part, the design is accompanied with configurable software modules and configuration guidelines for the chosen system platform. The automation subcontractor (in this case, a subcontractor in China) configures and tests the software modules according to the standards and hands over the software modules to the equipment subcontractor (in this case, a subcontractor in Italy). The equipment contractor that has implemented the process modules based on design specifications from the main contractor will then integrate the solution with the software modules and test the complete solution before shipping it to the customer in the United States. Parallel to this, the equipment subcontractor has implemented and installed the building modules, based on the design specifications from the main contractor. All modules can now be installed, integrated, and tested on-site by the customer. Due to the use of the modular engineering approach, it has been possible to run all necessary engineering activities in parallel at the most cost-efficient places in the world and to establish the new plant faster, cheaper, and with better quality, compared with the traditional sequential implementation of buildings, equipment, and automation solutions on-site.

A Practical Example

An example of state-of-the-art modular design and engineering is described in this section. In 18 months, it was possible to establish a new turnkey biotechnology plant for the production of FVIIa, a hemophilia medicine (Fig. 15.9). FVIIa enables the blood to coagulate. FVIIa has been approved for the treatment of bleeding in inhibitor patients (i.e., patients with hemophilia A or B who have developed antibodies against coagulation factors VIII or IX, which are normally used for treatment of hemophilia patients). The target group is patients for whom there are no alternative treatment possibilities.

The new plant is a self-supporting unit with own energy supplies. It consists of five building sections, comprising an administration building with a connecting corridor to the process building, utilities, and energy center. The process building has large glass facades, making the compact building seem light and transparent. Windows and roof sections can be easily dismounted to allow for the insertion of EMs. The classified clean rooms in the process building have been isolated in separate glass buildings. The quality level is set to comply with the requirements of the authorities and the client. The modular construction meets the demands of flexibility for a modern facility (Fig. 15.10).

Eighteen months is twice as fast as the normal rate for a pharmaceutical plant of this size and complexity. The key to building the plant in 18 months is the up-front modular design that makes use of known solutions. The modular design has enabled nearly all process and building units to be manufactured in complete modules. Based on the modular design, vendors have manufactured and pretested every single process module, including automation. The backbone and prequalified standard modules for the automation solution were provided by a system subcontractor. The process modules were transported to the production site and

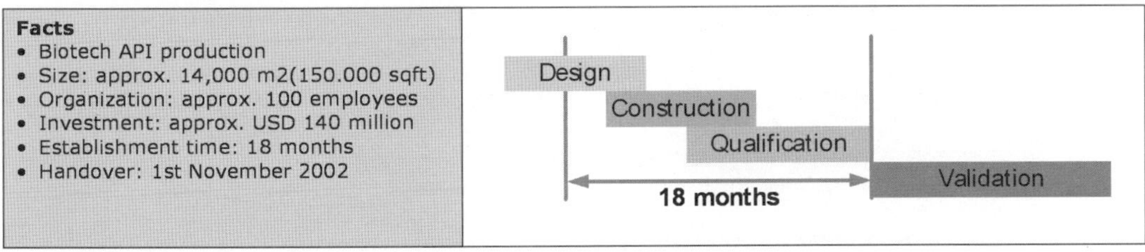

Figure 15.9. Overall facts and time frame for project implementation.

Figure 15.10. From a 3-D model to final implementation of the pharmaceutical plant.

lowered into the plant, where the final IQ, OQ, and PQ were performed. Full documentation, including test documentation, was prepared for each process module, as part of the vendor's delivery. The modular approach was taken up-front in the conceptual design phase, where relevant building and process modules were identified from the initial process flow diagrams (Fig. 15.11).

Requirements for each module were further detailed during the basic design phases. A full 3-D model of the complete plant that includes all process modules was developed in the detailed design phase (Fig. 15.12). The subcontractors implemented the process modules on various locations in Europe.

The development of the automation solutions was based on the structured approach described in ISA-88. As outlined in Figure 15.13, two separate models were developed top-down. The equipment model describes the hierarchical breakdown of the plant into process cells, units, EMs, and CMs. The procedural model describes the hierarchical breakdown of the overall manufacturing procedure into procedures, unit procedures, operations, and phases. The linking between the two models is indicated by the arrows in Figure 15.13. Based on these models, the automation solution was implemented and tested bottom-up (i.e., implementation, test,

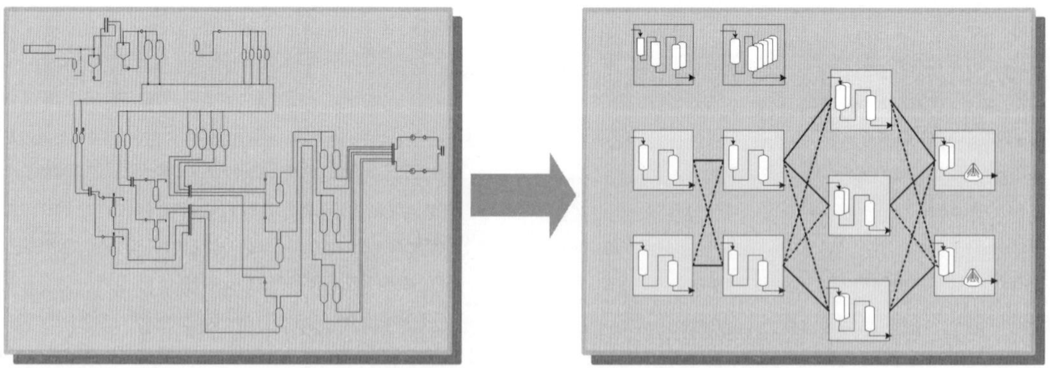

Figure 15.11. Process flow diagrams transferred into process modules.

Figure 15.12. Example of process module from 3-D design to installation.

and qualification of CMs, EMs, and process modules off-site and final integration, test, and qualification of the process modules on-site). The prequalification of all modules off-site was one of the major contributors to the short implementation time.

Conclusion

When a new drug is approved, the market demand for fast implementation of the necessary manufacturing facilities can be met by fast-track engineering based on configurable, standard modules.

A modular fast-track facility costs approximately the same as a traditional facility. Cost-effective production of modules off-site counterbalances shipping expenses, added structural steel, and so on. The structured work packages and streamlined qualification and validation save time and costs in the commissioning and qualification phase. The prime advantages of a modular fast-track approach are the following:

- Shorter time to market and faster Return On Investment (ROI)
- The ability to initiate projects later in the schedule, thereby reducing the risk for wrong decisions and scope changes
- More cost-effective consecutive projects

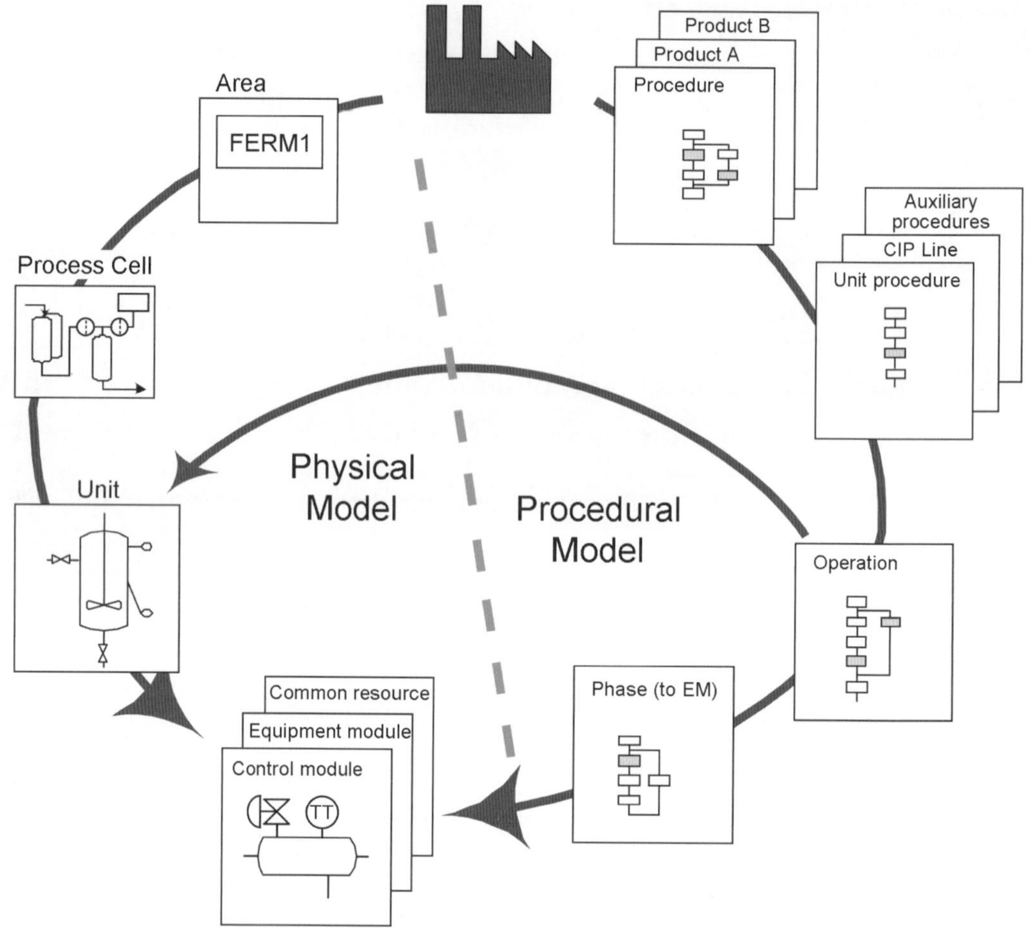

Figure 15.13. Modular engineering approach based on ISA-88.

However, it takes more than modularization to reduce time, risk, and cost. The following aspects are very important for a successful fast-track implementation:

- Clear strategy and goals
- A winning culture and team spirit
- Clear responsibilities for contractors
- Firm design packages (i.e., "approved for construction")
- Flexible building conditions
- Building and process modules engineered and established concurrently

Fast-track implementation also imposes a number of challenges for the user organization:

- Finalize process design in the short design phase.
- Identify known process uncertainties.
- Create, hire, and train a highly professional team to run the plant in less than 18 months.
- Create validation protocols and a quality system that include instructions.

Our experience with fast-track implementation based on modular engineering has shown remarkable results. Step by step, we have approached an overall goal that was defined 5 years ago: establish a new pharmaceutical plant within 12 months. The example described in this chapter cut down the implementation time from 36 to 18 months, and recently, we have delivered a complete new pharmaceutical plant in 12 months. The goal was reached.

Acknowledgments

I would like to thank my colleagues Kasper Bonnevie and Klaus Illum for their valuable input to this chapter.

Abbreviations

A-PAM	Automation-Project Activity Model
API	Active Pharmaceutical Ingredient
CIP	Clean In Place
IQ	Installation Qualification
ISA	International Society for Automation (http://www.isa.org)
ISPE	International Society for Pharmaceutical Engineering
MES	Manufacturing Execution System
OQ	Operational Qualification
PAM	Project Activity Model
PQ	Process Qualification
ROI	Return On Investment

Further Reading

International Society for Pharmaceutical Engineering. 2001. *GAMP4: Good automated manufacturing practice guide for validation of automated systems*. Ed. Sion Wyn. Tampa, FL: ISPE.

Instrumentation, Systems, and Automation Society. 1995. *ANSI/ISA-ISA-88.01-1995: Batch control part 1: Models and terminology*. Research Triangle Park, NC: ISA.

———. 2010. *ANSI/ISA-95.01-2010 (IEC 62264-1 Mod): Enterprise-control system integration part 1: Models and terminology*. Research Triangle Park, NC: ISA.

CHAPTER 16

Risk-based Engineering Assessment and Qualification: A Case Study

Presented at the WBF North American Conference, March 5–8, 2006, by

Chinmoy Roy
Senior Automation Engineer
croy@gene.com
Genentech Inc., 1 DNA Way, MS 252A,
South San Francisco, CA 94080 USA

Tom Johannessen
Automation System Administrator
hemnes@gene.com
Genentech Inc., 1000 New Horizons
Way, Vacaville, CA 95688 USA

Abstract

A recent spate of life-saving drug discoveries is requiring drug manufacturers to bring these drugs to market faster, at a lower cost. These challenges need to be met while maintaining a high level of drug manufacturing safety. Besides the challenge of reduced time to market, there is also an increasing realization that drug costs can be lowered through an environment that is conducive to innovation and continuous process improvement in order to realize process optimization and elimination of wasted efforts in production. To support this environment, a flexible regulatory system that focuses on risk-based approaches to critical areas of the manufacturing process is gaining industry-wide acceptance.

This chapter outlines a risk-based approach to the validation of a plant utilities and facility management automation systems in a bulk-manufacturing biotech facility. The facility management automation system is one of the four key subsystems that regulatory agencies consider to represent the largest risk, and if this is under control, then the manufacturing process is considered to be in compliance. This chapter describes how existing techniques such as Failure Mode Effects Analysis (FMEA) and Fault Tree Analysis (FTA) were tailored and applied to engineering principles in order to develop an integrated approach to good engineering practice and validation testing, without compromising production safety. Business process benefits derived from adopting such a risk-based strategy are also addressed.

Introduction

The past few years have witnessed a spate of new, life-saving drug discoveries. These drugs are required to be manufactured at a lower cost so as to make them more affordable for all. Lower manufacturing costs can be realized through efficiency improvements obtained through a scientific understanding of how formulation and process factors affect product quality. Additionally, a mechanistic understanding of the process results in reduced downtime caused by process and equipment failures.

The drive to lower manufacturing costs through a better understanding of the drug manufacturing process presents quality and validation groups with an opportunity to develop qualification strategies that are not only cost-effective but also regulatory compliant. Developing such strategies provides the added benefit of being less rigid and more conducive to scientific innovation.

The effort to better understand the process provides insight into the technical aspects of the manufacturing process, including the stratification of process risk factors. For factors that present a higher process risk, a detailed functional and area risk assessment is conducted to determine where to focus validation activities. The others are subjected to good-engineering-practice testing. This risk-based approach to validation provides for a safe and affordable drug product. It also provides a level of assurance that validation strategies, developed from an approach based on a level of process understanding, will survive regulatory scrutiny.

Improving Inspection

Risk can be defined in many ways. Before establishing a risk-based validation strategy, it is important to look at the Food and Drug Administration's (FDA) risk-based systems approach to inspection of facilities. This provides insight into one of the FDA's inspection strategies, which could be helpful during evaluations of risk assessment.

Of interest is the FDA's Quality System Inspection Technique (QSIT), which focuses on key manufacturing and quality areas during inspections. The QSIT approach is illustrated in Figure 16.1.

Under this approach, the inspection focuses on four primary areas: Management, Design Controls, Corrective And Preventive Actions (CAPA), and Production and Process Controls. There are three other subsystems (sometimes referred to as secondary systems) that the FDA may opt to inspect at its discretion. These subsystems include Material Controls; Records, Documents, and Change Controls; and Equipment and Facility Controls.

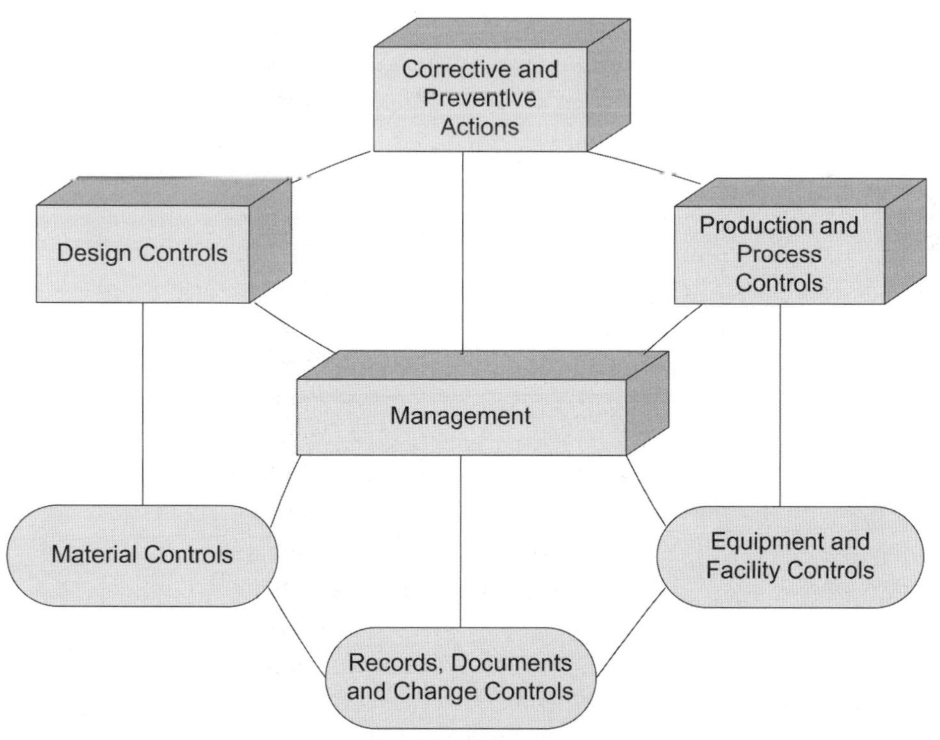

Figure 16.1. QSIT.

While assuming that product quality is significantly dependent on the four primary areas mentioned previously, it is also recognized that the three other subsystems also play a major role in ensuring product quality. The interconnections among the systems provide the forward and backward traceability perspective, which could produce an audit trail across the primary and secondary systems. Hence, the validation of all subsystems is equally important. With the new FDA guidance that validation decisions can be based on risk assessment, the risk-based validation decision-making process for the seven subsystems is subject to scrutiny during the FDA inspection process.

Automation System Overview

Plant utilities and air-handling units at Genentech's bulk-manufacturing site in Vacaville, California, are controlled by a Building Automation System (BAS). The BAS is a centralized client server system. A central server communicates with multiple field-based controllers over the corporate network. Operator terminals communicate with the central server to interact with and monitor the process. The BAS also provides the monitoring infrastructure to monitor cold room temperatures, where product is stored. Figure 16.2 is a representative architectural block diagram of the BAS.

The BAS monitors and controls the following systems:

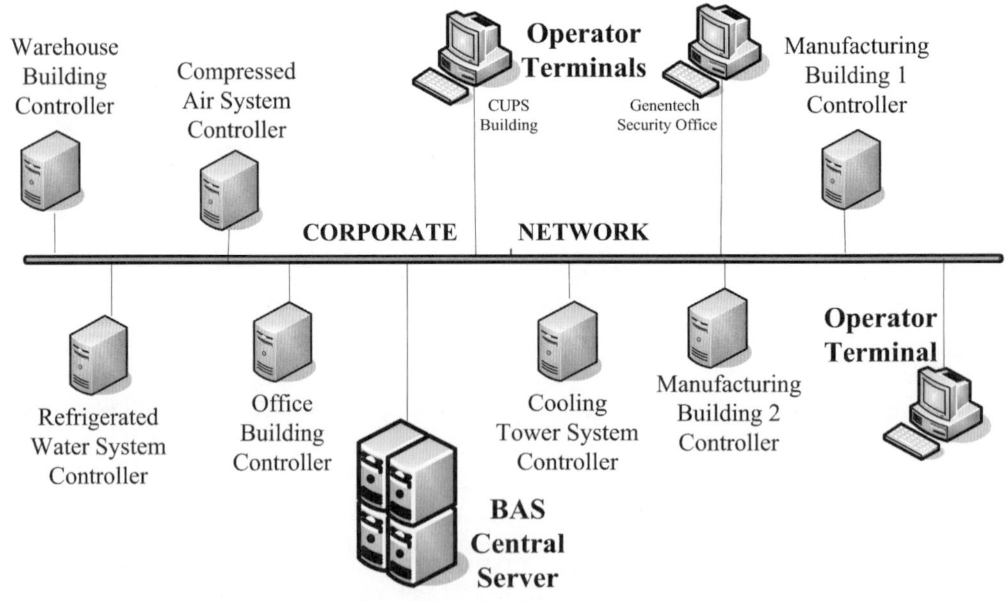

Figure 16.2. Architectural block diagram of the BAS.

1. Plant Refrigerated Water (RW) system
2. Cooling tower system
3. Process RW system
4. Compressed air system
5. Plant steam system
6. Air-handling units
7. Water heating units
8. Facility alarm monitoring system

Risk Assessment Approach

The engineering assessment's objective is to understand BAS-controlled processes and systems and focus qualification efforts on only those factors that can affect the quality of the product. The assessment is done in three steps:

1. Identification
2. Impact assessment
3. Criticality assessment

During identification, a team of subject-matter experts identify all BAS-controlled systems, along with their primary function. For each system, the component subsystems are also identified, as shown in Figure 16.2.

The impact assessment follows after systems and their functions are defined. During this step, the team assesses each system's impact on overall product quality. Based on their expertise, they provide "yes" or "no" answers to a series of questions pertaining to each system's impact on product quality. The answers form the basis for each system's categorization as a direct-impact, indirect-impact, or no-impact system.

During the criticality assessment, subsystem components and subcomponents of direct and indirect impact systems are ranked to reflect the magnitude of their failure consequence. The ranking provides the basis for identifying the critical subsystems.

The criticality assessment adopts a top-down and a bottom-up approach. The top-down approach is a functional or equipment-level assessment in which the system function is viewed as a collective function of its subsystems. The critical effects of each subsystem's potential failure to achieve product quality are assessed. Typically, each subsystem's failure consequence is examined by assessing its severity, along with the likelihood that failure can occur. Based on the probability of these occurrences and their effects on product quality, the subsystem is given a criticality rating. Those with the highest criticality rating or with the highest impact to process are identified for validation testing. The remaining subsystems are tested using good engineering practices.

Process-control assessment constitutes the bottom-up approach to criticality assessment. Process parameters whose control or monitoring ensures proper operation of other (subordinate) process parameters are identified as critical process parameters. Control elements that directly impact the measurement, control, and excursion detection of these parameters are identified as requiring validated level testing. The remaining components are tested using good engineering practices.

Case Study of a Plant Utility RW System

The RW system is one of the BAS-controlled plant utility systems. Its key function is to produce and supply RW to all buildings on-site. Figure 16.3 is a representative block diagram of a typical RW system.

The identification step identified the RW system's primary function to be the production and supply of RW at a fixed temperature and pressure differential for varying loads in all buildings on-site. The subsystems were also identified, and their results can be found in Table 16.1. During impact assessment, the RW system was deemed to be an indirect-impact system, based on the analysis presented in Table 16.2.

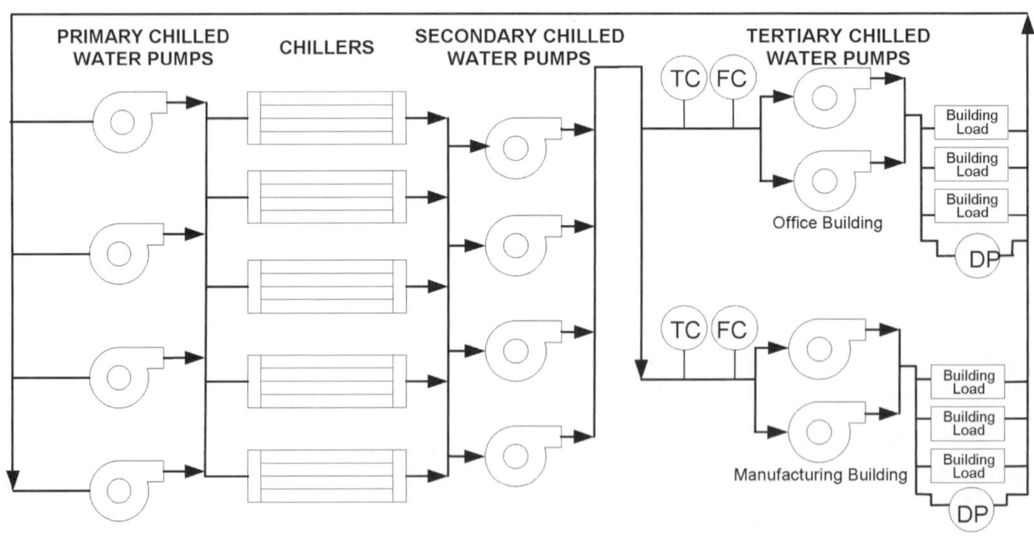

Figure 16.3. RW system.

Table 16.1. RW subsystems		
System	*Subsystems*	*Operational function*
RW	Chillers	Produce and provide RW at a preset temperature differential, flow rate, and differential pressure.
	Primary RW pumps	
	Secondary RW pumps	
	Tertiary RW pumps	

Table 16.2. RW system analysis	
	RW system
Does the system have direct product contact?	No
Does the system provide an excipient or produce an ingredient of solvent?	No
Is the system used for cleaning and sterilizing?	No
Does the system preserve or monitor product status?	No
Does the system produce data that is used to accept or reject product?	No

Since RW does not contact the product, it does not have a direct impact on product quality. However, the systems that receive RW do impact product quality (e.g., air-handling units, jacket coolers). Consequently, RW quality is somewhat critical. Based on this conclusion, subsequent assessment efforts shifted their focus from the RW system's impact on the plant product's quality to the impact of RW subsystems on the RW system's product (i.e., the production and supply of RW to the plant).

Functional criticality was determined using FMEA techniques. The team of experts identified various subsystems. For each subsystem, failure modes, failure impact severity, and the likelihood of failure were assessed. The FMEA analytical technique addressed the following questions:

1. How can each subsystem fail?
2. What might cause these modes of failure?
3. What could the effects be if the failures did occur?
4. How serious are these failure modes?
5. How is each failure mode detected?

A chiller failure was deemed as having the highest impact to RW quality. Consequently, chiller failure detection and alarming, along with failure recovery, were validated. Components such as RW pumps were not deemed critical and were tested per general engineering practice.

Operational criticality was assessed through a review and analysis of the system's Control Modules (CMs), control sequence, or procedure and alarms. CMs and procedures that directly impacted the control of the system's primary function were validated. Since the RW system's primary function was to produce RW at a desired temperature and supply it at a desired pressure, the following control entities were validated:

1. Proportional Integral Derivative (PID) temperature control
2. PID flow control
3. PID control of pump speed (for secondary pumps only)

Conclusion

The conceptual understanding and perception of risk is expected to differ, depending on an individual's pertinent training, prior beliefs about the risks, and other factors, such as the vulnerability of the human mind to fallacies in human reasoning. These vulnerabilities may be partially overcome through an open-ended brainstorming approach to identify hazards that will generate an abundance of factors believed by individuals or groups of individuals to contribute to risk. Some factors are likely to be objective and qualitatively supported, while others are likely to be subjective and value based.

While cognizant of these vulnerabilities, regulatory agencies are cautiously embracing risk assessment and mitigation initiatives by the industry. They recognize that such initiatives are based on a better process understanding that leads to built-in quality by design. These initiatives also lead to the development of better qualification efforts, which are both cost-effective and regulatory compliant.

Significant challenges lie ahead for the pharmaceutical community to move to the "desired state" for pharmaceutical manufacturing in the 21st century. Nevertheless, important steps have already been taken. Industry efforts to develop a structured regulatory format for risk assessment and management are currently in progress. Risk-management principles and tools will be necessary to describe and communicate the level of risk mitigation achieved through quality by design and process understanding. Current efforts based on initiatives such as Process Analytical Technologies (PAT) and the International Conference on Harmonization (ICH) and committees such as the American Society of Testing and Materials' (ASTM) E55.03 will provide a basis for risk mitigation through the development tools and risk management principles.

Further Reading

IEEE. *Risk and hazard analysis: Risk assessment techniques, part 1—FMEA. Briefing 26a*. New York: IEEE.

———. *Risk and hazard analysis: Risk assessment techniques, part 2—ETA. Briefing 26b*. New York: IEEE.

———. *Risk and hazard analysis: Risk assessment techniques, part 3—FTA. Briefing 26c*. New York: IEEE.

Sondalini, Mike. 2002. *Production risk management using equipment criticality analysis*. Saint Louis, MO: Business Industrial Network. E-book available at http://www.feedforward.com.au/criticality_definition.htm?source=froogal.

U.S. Food and Drug Administration. 2004. *Risk-based method for prioritizing CGMP inspections of pharmaceutical manufacturing sites: A pilot risk ranking model*. Washington, DC: Government Printing Office.

CHAPTER 17

Lean Computer Validation through a Risk-based Approach: A Case Study

Presented at the WBF North American Conference, March 5–8, 2006, by

Peter Werner Christensen
Quality Professional
pwch@nne.dk
NNE A/S, Gladsaxevej 363, DK- 2860
Soeborg, Denmark

Abstract

In the past, a substantial Supervisory Control And Data Acquisition (SCADA) upgrade would have caused full revalidation, including updates of basically all documentation and verification of a major part of the original functionality. This activity is costly and time-consuming. This chapter describes how to save up to 75% of the expected cost by applying risk assessment to the changes to the system so that risks are identified and eliminated, a feasible level of revalidation is defined, and all existing validation reused to the highest degree possible.

Using a case study from a SCADA upgrade project, this chapter will describe how to structure the project and how to define the system model that is a necessary input to the risk assessment. It describes how to carry out a series of risk assessments and subsequently eliminate the unacceptable risks by changing key concepts or by adding controls and barriers. A very large number of possible test combinations were initially defined, but a risk assessment analysis reduced this to less than ten situations to be tested. The result was a more controlled upgrade project, fewer test protocols, and considerable savings in time and cost. Key aspects of our analysis include the following:

- Relevant system model used
- The risk analysis influence on the validation planning
- Lessons learned

Project Background

In the spring of 2005, our client wanted to renew their SCADA system. The factory was built and initially validated in 1996, and all hardware in the SCADA system had the same age. During the intervening years, all software was continuously updated, and the hardware was kept alive. The validation status was maintained using a strict change procedure.

It became more and more difficult to get the necessary spare parts. The operating system on the computers (Windows NT) was no longer supported by Microsoft. The up time on the old equipment was no longer as good as it was initially.

The client was therefore forced to take corrective action. A decision was made to replace all hardware with the newest products possible and to upgrade to the latest versions of the operating system. The decision was made to move the servers from a Windows NT platform to a Windows 2003 platform and change the clients from Windows 98 to Windows XP. There was also a wish to migrate from an older SCADA version to a newer version.

The total SCADA upgrade was to be done with a focus on reducing the amount and duration of noncompliance situations for the entire plant. The goal was to be able to produce product while being in compliance all the time, except for the small window necessary to make the physical change of equipment and the time subsequently needed to get the system validated and released for production again.

The layout of the factory consisted of several different cells that, in principle, worked independently of each other, with their only communication occurring through the overlying Manufacturing Execution System (MES). Figure 17.1 shows a schematic system architecture where the SCADA only has interfaces upward to MES and downward to Programmable Logic Controller (PLC) systems. This is used as an important input to the risk-evaluation study later in the project.

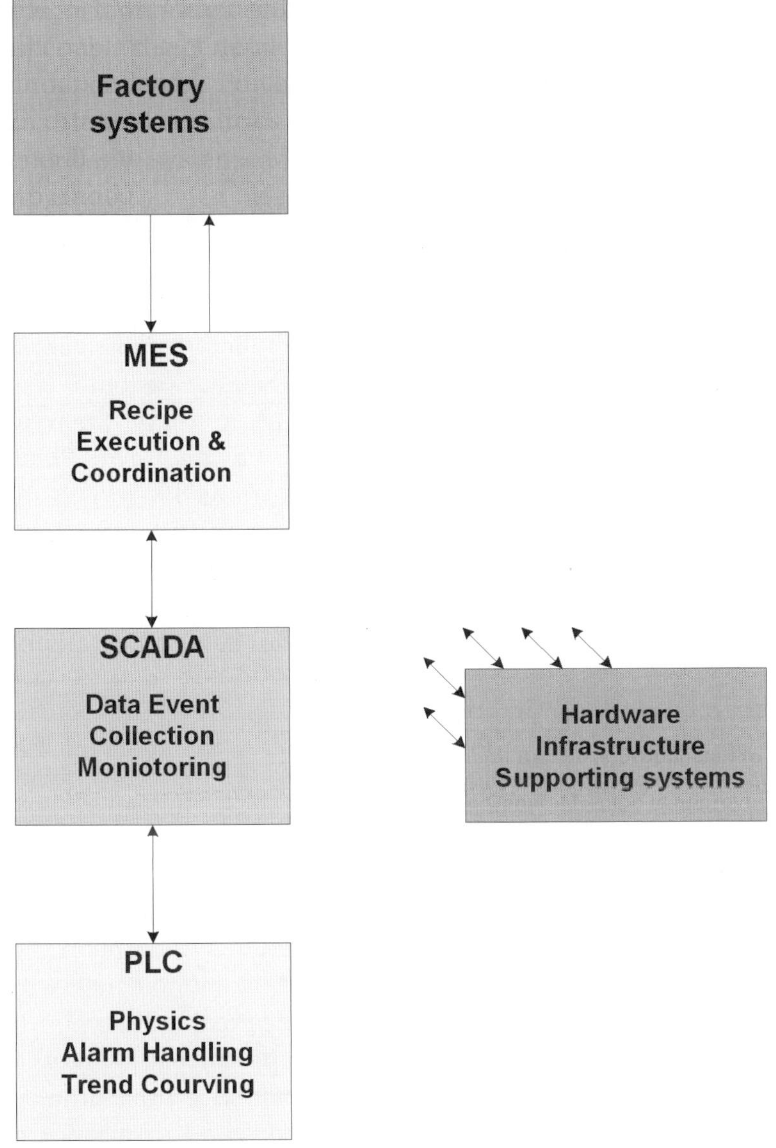

Figure 17.1. Interfaces to SCADA.

Validation Challenge

The existing validation documentation was huge and filled many binders. Furthermore, many change requests supplemented the original validation. When the SCADA system was being updated, large volumes of old validation documents and protocols needed to be found, opened, and updated. In the case described here, it would normally have been necessary to perform a full, new revalidation, as the change of the system would have a direct impact on the product.

From the beginning of the project, the client wanted to make the necessary revalidation as Lean as possible. The whole system was already in full current Good Manufacturing Practice (cGMP) compliance, and a detailed record of the system's performance was available. A key issue for the validation challenge was to cause the smallest possible interruption of the running production.

The upgrade project was not allowed to change any of the functions of the system, and the application software was to remain unchanged. The supplier of the SCADA software had stated that existing application software from the old system could be reused without any changes in the new system.

Very early in the project, it was decided to try to follow the new Risk-Based Approach strategy, described in *Good Automation Manufacturing Practices (GAMP) 4*. This strategy helped us choose the necessary efforts to put into the update project and only document what was really needed. It was also decided that the upgrade of the individual servers should be made independently of each other so that they could be planned into the natural slots for maintenance during production.

Our Interpretation of the FDA Risk-based Approach

Since the fall of 2002, the U.S. Food and Drug Administration (FDA) has worked on a new strategy, which was launched under the headline "A Risk-Based Approach." In September 2004, the FDA issued their new thoughts in the paper "Pharmaceutical cGMPs for the Twenty-First Century—A Risk-Based Approach." In November 2005, the FDA consistently promoted this thinking in statements and conferences. These new FDA signals have opened up a whole new way of thinking about upgrades and new projects in the pharmaceutical industry.

The validation process should no longer produce a "Great Mountain of Paper" ("GMP") but, instead, produce proven evidence providing a high degree of assurance that a specific process will consistently produce a product meeting its predetermined specifications and quality attributes, as already stated by the FDA in 1987. As the FDA stated again in November 2005, we must all go back to establishing only the documents that really produce the stated evidence, and they must be based on scientific judgment.

One of the FDA's goals for this approach is to get more and cheaper medicine to the public of better quality than it is today. The FDA has seen that the pharmaceutical industry today exhibits a very high degree of variation in the medicine that it produces. It is also acknowledged that the industry has low utilization rates and high scrap percentages during production. The FDA has stated that their old way of thinking may have prevented the industry from making the necessary changes.

The FDA encourages the industry to make quality improvements, and they are prepared to make it easier to do this. A means to do this is to implement a risk-based approach by defining the need for validation efforts. That is what we did in this project. We decided to perform risk management according to the most-used guidelines and standards in the industry today; thus we used the principles laid out in ISO 14971 and ICH Q9. These two documents provide good inspiration for the work.

Planning the Project

Before the SCADA upgrade project could be started, it was necessary to define an appropriate system model for the upgrade of the SCADA system in question. A team of experts was established to analyze the contents of the upgrade. The GAMP 4 software categories were used as guidance. These categories are defined briefly in Table 17.1.

The SCADA system was then divided into individual software components that all were all evaluated against the GAMP 4 software categories to identify the validation approach. It ended up with a system model as shown in Figure 17.2. Focus was put on configuration, setup, drivers, and supporting software. The hardware and operating system were also included.

Figure 17.2 illustrates how the system model was defined. The white boxes show the known components that were to be upgraded. It was important to keep this abstraction very simple, in order to be able to use it in the forthcoming work. To simplify the light gray and white "PLC drivers" and "SCADA application," the dark gray box on the left with driver configuration was drawn out. The setup activities were also separated and illustrated as two dark gray setup boxes. The light gray boxes illustrate components that were being reused.

Table 17.1. GAMP 4 software categories

Category	*Software type*	*Validation approach*
1	Operating systems	Record version; check applications.
2	Firmware	Record version, configuration, and calibration. Verify requirements and test functionality.
3	Standard software packages	Record version (and configuration of environment). Verify requirements. Consider auditing supplier.
4	Configurable software packages	Record version and configuration. Verify requirements. Validate any bespoke code. Audit supplier.
5	Bespoke systems	Audit supplier and validate complete system.

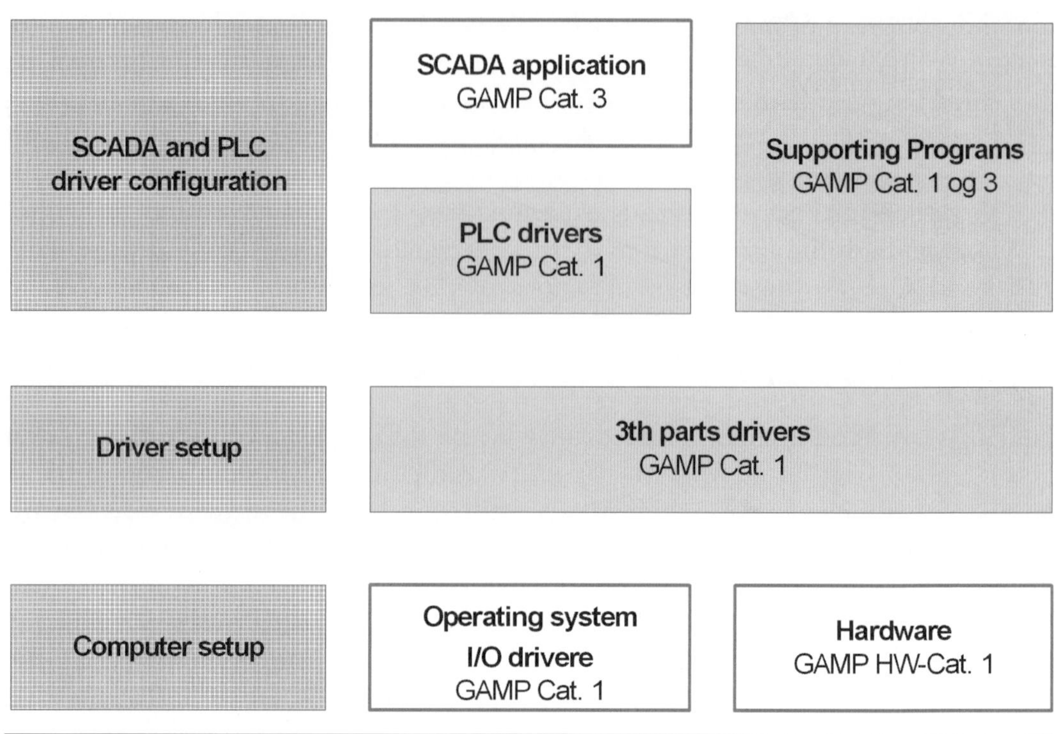

Figure 17.2. System model.

The SCADA system software itself was categorized as GAMP software category 3 and the supporting programs were categorized as 1 to 3 depending on their respective types. The rest were all software category 1. The validation efforts of this software were used as an input for the risk assessment later on.

Basic Validation Concept

It was decided to minimize the validation efforts to the lowest possible level while still achieving full compliance with cGMP and leaving the SCADA system in a new validated state after the upgrade. Therefore, Personal Computer (PC) clones were used as frequently as possible.

Using cloned images of a PC means that once one PC is validated, the rest are treated as mere images of the first one; the validation efforts of the rest will only serve to ensure that the image is restored successfully without failure. The validation documentation for the rest of the group will, therefore, only be a test plan to show that this situation is established. We decided to rely on the results of the risk assessment to define the additional validation efforts.

Risk Assessment and the Validation Planning

As mentioned earlier, it was decided to use a risk method that was inspired by ISO 14971. Therefore, it was necessary first to define an As-Low-As-Reasonably-Possible (ALARP) distribution. The concept of ALARP is a key element of the ISO 14971 standard. This is something that should be agreed upon before starting a risk assessment.

Before the risk assessment was carried out, it was decided that the User Requirement Specification (URS) for the SCADA upgrade project should form the basis for the analysis and that the validation plan should be directly affected by the result of the risk assessment.

A risk team was established. Members of this team consisted of a facilitator who was an expert within the risk assessment work and a quality expert who knew the procedures to follow. The rest of the team were experts in the specific systems. The risk team agreed upon an ALARP scheme with logarithmic values and three levels for probability (P) and severity (S). It was used in the further analysis work (Fig. 17.3). Furthermore, a three-level evaluation of the detection was used.

Figure 17.3 shows the chosen ALARP scheme. As it is shown, a logarithmic scale was used and only a simple band of ALARP is accepted. During the risk evaluation, it was decided that all black and white hazards should be mitigated.

The first task for the team was to go through a brainstorming session regarding possible hazards in relation to the coming upgrade project. The Preliminary Hazard Analysis (PHA) method was used for this purpose. It ended up with a list of hazards that needed to be grouped to continue the work. The time used for this first brainstorming meeting was fixed to only 2 hours. It is our experience that these meetings should not take longer in order to make sure that they remain effective (Table 17.2).

P_{harm}				
P3	1,00E-02	1,0000	10,0000	100,0000
P2	1,00E-03	0,1000	1,0000	10,0000
P1	1,00E-04	0,0100	0,1000	1,0000
P0	1,00E-05	0,0010	0,0100	0,1000
P	1,00E-06	0,0001	0,0010	0,0100
		1,00E+02	1,00E+03	1,00E+04
Severity		S1	S2	S3

Figure 17.3. Logarithmic ALARP scheme.

Table 17.2. Summary of the risk assessment flow—step by step		
Step	*Action*	*Result*
1	Perform a PHA (brainstorming).	List of possible hazards in relation to the upgrade project
2	Define a suitable detail level for the analysis (use of the system model).	Three possible hazards and six root causes
3	Create an Excel worksheet.	Overview of possible risk scenarios
4	Calculate the risk and evaluate possible mitigations for all priority risk scenarios.	Excel worksheet with calculation of all scenarios. Proposal for mitigations
5	Review the final analysis.	Update of Excel worksheet

After the first PHA, we ended up with three possible hazards that could be introduced by the SCADA upgrade project. The three hazards were as follows:

- Wrong data in the batch report
- Erroneous interface to the operator
- Wrong control of equipment

Things that could cause these hazards were also identified. This was limited to only six different causes (i.e., scenarios):

- The installation and configuration of the new operating system (i.e., Microsoft Windows)
- The installation of the new SCADA software
- The installation of the old supporting software
- The conversion of files
- The configuration of the software
- Changed functionality of the new software

The Risk Estimation

After identifying the possible hazards and their root causes, we needed to evaluate the probability and the severity of the hazards for each of the components from the system model in Figure 17.2. See Table 17.3. For each of the analyzed situations, the possibility of detection was evaluated. In many cases, the system would not

Table 17.3. Definitions of severities used in the risk assessment			
Severity	*Wrong data in batch report*	*Erroneous interface to the operator*	*Wrong control of equipment*
High S3	The batch report contains critical errors (e.g., wrong or missing data).	The operator interface contains critical errors (e.g., wrong or missing information on screen), or the operator cannot send required acknowledgments.	The system does not react or reacts incorrectly, and feedback from equipment is wrong or missing.
Medium S2	The batch report contains insignificant errors (e.g., layout or wrong date and time format).	The operator interface contains insignificant errors (e.g., wrong format of date and time on screen).	The system reacts incorrectly or not as intended.
Low S1	The errors are of no significant importance or not relevant.	The errors are of no significant importance or not relevant.	The errors are of no significant importance or not relevant.

be able to run, and therefore the detection was obvious, as nothing would happen and thereby no risk would be possible.

Many other situations were more difficult, and it was necessary for the experts on the risk team to make science-based decisions to secure a rational result. It was often a balance between the degree of detection and the related probability. When this risk estimation was complete, it was time to make the risk evaluation and find proper mitigation for nonacceptable risk scenarios.

The Risk Evaluation

With the ALARP scheme in Figure 17.3 in place, we ended up, as expected, with many potential risk scenarios. The risk team then started new working sessions for identifying acceptable mitigation for all substantial cases. This work led directly to the validation effort in the form of necessary Installation Qualification (IQ) and Operational Qualification (OQ) validation activities.

The process itself turned out to be very important, as a lot of discussions were carried out during the evaluation. The gathered information was of great importance for the upcoming validation planning. A risk report that contained guidance for the validation activities, as well as a checklist for specific test items in the validation protocols, concluded the risk assessment.

The whole effort of performing the risk assessment ended up with the use of only 33 hours in total. During the project, we often ran into situations where we believed that we needed to reconsider the risk assessment again, but every time, it turned out that we had already dealt with all relevant and necessary decisions.

Preparing the Documentation

The supplier of the SCADA system had prepared a complete migration document that in detail explained how the upgrade from the old version to the new version could be done. This document was used as a key guidance for the upgrade project. A special IQ test was prepared to secure correct use of this guidance document.

The risk report concluded (among other things) that after successful upgrade, configuration, and approved IQ off-line testing, each SCADA cell should be tested with a reference batch, on-line in a production environment. To get a qualified input for these tests, a meeting was set up with selected operators for each of the changed SCADA cells.

It turned out to be a very good investment, as many interesting issues were raised during this process. Input from the operators, in combination with the checklist from the risk report, constituted adherence to the relevant OQ protocols. In full agreement with the validation plan, this approach was then deemed sufficient, and no other validation tests were conducted.

Key Documents

The key documents of the SCADA upgrade project were the URS, risk report, and the validation plan. They all acted as the guidance for the project and were used again and again as reference documents.

Only one validation report was prepared for each kind of IQ or OQ activity (even though there were several cells with servers and clients throughout the entire SCADA system). The reason for this decision was simple: the risk assessment had identified the necessary tests and then the conducted, completed, approved, and reviewed test plans were sufficient to establish the documented evidence, as required by the FDA.

The client's change request system was used during the implementation to keep track of the status and progress of the work. Following that system, all the work was carried out, and all deviations were closed and approved. A final validation report was written, containing a summary of the entire SCADA upgrade project. Furthermore, a new baseline was established and documented.

Lessons Learned

In order to control the project and validation impact on the whole system, it is very important to stick to the existing functionality and not to be tempted to make any changes during the project phase. It is crucial for the project's success for it to be based on a validated platform with a well-known performance because the upgrade project can then be considered an incremental change.

This project showed us that it is very important to define a sustainable system model that can be used again and again during the many decisions that need to be made when preparing the IQ protocols and defining the necessary OQ tests. A system model is also necessary for settling risk evaluations in analyzed hazard scenarios.

During the risk assessment, it was recognized that the discussion process itself was very important. It is also important to select the right team of experts to participate in the risk process. A special focus must be made on the probability of human errors, which must be considered in more depth than an equivalent error made by machines.

When preparing the contents of the test batch that was used during the OQ activities, it was important to involve the operators. They knew many important things about the system and the "intended use," which is a critical focus point of the FDA. The OQ test was defined in a way so all critical items were included; at the same time, it was defined in a way that allowed it to be verified in practice.

The FDA requirement to use science-based methods should be adhered to in order to avoid duplicated works in the validation activities. Use of computer "images" to make copies of computers turned out to be a very good investment. It saved us from a lot of boring and repetative tests. In addition, there were lots of hours saved in preparing the computers for the different SCADA cells.

The amount of paper used in this project was kept to a minimum, when compared with the traditional way of performing an upgrade project. It is estimated that there has been a savings of approximately 75% of the validation effort. The goal of minimum disturbance to production in the form of wasted production hours was met. In one case, changing the SCADA system in one cell, including the final part of the validation activity, took less than 4 hours.

CHAPTER 18

Multiple Products in a Monoclonal Antibody ISA-88.01 Batch Plant

Presented at the WBF North American Conference, May 16–19, 2004, by

Mahasti Kheradmand
Automation Supervisor
mkheradm@gene.com
Genentech, 1000 New Horizons Way, Vacaville, CA 95688 USA

John O'Connell
Automation Supervisor
johno@gene.com
Genentech, 1000 New Horizons Way, Vacaville, CA 95688 USA

Abstract

The most successful biotech companies have multiple products approved for market and must make use of existing manufacturing capacity to produce them. Oftentimes, this requires rapid changeover from one product to the next. This chapter addresses the challenges of bringing new products into an existing ISA-88.01 facility and the challenges involved with implementing a change in an operating plant, with a minimum of downtime. It was found that the ISA-88 concepts are enormously helpful for implementing changes within the equipment capability, but challenges arise when the equipment capability must also be changed. Our recommendation is to standardize the manufacturing process and build the necessary capability into the plant up front to avoid costly downtime during product changeover.

Introduction

The ISA-88 batch standards have been a tremendous benefit to the pharmaceutical industry. The standardized approach in the design of process equipment and control-system software has dramatically reduced the effort required to design, construct, and validate a batch manufacturing facility. With the increased number of marketed products, the market leaders in biotechnology are approaching the limits of their in-house manufacturing capacity. The net result is that existing plants are being required to produce multiple products and to switch as quickly as possible between production campaigns. For those companies fortunate enough to have newly approved products, there is an additional demand to quickly introduce the manufacturing of new products into existing facilities. This chapter explores some of the major aspects related to bringing a new manufacturing process on-line in an existing multiproduct monoclonal antibody facility already automated to the ISA-88 standard. Some thoughts on future directions for easing these difficulties in new plants are also offered.

Process Background

The typical monoclonal antibody facility includes a fermentation train that ranges from laboratory-scale fermenters, in the 10-to-20-liter scale, to large-scale cell culture vessels, in the range of 10,000 to 20,000 liters. The cell culture vessels are held under controlled conditions of level, pH, agitation, and dissolved oxygen in order to provide the optimal conditions for production of the molecule of interest by genetically engineered organisms. The facilities also include a harvest system that is used to clarify the cell culture fluid as it is removed from large-scale cell culture vessels at the end of a batch (Fig. 18.1).

The clarified harvest is typically processed further in a product recovery suite that includes one or more chromatography steps and final product formulation using ultrafiltration. Chromatography is used to purify the product by separating it from the clarified harvest under carefully controlled conditions. Both the cell culture and protein areas are serviced by preparation and storage tanks that provide and store growth media for the cell culture vessels and buffer solutions for the chromatography and ultrafiltration units.

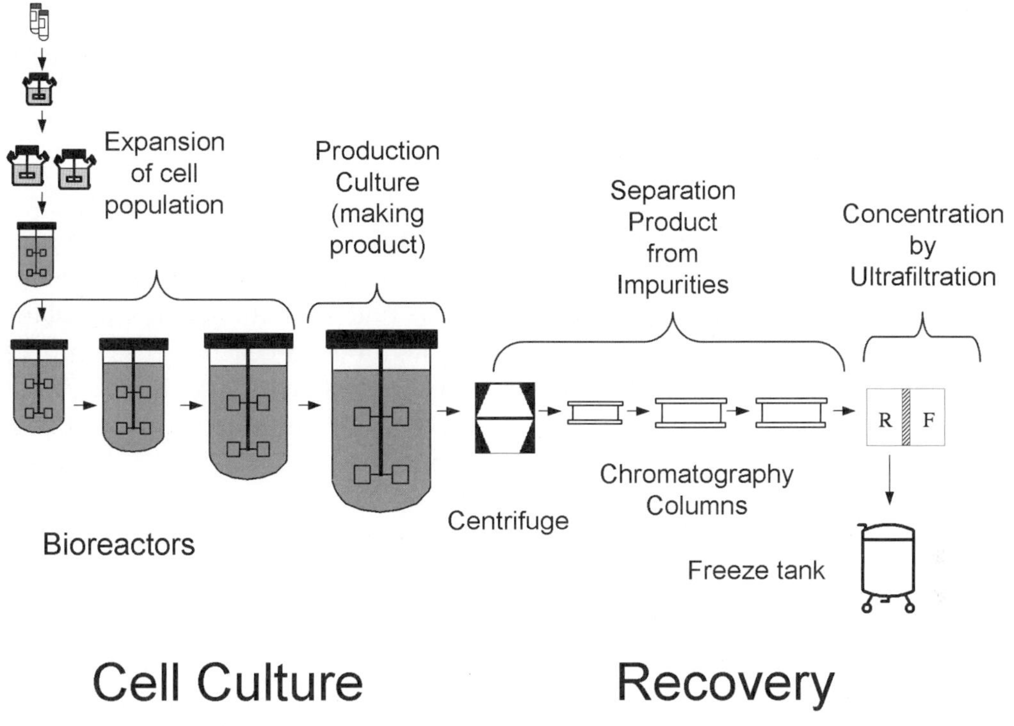

Figure 18.1. Process overview.

Most Changes Happen in Product Recovery

Most of the changes from product to product occur in the recovery area, as this is where the manufacturing process for monoclonal antibodies is the most variable. It is very rare that major equipment changes need to be made in cell culture, since this portion of the process has been somewhat standardized across the industry. The process variation can take the form of new production steps, such as replacement of a chromatography step with a filtration step; new capability requirements, such as in-line dilution as part of a chromatography step; or increased yield. Variations in yield can render the existing recovery process equipment too small or too large for the intended purpose. New process steps or equipment capabilities may require the modification of existing recovery equipment. These process changes may require plant downtime between products, as both the equipment and the associated software will need to be modified and fully tested before the next production campaign can begin.

Production campaign changeovers are usually much less trouble in the cell culture area. Since it is rare that new steps or capabilities need to be added, most new products can be introduced with recipe-level changes only. This dramatically reduces the time spent defining requirements and developing software in cell culture, as compared to recovery.

More Products Make Changes More Difficult

The complexity of a multiproduct monoclonal antibody plant tends to increase over time. For a single-product plant, only one set of conditions needs to be considered for every phase or Control Module (CM) that is added or changed in the plant. In a plant that supports six different products, the impact on each of these processes must be considered for every change. The level of design effort is increased as the changes become more complex and the functional requirements differ between products. Piping and control schemes that would be relatively straightforward for a single product become more involved as the process requirements for each of the various products are built into the design. The net result is that, as more capability and flexibility is added to the plant, the design and implementation cycles will lengthen, since any changes must preserve existing functionality while meeting the requirements of the new product.

Overhead Required to Maintain the Master Recipes

The overhead required to maintain a library of master recipes increases with the number of supported products. All products have a particular set of processing requirements that are defined in their set of master recipes. As the number of products supported by a particular plant increases, the number of master recipes associated with the plant also increases. Process-model changes for any new product may invalidate existing master recipes, requiring them to be revised before they are used again and therefore extending the project schedule. Depending on the complexity of the changes and the level of testing required, it may take even longer to revise the existing product recipes than it did to develop recipes for the new product or process.

A large library of master recipes also tends to increase the time required to make routine changes to the plant (Fig. 18.2). For a facility that supports multiple products, it becomes more difficult to assess the impact of a particular change on all the recipes. For example, a recent project to introduce a new product required the development or modification of over 171 recipes, 95 of which were existing product recipes. In addition, a process improvement change that is common for all products will require multiple edits, since each product will have its own set of master recipes.

First Product

Requirements Definition	Develop and Test

Second Product

Requirements Definition	Develop and Test	Revise Software For Other Products

Third Product

Requirements Definition	Develop and Test	Revise Software For Other Products

Figure 18.2. Impact of multiple products.

Transferring a Process from One Site to the Next

For organizations that have multiple sites, transferring processes between facilities can be extremely challenging (Fig. 18.3). This is particularly true if the two facilities have different control systems, equipment capabilities, or business systems. In particular, process transfer between paper-based and paperless facilities is very labor intensive without general or site recipes. In the absence of an enterprise-wide information structure, such as the one defined in ISA-88.03 and ISA-95, it is necessary to convert the words and steps from paper-based operator instructions to recipes, phase logic, and Control Modules (CMs). An exhaustive verification is required to ensure that the critical process parameters and steps have been correctly coded into the recipes. For a complex process transfer, this verification can add several weeks to the design effort and require many hours of cross-functional team meetings. Adopting the common data structure of a general recipe would speed this process up tremendously by quickly allowing the identification of the critical steps, equipment capabilities, process parameters, and material requirements. Implementation of the ISA-88.03 standard for general and site recipes requires an enterprise-wide commitment, and there may be some institutional resistance to expend the time and capital required to implement the concept. It would, however, be worth the investment, since it reduces the time it takes to move a production process between sites and, as a result, makes the manufacturing capacity of the organization more responsive and flexible.

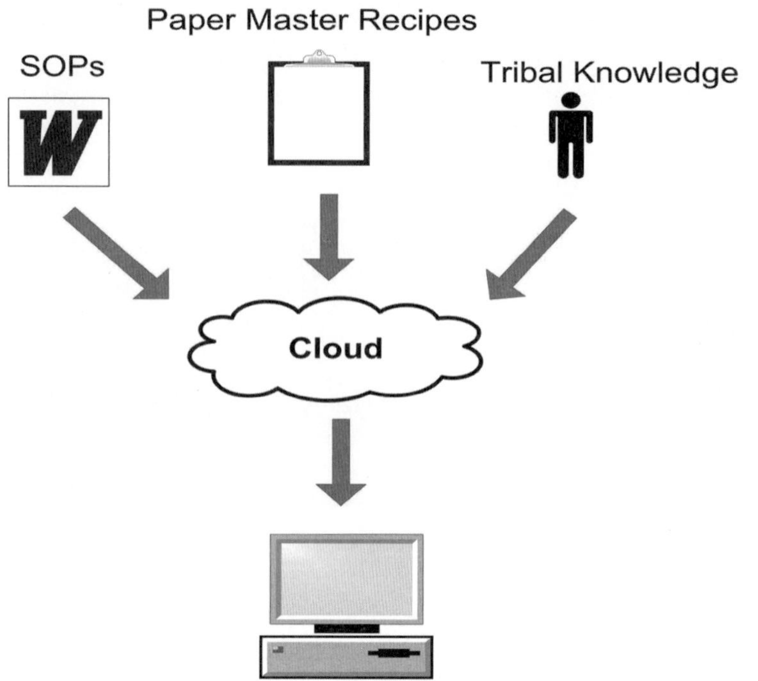

Figure 18.3. Process transfer.

Planning Is Essential for a Smooth Product Transfer

It is essential that monoclonal antibody manufacturing firms adopt a plan to manage process-to-process variability, if they are to maximize the capability of existing facilities. Since it would be prohibitively expensive to build a plant that can run all current manufacturing processes for monoclonal antibodies, most manufacturing plants can support a limited subset of these processes. Information about the existing manufacturing capabilities must be communicated to the process development staff. Oftentimes, the process developers have choices in the methods they select to manufacture and purify a new molecule. If they are aware of downstream constraints, then they are more likely to develop a method that will fit within the existing capabilities. At the very least, they will be able to provide advance warning when changes are required for an existing facility.

The process development scientist must communicate the required manufacturing capabilities of the process equipment early in the process transfer effort. This is most conveniently done in a User Requirements Specification (URS) that forms the basis for both the process and control system design. Having well-defined user requirements up front allows the project team to quickly assess whether a manufacturing site has the necessary capability to run a process and determine what changes will be needed to support the new product. Planning for product-to-product process variation will allow the organization to more rapidly respond to required changes and maximize the use of existing manufacturing capacity.

Minimizing Downtime

Keeping plant downtime to a minimum is mandatory during a campaign changeover. Downtime in a manufacturing facility is never a good thing, and since most monoclonal antibody plants run on a continuous basis, it is also difficult to schedule. A successful product changeover with minimum downtime requires careful planning. Sufficient lead time should be allowed for revising the requirements documents and developing and testing most of the software changes in an off-line environment. Since downtime comes at such a premium, early involvement and training of persons performing the activities is a must. Adequate staffing should be provided across all disciplines to allow around-the-clock coverage during the plant shutdown for all installation, debug, and validation activities. Although detailed planning is a required, the project team must also be flexible enough to take advantage of opportunities to complete tasks ahead of schedule when debugging and testing efforts go well, in order to compensate for the time lost when these efforts do not go as expected.

The ISA-88 standard is both a help and hindrance in the effort to minimize plant downtimes during campaign changeovers. The modular software structure of an ISA-88 plant allows changeover lead times to be reduced because software objects such as phases and CMs that are used in multiple units can be modified and tested once in an off-line environment and then copied to where they are used. For example, in a recent project, over thirty-one CMs needed to be added to the configuration, but since the modules were designed to be portable between multiple units, only one new module needed to be created. This reduced the level of effort required for software engineering of CMs by at least two orders of magnitude, since there was no need to specify, develop, and test the existing CMs that were being reused.

For product changeovers that do not require unit level changes, software development is limited to the creation of new master recipes. Designing software according to the ISA-88 standard reduces the downtime required to change from product to product by allowing the same code to be used on multiple units and multiple products with simple recipe parameter changes.

The ISA-88 standard can become a hindrance in cases where following the standard will require unusual effort. For example, it may be much quicker to solve a problem found during debug by modifying a CM on a single unit so that it is no longer portable to other similar units. In almost every case, it is best to make the extra effort to comply with the standard, since nonstandard code will make the plant harder to maintain in the long term.

Stay on Top of Change Control

An important factor in minimizing plant downtime is managing the change control process. After requirements definition, most of the hours in a software development effort are spent developing and executing test plans. Doing this work correctly is essential for project success. Executing the change control process poorly by not following good documentation procedures, making unauthorized changes, or taking shortcuts on testing will increase the number of hours required to complete a project. In a regulated industry, if the software development and testing is not done under proper change control, it will normally require the involvement of department management and quality assurance to justify the deviation from the process. In some cases, repeating the work may also be required, and in all cases, it will take much longer to correct an error than to follow the correct procedures in the first place. Doing it right the first time is definitely worth the effort in the final quality of the software and the overall productivity of the project.

Before a large implementation project begins, it is best to prepare to work under the change control system. The following steps will go a long way when preparing your team:

- Review the change control process, and streamline it, wherever possible, prior to the start of software development.
- Refresh the team members' understanding by conducting change control and configuration management training.
- Put an administrative assistant on the project to assist with managing and organizing the large volume of paperwork associated with software testing.
- Have a written configuration management plan and all the necessary tools in place to manage the software objects through development, testing, and implementation.
- Review and close out tests and testing deviations as soon as possible, following test execution. Doing this rapidly will allow testing issues to be documented and dealt with while the events are fresh in the memory of the parties involved.

Applying Lessons Learned to a New Facility

Designing a new facility presents the opportunity to avoid using the limitations inherent in older concepts. With the pace of technological change in monoclonal antibody manufacturing, a new facility must have the ability to deal with increase yields and process changes via a scalable and flexible plant. This is especially true in the recovery area, which is extremely sensitive to yield improvements and is the area with the greatest product-to-product process variability. Consideration should be given to technologies that increase the capacity of processing suites, such as buffer concentrates, and skid-mounted recovery equipment that can be changed out quickly for a product changeover. To help build in the required capabilities, a URS should be developed for each process unit. The URS should specify the equipment capabilities that will be required to support all the products that the plant is intended to produce, not just the product that the plant will begin to manufacture immediately after startup. The process development staff needs to be fully engaged with engineering and manufacturing in the requirements development process because they are most fully aware of the manufacturing processes being considered for all the products in the pipeline. Involving them in the requirements definition document also makes them aware of the capabilities of the new plant. Having this document available will allow them to make more informed choices as they design the process for future products. Finally, there should be a formal risk assessment to analyze the impact of emerging technologies related to cell culture yield improvements and protein-recovery operations. Asking the "what if" questions up front can allow some level of risk mitigation to be built into the plant and reduce costly downtime and changes over the life cycle of the plant, as new technologies are employed.

What's Ahead in Monoclonal Antibody Manufacturing?

According to the "BIO Editor's and Reporter's Guide 2003–2004," published by the Biotechnology Industry Organization,[1] there were 35 new biotech drugs and vaccines approved for market in 2002, and there were more than 370 new biotech drugs in clinical trials. It has also been speculated that the current industry standard yield for a monoclonal antibody cell culture process may triple from 1 g/L to 3 g/L; this will require continued innovation in the product recovery technology downstream of cell culture.[2] Plants designed for the 1 g/L scale will be challenged to take full benefit of these yield improvements, and new facilities must be both scalable and flexible to accommodate product-to-product yield changes and evolving technology. With the large number of products in the pipeline, those monoclonal antibody manufacturing firms that can deal best with multiple products, fast product changeovers, improved yields, and evolving technology will have a key advantage over their competitors.

References

1. Biotechnology Industry Association. 2004. Bio editor's and reporter's guide 2003–2004. http://www.bio.org/er/BiotechGuide.pdf (site discontinued).
2. Gottshalk, Uwe. 2003. Biotech manufacturing is coming of age. *Bioprocess International,* April 10. http://www.pharma-und-chemiepark.de/en/pub_event_report_biotech_manuf.pdf.

CHAPTER 19

Considerations for Managing Global Recipe Development

Presented at the WBF North American Conference, May 15–18, 2005, by

Tom Farenholtz
Principal Consultant
tom.farenholtz@honeywell.com
Honeywell International, 13655 Dulles Technology Drive, Leesburg, VA 20171 USA

Abstract

Batch control systems have grown from single proprietary systems to multiple open systems that can be networked within the plant and also expanded with links throughout the enterprise. As more vendors adopt Batch Markup Language (BatchML) as an Extensible Markup Language (XML) format for publishing recipes and ISA-95 standards for organizing recipe content, the ability to build recipes that can be managed globally and then distributed to local batch execution systems is emerging as a requirement for enterprise batch management. The need for flexibility in managing data for different target systems and for the collaboration of authors during recipe development, version management, and workflow for approvals are creating new challenges for recipe authoring systems. This chapter explores several issues related to the centralized management of batch recipes and recipe management architectures and methods for distributing recipes to batch execution systems.

Overview

Batch and other process control systems are being deployed and upgraded to provide greater flexibility and standardized process models, in order to keep up with demands for global manufacturing capacity. This trend is being driven by high market demand for blockbuster products, mergers, the economics of offshore manufacturing, and continued pressures to reduce manufacturing and construction costs.

Various standards and committees (e.g., ISA-88, ISA-95, WBF) have been making progress to establish standards, guidelines, terminology, and models for describing, documenting, and exchanging batch information. As systems are developed using the standards and become available for use, there are a number of questions that need to be considered when deploying a multisite or global solution.

This chapter will explore some topics related to the authoring of recipes to support multisite and global batch manufacturing. These factors include organizational structure, scalability, data segmentation, collaboration, version control, and system architecture.

Local or Central Engineering?

Over time, engineering tends to shift between plant- and central-engineering groups. When a shift in ownership occurs, it is disruptive to manufacturing operations that are required to continue producing quality product on time and meet production goals. A shift in focus may occur in response to either a corporate reorganization or a special project (e.g., a new plant or changing business priorities). Neither approach is perfect, with each having its own advantages and disadvantages, as shown in Table 19.1.

A Global Approach

The goal of global recipe development is to take advantage of the best practices of the central and site organizations, while minimizing the weaknesses of each organization. Although it is an organizational challenge, the goal cannot be achieved for batch and other recipe driven processes without the help of underlying enabling technology.

Global recipe development requires a common recipe authoring model that is independent of the target system. Even if the authoring system uses a distributed

Table 19.1. Comparing central and local engineering		
	Central engineering	*Local engineering*
Location of engineering staff	Corporate engineering group	Separate groups in each plant
Funding	Corporate	Plant Operations
Responsiveness	Good for large projects	May have long delays for small problems
	Responsive for critical fixes	Resource constrained for large projects
Process knowledge	General knowledge is shared by many; backup is usually available if a specific person is absent.	Plant-specific knowledge is weak.
	Details tend to be "owned" by individuals, and problems occur if a person is absent or leaves the company.	Plant-specific knowledge is strong.
Reusability of process engineering	Recipes and processes able to be shared across multiple plants if the batch systems are similar and standards are used	Recipes and processes are specific to the plant and are not easily shared between sites

architecture, the system behaves as a single logical system from the user's perspective. This has the effect of creating a virtual organization co-located in the same office.

With the global model, central engineering can manage the General Recipe (GR), provide continuous support for the plant, and supply engineering services to the plant during peak periods. Plant engineering can manage the site and master recipes, perform day-to-day maintenance on the system, and maintain plant-specific process knowledge. The plant can maintain a stable staff supplemented by the central group, as needed. One additional benefit resulting from the global model is that all process engineers will use the same system, thus reducing the common complaint, "they always get the new technology and tell us how to do our jobs." The remainder of this chapter will discuss the technology behind the global recipe model and discuss some practical applications.

GR Considerations

GRs have become a topic of discussion at most meetings where ISA-88 is discussed. A GR is defined in ISA-88.03 as "a type of recipe that expresses equipment and site independent processing requirements." The boundaries between general, site, and master recipes are described in the standard. These boundaries also imply that a central organization exists to manage the GR.

GRs do not need to contain equipment or batch system specific information. This allows a GR to be authored on a system other than the final batch system or even a site system. In fact, several ISA-95 level 3 and 4 systems such as Enterprise Resource Planning (ERP), scheduling, and Manufacturing Execution Systems (MES) now use GR data for their own needs. ISA-95 and Business To Manufacturing Markup Language (B2MML) define formats that facilitate the exchange of data between these systems.

Organizations need to decide if they will implement a single GR authoring system that contains the recipe information required by all systems or if each system will maintain additional information unique to that system. In practice, each type of data should be maintained in a single master system and then published to any system that needs the data. For example, the material specifications could be maintained in the ERP system and published to the recipe authoring system, while process routings are maintained in the recipe authoring system and uploaded to the ERP system. In all these cases, the B2MML definitions facilitate the exchange of data among systems at the GR level.

Propagation of Recipes to Lower Levels

Assuming that a GR authoring capability exists, the GR definition needs to be propagated to a site recipe. The first consideration is the site authoring system: will the site authoring system be the same system as the GR authoring system, or will it be supplied by a different vendor?

If the general and site systems are provided by different vendors, then the GR needs to be published in a standard format (e.g., BatchML, B2MML) from the general authoring system to the site authoring system. Prior to the establishment of standards to define the components of a recipe, a custom integration effort was required to accomplish the migration. Custom integration was rarely, if ever, implemented because the costs outweigh the benefits.

If the general and site authoring systems are provided by the same vendor, then the propagation of GR to site recipe is usually embedded in the vendor's package. This raises the question of integration with other systems that need to share the GR or some of its components: if the general and site authoring systems are the same, will they be installed on the same instance, or will there be separate instances of the general and site systems? Again, the authoring package should contain a standard interface such as B2MML for exchange of data with other systems.

For performance and reliability, companies may decide to install a separate instance of the site and master authoring system at each plant or in each region.

When multiple instances are used, the transfer of data should be in a standard format. This will enable the general and site authoring systems to be decoupled and maintained independently. For example, different site authoring systems could be installed in different countries but could share a common GR. Proprietary integration of general-site systems would create a support issue if either system was replaced or upgraded.

Are we done once the system architecture has been determined and the standards for transferring the GR to the site recipe have been defined? The answer is no. We have yet to encounter the hard part. If the GR is for a new product or an entirely new recipe for a given product, then a new site recipe can be established based on the GR received. But the problem is much more complex when a change to an existing recipe is received. How is the change identified, and how can the change be propagated to site recipes that were derived from the GR? There are currently no standards for updates, and many implementations reach an impasse at this point.

As shown in Figure 19.1, versioning of recipes and recipe components provides a mechanism for keeping track of updates. The WBF's BatchML defines XML schemas that can be used to exchange information about recipes, equipment, and batch lists. BatchML provides for version tracing as part of the recipe header, but

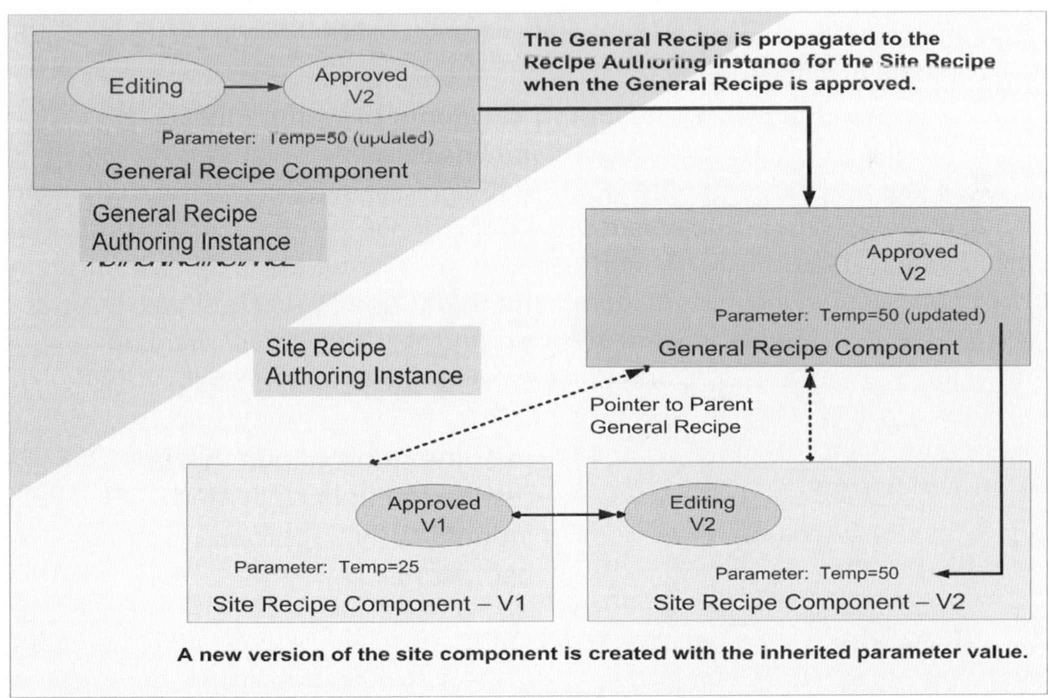

Figure 19.1. Versioning of recipes.

it does not specify how different versions are stored in an authoring environment. Business-system integration uses B2MML, which is based on the data models and attributes defined in ISA-95. B2MML does not contain a version field in any of the definitions. If a recipe or component has the same name and different versions, then an implementation scheme must be developed to keep more than one version of the object in the authoring system.

Versioning becomes very important if the recipe is built using reusable components. Many commercially available recipe and batch execution systems implement a reusable component model (Fig. 19.2). This means that a change made to any component (e.g., a parameter in a recipe operation) will be applied to all recipes using that component. This is problematic in regulated industries that require change control on all recipes. An alternative to this is to provide different versions of each component and coordinate the updates to the recipes that will use the updated component, as shown in Figure 19.3.

The use of the global recipe authoring system to manage the versioning of objects provides a way to allow existing systems to use their current component models. This can be accomplished by transforming the versioned recipe components into uniquely identified components as the components are transferred from the recipe authoring system to the batch server. The batch server would then see a uniquely named component for each version. The higher-level components (e.g., a recipe unit procedure using a versioned recipe operation) would then refer to the new component name.

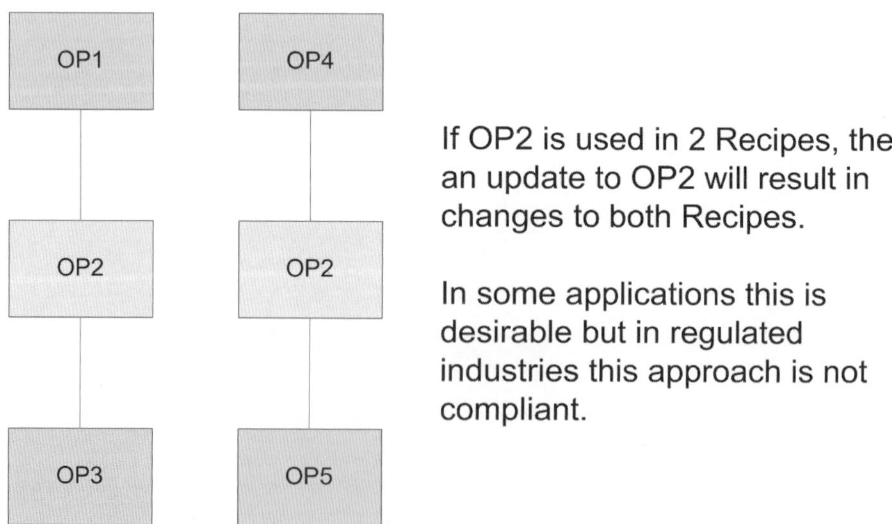

Figure 19.2. Reusable components.

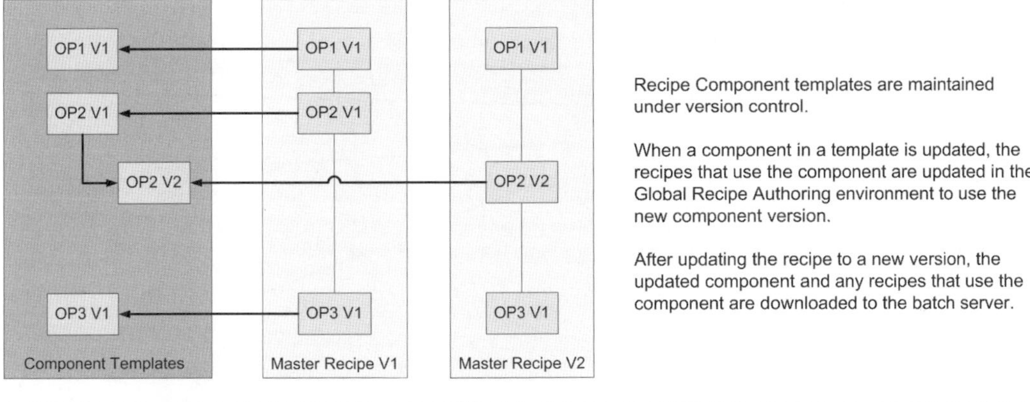

Figure 19.3. Different versions for each component.

The mechanism used to update components to new versions needs to compensate for several additional factors. Dealing with each of these factors in detail is not possible here, but some of the factors to consider include the following:

- Identify all the higher-level recipe components that refer to the modified component (i.e., "where used" capability). Will the change be applied to all the referencing components or to each component individually?
- Concisely identify the changes made to a recipe component (i.e., the difference tool).
- Add a parameter or other element to a component. Will a default value of the parameter propagate, or will the process engineer need to add a value for each instance of the component?
- Will deleted parameters break other parts of the recipe?

Global Authoring Architecture

There are two basic architectures for global authoring systems: The first architecture (Fig. 19.4) could be considered an all-in-one architecture. This architecture provides all levels of the recipe in the same instance and allows both plant and central engineering access to the same data. It has the advantage of a single system to maintain and fewer integration issues. It has the disadvantage of a single point of failure and scalability.

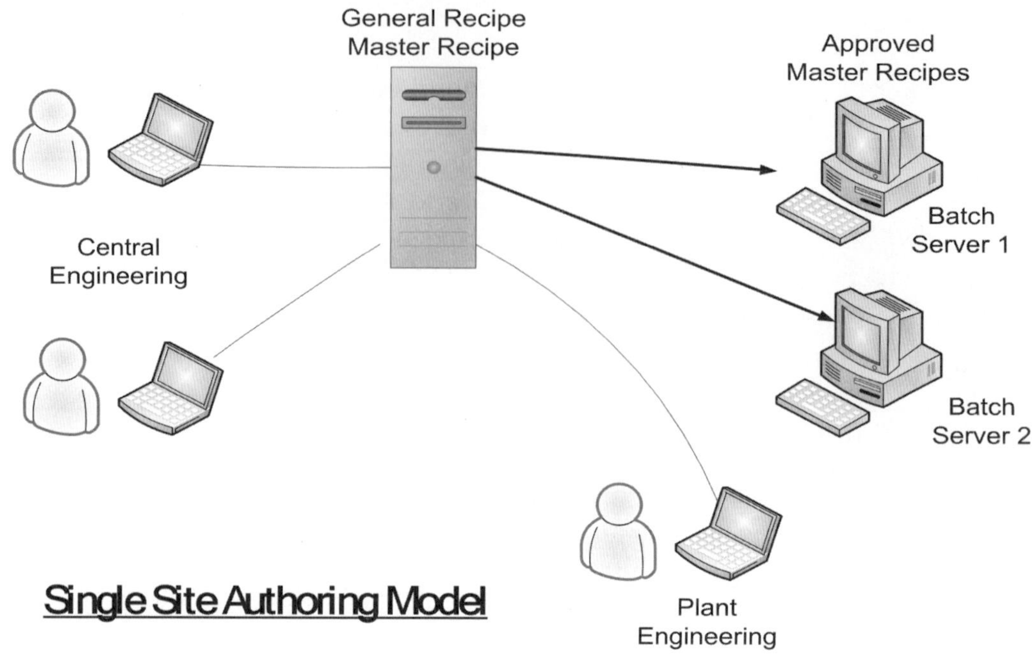

Figure 19.4. All-in-one architecture.

The second architecture (Fig. 19.5) is suited to larger corporations operating several plants that may have different products or processing steps for the same product at different sites. This architecture employs multiple instances of the recipe authoring system. It is important that the different instances exchange data using standards such as B2MML. Using a standard to exchange recipe data allows the instances to be updated independently or even to be provided by different vendors. A distributed architecture requires an integration layer but provides better scalability and fault tolerance than a single-instance solution.

Globalization Issues

In conclusion, a few remarks on culture in a global recipe authoring environment are in order. Different countries have different standards for time, date, currency, numeric formats, and so on. The challenge of a global authoring system will be to present these data types to engineers in formats that are acceptable for their respective cultures.

Readers of this chapter are encouraged to look into the new features provided through new operating systems. Features are being included at the operating-system level that will permit users to select the culture most familiar to them. Global recipe authoring systems will need to take advantage of this type of technology in order to be useful across national boundaries.

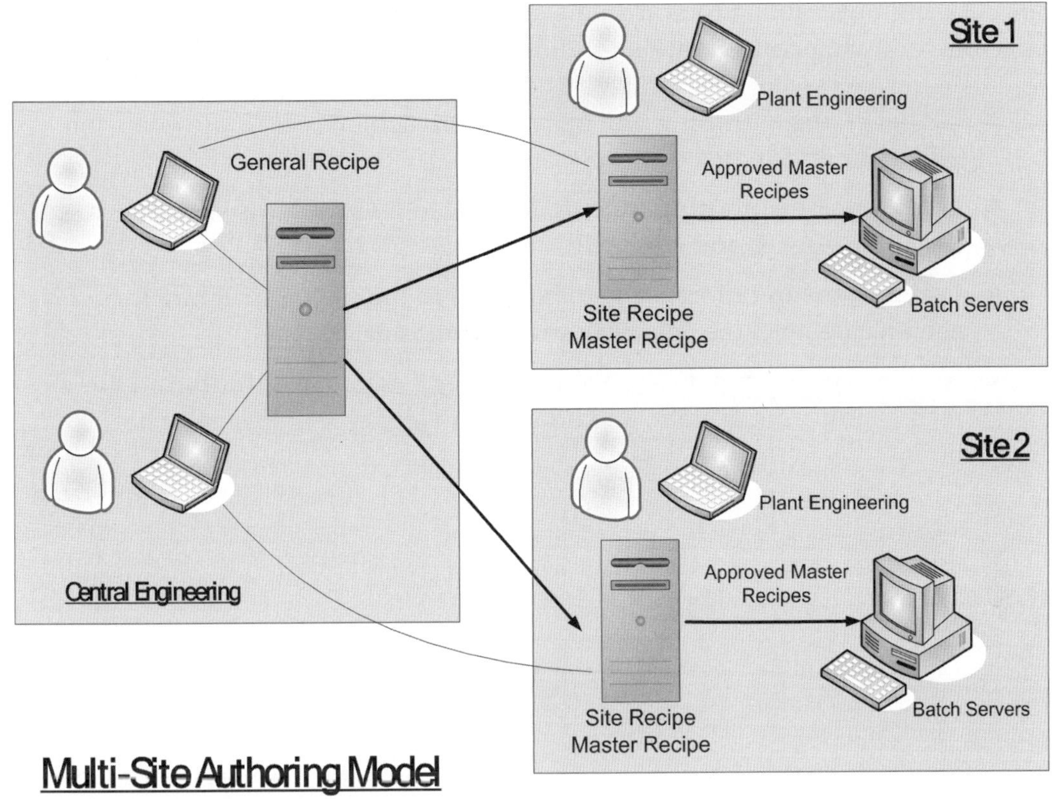

Figure 19.5. Distributed architecture.

Summary

Global recipe authoring can provide several advantages over stand-alone authoring environments. These advantages become more compelling as the global manufacturing enterprise evolves and include the following:

- A common authoring environment for the corporation, independent of individual batch systems
- Interorganizational collaboration on the same recipe that helps avoid the "local-central" dilemma
- Training that is the same for all engineers
- More efficient and easier staffing for large projects
- A higher degree of reusability of recipes across sites

As companies move forward and plan for the next generation of batch systems, the following considerations should be factored into deployment plans:

- Global recipe authoring will be needed to achieve corporate goals for efficiency and reduced engineering effort.
- Global recipe authoring is being facilitated by standards such as B2MML, ISA-95, and BatchML.
- Careful consideration needs to be given to the global structure for each recipe level (i.e., general, site, and master).
- New batch systems must be able to accept standard output from the recipe authoring system, and adapters to legacy systems must be built.

CHAPTER 20

General Recipes as Contracts with Manufacturing

Presented at the WBF North American Conference, April, 7–10, 2002 by

Dennis Brandl
BR&L Consulting
dnbrandl@brlconsulting.com
208 Townsend Ct, Suite 200,
Cary, NC 27511 USA

Abstract

General Recipes (GRs) are ISA-88's definition of how to document the way a product is manufactured without specifying the exact production equipment. GRs are transformed into master recipes through an engineering process, and they can often map to a wide range of physical equipment layouts. The target equipment varies in unit layout, level of automation, physical properties, and process control capability. Usually, local engineers with a deep understanding of the target process cell layout do the transformation from a general or site recipe to a master recipe. The GR thus becomes the controlling document that is exchanged between sites. The GR can be considered a "contract" document between Research and Development (R&D) and manufacturing, defining the chemistry and physics that need to be performed to manufacture the product. Because this document must be unambiguous, the elements that make up the GR must also have an unambiguous definition. In ISA-88, these elements are *process actions* and *equipment constraints*. The actions and constraints are used during master recipe creation to identify master recipe unit procedures and operations and to bind the master recipe unit procedures to equipment. This chapter defines some rules and considerations for defining corporate-wide process actions and equipment constraints, so that they can become the complete and unambiguous definitions required by manufacturing.

Introduction

One of the final but most important tasks that an R&D organization must do is to help make a product ready for manufacturing. This is often referred to as the "technology transfer" stage in new product development. Information about how to make a product in laboratory systems using test tubes must be scaled up to production-scale chemistry. There are often two scale-up efforts: one from laboratory scale (i.e., test tube) to pilot scale (i.e, drums) and then a second from pilot scale to full production scale (i.e., units). In some industries (e.g., the pharmaceutical industry), there may be many scales of production. These scales can vary from micrograms to metric tons, with small samples being manufactured for trials or marketing tests and large production runs after government approval. Often, the different production scales are performed at different facilities or laboratories. These issues make it vital for a company to be able to define how to make a product in a comprehensive and unambiguous manner.

The ISA-88.01 and IEC 61512-01 standards introduce the concept of GRs as the repository for product manufacturing definitions. A GR is intended to be the primary document used by process engineers in the generation of equipment-specific master recipes. Site recipes are local versions of the GR, containing the elements to be produced at the site, local languages, local units of measure, and variations for local material availability. GRs and site recipes can be represented as process and material dependency diagrams, as shown in Figure 20.1.

Usually, engineers with a deep understanding of the target process cell layout do the transformation from the general or site recipe to master recipes for each process cell. The target process cell equipment can vary in unit layout, level of automation, physical properties, and process control capability, so there can be considerable variation in different generated master recipes. Because of this variation, master recipes are not the best way to exchange manufacturing definitions between sites. GRs, which are independent of specific unit layout, automation levels, and equipment specifics, are a better way to describe manufacturing definitions. GRs can be used as a controlling document across sites, defining the chemistry and physics that must be accomplished in order to manufacture a product.

Another way to view a GR is as a "contract" between R&D and manufacturing. A contract is a binding agreement between two or more parties. In this case,

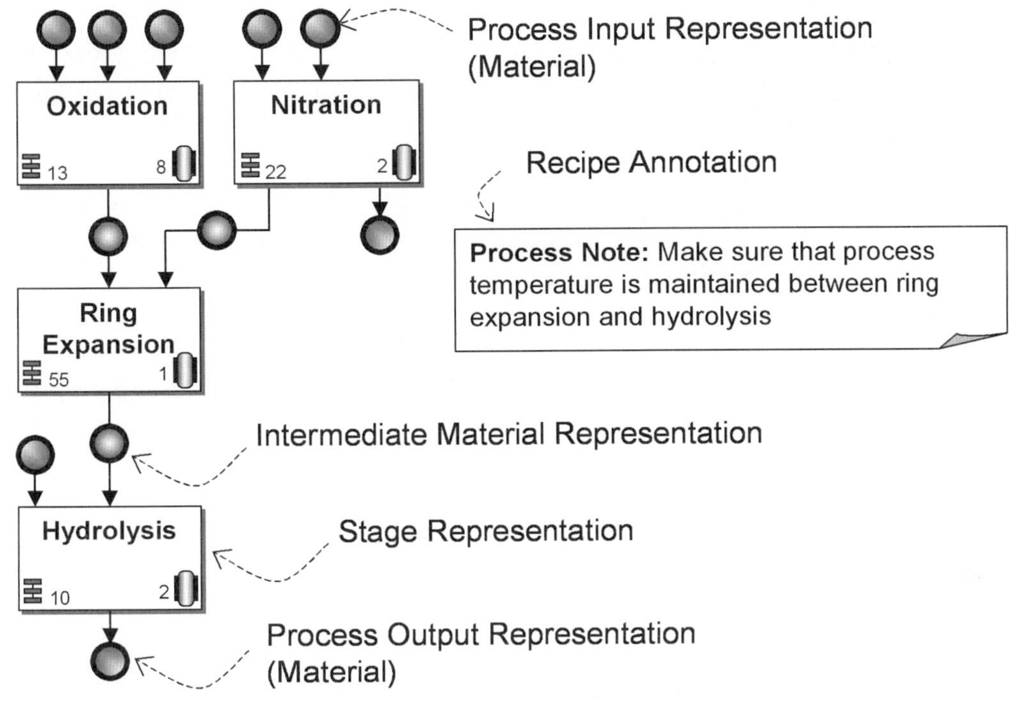

Figure 20.1. A GR representation.

it is an agreement on how to produce a product. In order to be useful, the contract must be both comprehensive and unambiguous. This is accomplished in GRs through a variety of elements, including the process definition, process stages, process operations, process actions, and equipment constraints. Figure 20.2 illustrates the ISA-88 model for the process definition. A GR process is defined using process stages. Process stages contain one or more process operations, and process operations are accomplished using one or more process actions.

Figure 20.3 illustrates a method for documenting the process operations, process actions, and equipment constraints of a GR's process stage. The purpose of this documentation is to provide the master recipe author with a sufficient definition of the manufacturing process, so that it can be converted into a master recipe procedure that is bound to the specific units or class of units. In this example, the equipment constraints are associated with a process stage definition, and the process operations are defined as an organizing structure of process actions.

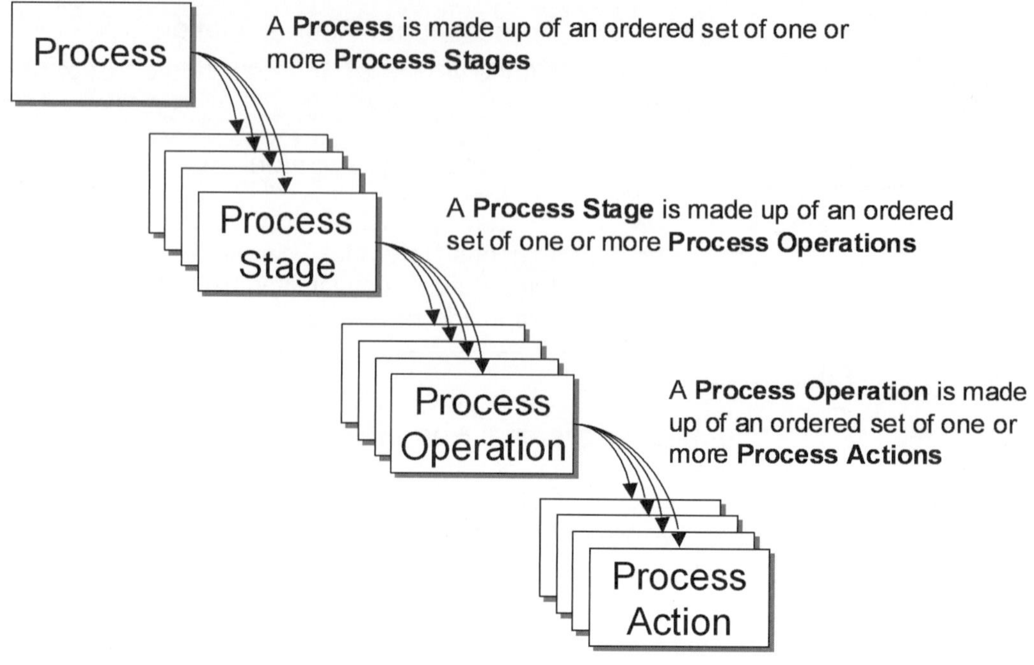

Figure 20.2. Process definition within a GR.

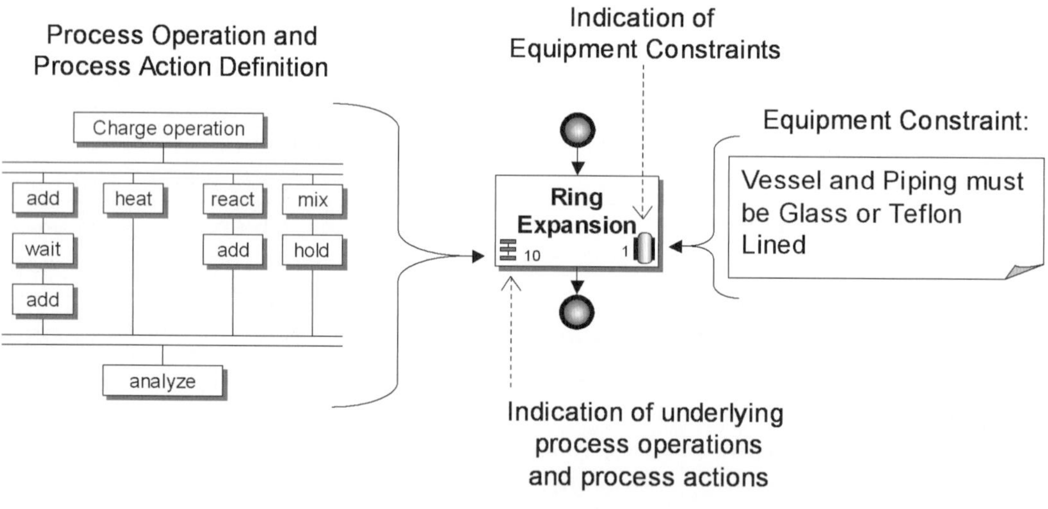

Figure 20.3. Elements of a GR.

Process Actions

A key element of the contract with manufacturing is the definition of the process actions that are used to make up GRs. The process stages and process operations will vary based on products, but they should be constructed from a standardized company- or division-wide library of process action definitions. The process actions define the basic production process capabilities of the company (or division) in an abstract manner. They define the processes that manufacturing can perform to make products. Specific GRs describe the order and timing of the processes to make specific products.

As with any good contract, a process action definition must define all the assumptions and contingencies associated with its use. The elements of a process action include the following:

- A description of the basic functionality provided by the process action
- A definition of any parameters that may be passed to the process action to modify or characterize its behavior
- A definition of any information that may be returned as a result of the process action execution
- A definition of any assumptions about the execution environment of the process actions,
- A definition of any required preconditions that must be met for the process action to execute correctly
- A definition of any expected postconditions that must be handled as a result of the processes action execution
- A definition of known exception conditions and how they should be handled

Experience has shown that most companies (or product line divisions) can define their basic manufacturing process capabilities by using between thirty to fifty process action definitions. These range from simple capabilities (e.g., adding and mixing materials) to industry- or process-specific capabilities, related to separation, extraction, and final packaging of materials.

Process actions can generally be divided into two subtypes—general and company specific. General actions deal with (1) adding materials, (2) removing materials, (3) adding energy, (4) removing energy, and (5) setting material environments (e.g., pressure, agitation). Company-specific actions may deal with unique

(6) material preparation, (7) material extraction, (8) material shaping, (9) material packaging, and (10) material testing. Figure 20.4 illustrates this concept. These ten subtypes usually each have additional variations on the methods that bring the total number of process actions to between thirty and fifty.

Table 20.1 is a short list of sample process actions that can be used as the starting point for the development of a company specific library of process actions. Each process action has multiple parameters that can be used to characterize each instance of use.

Because all general and site recipes are built from a basic set of thirty to fifty process action definitions, the task for process engineers constructing master recipes is made easier. Local site engineers need only determine the best methods for implementing the process actions on their local equipment. The engineers can then use these best methods when constructing master recipes. Using a well-defined library of process actions and a well-defined library of site-specific implementations can significantly reduce the amount of work required to construct master recipes. It can also reduce the variations in recipe quality due to different authors.

The definition of the process action library should be worked out between R&D and manufacturing. A well-documented library of process actions will reduce the ambiguity of the GR definition and make technology transfers faster

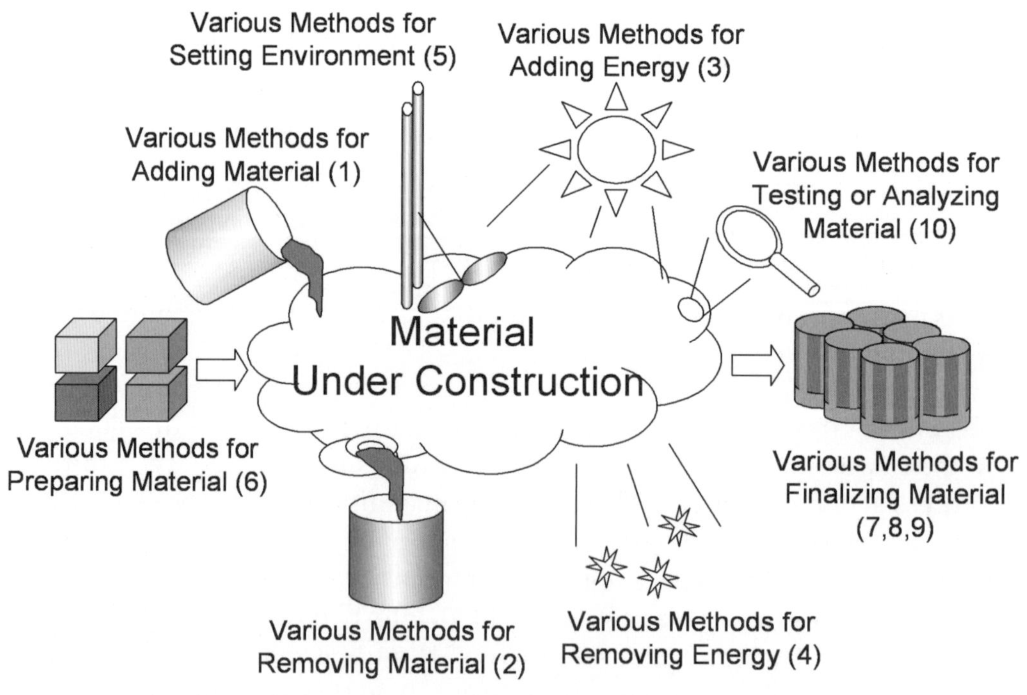

Figure 20.4. Basic processes generally used for process actions.

Table 20.1. Sample process actions		
Process action name	*Functional description*	*Parameters*
Charge	Adds the specified material to the material under construction. There is no rate constraint on the material addition. Usually used where there is no expected chemical reaction.	Material to add Amount to add
ChargeAtRate	Adds the specified material to the material under construction at the specified rate and tolerance. Usually used when mixing is required or too fast a rate will cause an undesired chemical reaction.	Material to add Amount to add Percent per minute Percent per minute tolerance
ChargeAtTemperature	Adds the specified material to the material under construction so the temperature stays within the specified value. This may require heating or cooling capability. Usually used when an exothermic or endothermic chemical reaction will occur.	Material to add Amount to add Maximum temperature Minimum temperature Temperature tolerance
Dry	Dries the material to remove any water or other safely evaporated materials.	Material to remove Expected amount removed Minimum temperature
EvaporateSolvent	Removes a solvent through evaporation. The solvent must be retained and not dispersed into the atmosphere.	Material to remove Expected amount removed Evaporation temperature
FilterSolids	Removes solids in the material under construction.	Material to remove Expected amount removed
HoldTemperature	Adds heat to or removes heat from the material under construction so the temperature stays at a specified value. Usually used to cause a chemical or physical change in the material.	Maximum temperature Minimum temperature Temperature tolerance Time to hold temperature
TemperatureProfile	Adds heat to or removes heat from the material under construction so the temperature follows a defined temperature profile. Usually used to control a chemical or physical change in the material.	Temperature profile
SetPressure	Sets the pressure of the material under construction. Usually used to control boiling or evaporation of materials or to force a chemical or physical change to the material under construction.	Maximum pressure Minimum pressure Pressure tolerance

with fewer errors. The library of process actions should be tightly managed, with controls in place to eliminate unintentional changes. Any change to the library of process actions could impact hundreds to thousands of recipes and require reengineering in all production plants; changes must be tightly controlled because of their potential impact.

Equipment Constraints

The second main element of a GR is the definition of equipment constraints. These are specifications for constraints that must be met by production equipment in order for the material to be correctly produced. Examples of these include constraints on the type of materials used in the production equipment and constraints on the type of heating or cooling used for the material. Equipment constraints are utitlized when the equipment used for production may affect the chemistry or physics of the process. For example, the material under construction may react with the vessel, thus producing a contaminated product or even preventing the desired reaction. Experience has shown that most companies can characterize their equipment using ten to twenty equipment constraints.

Process engineers use equipment constraints to select the appropriate equipment for the manufacture of a product. The GRs constrain the target equipment through its properties without specifying the exact equipment. Both equipment constraints and the ability to perform process actions on equipment enable the process engineer to select the best manufacturing equipment. For example, the requirement that a material must be heated, combined with the requirement that it must be mixed at the same time in a copper-free environment, would eliminate any nonheating, nonmixing, or copper vessels from consideration.

Equipment constraints may also define "hints" for production (e.g., indicating that short transfer lines are needed because the material tends to clog lines or react during transfer). Other hints may involve company-wide definitions of classes of equipment that are appropriate. For example, a company may characterize equipment as "Type 1—Safe for Color Film" or "Type 2—Not Safe for Color Film."

Table 20.2 lists example equipment constraints that a company may define. User experience has shown that equipment constraints should be defined as either a class of equipment or a set of Boolean (i.e., true or false) choices. These methods provide the most flexibility and extensibility and are illustrated in Table 20.2.

Library Control and Workflows

Equipment constraint types must be controlled and managed in the same way that process actions are managed. Because equipment constraints are used by

Table 20.2. Sample equipment constraints.		
Equipment constraint	*Functional description*	*Parameters*
SafeTemperature	Defines the required safe temperature of any containing vessel	Temperature; Measured in degrees
ExternalHeatingCoils	Specifies if any heating coils are external to the containing vessel, so they cannot be fouled by the batch material	Boolean
LiningGlass	Specifies if the containing vessel is glass lined	Boolean
LiningNickelFree	Specifies if the containing vessel and pipes are free of any elements containing nickel	Boolean
LiningCopperFree	Specifies if the containing vessel and pipes are free of any elements containing copper	Boolean
TransfersShort	Specifies if any transfers between vessels are short, so that unintended reactions (such as crystallization) will not occur in the piping	Boolean
TransfersStraight	Specifies if any transfers between vessels are relatively straight, so that highly viscous or material prone to plugging may be transferred	Boolean
AgitationNonSheared	Specifies that the agitation method does not shear the batch material	True or false

manufacturing to select equipment, they must be well understood and contain unambiguous definitions. Equipment constraint types should also be maintained in a library so that they are available to general and site recipe authors. One method to manage process actions and equipment constraints is to apply configuration management to the definitions. There can be versions for each definition and states for each definition. Table 20.3 lists example states that could be applied. The states should be supported by a workflow process that requires sign-off approval to change the state of the object and a workflow to ensure that all parties are informed of any changes.

Summary

GRs and site recipes define a contract between manufacturing and R&D. They define the chemical and physical rules on how to make a product. They also define constraints for the target manufacturing equipment. The GR can also be a repository for other information about the product. This may include pictures, diagrams, chemical drawing, or even laboratory notes. The main constituents of a GR are process actions and equipment constraints. Experience has shown that the best

Table 20.3. Example definition states	
State name	*State description*
Approved	Indicates that the process action or equipment constraint definition is available for use in general or site recipes
Obsolete	Indicates that the process action or equipment constraint definition is no longer available for use in general or site recipes
Preliminary	Indicates that the process action or equipment constraint definition is complete and may be used in general or site recipes for testing but is not yet available for use in normal production recipes
InWork	Indicates that the process action or equipment constraint definition is not yet complete, and the definition is not available for use in general or site recipes

way to use process actions and equipment constraints is to construct a managed library of definitions. GR authors can then use these definitions. The library allows them to unambiguously define how to manufacture a product, ensuring that there are places to put all pertinent information. Process engineers can use the process action library to define a set of "best practices" to implement the actions on their equipment. This allows them to build consistent and predicable recipes. A controlled library of equipment constraints provides similar advantages to recipe authors and process engineers.

Using General Recipes for Standardized Multiple Plant Manufacturing Science

Presented at the WBF North American Conference, March 5–8, 2006, by

Velumani Pillai
velumani.pillai@pfizer.com
Pfizer, MSC 724, 100 Route 206 North
Peapack, NJ 07977 USA

Rob Burrows
rob.burrows@pfizer.com
Pfizer, MSC 420, 100 Route 206 North
Peapack, NJ 07977 USA

Dennis Brandl
dnbrandl@brlconsulting.com
BR&L Consulting, 208 Townsend Ct.
Cary, NC 27511 USA

Abstract

Innovation in pharmaceutical manufacturing with faster development, faster new product release, robust process understanding, and vastly improved quality by design is a key component of the FDA's 21st Century manufacturing initiative. A pioneering approach that uses General Recipes (GR) as a keystone element is accelerating our understanding of manufacturing science at Pfizer to support standardized multiple-plant investigations.

Manufacturing science is the body of scientific knowledge, regulations, and principles involved in the transformation of materials and information into products. Applying manufacturing science across multiple sites has always been daunting. Information that may take months to collect at one site cannot easily be compared against information from other sites, due to differences in equipment, unit layouts, and master recipes.

The ISA-88.03 GR and site recipe standard provides a method to document standard manufacturing science information. This chapter describes an innovative way to extend GRs to document critical and key process parameters and a method to convert that information into data collection requirements for plant-specific master recipes. Process knowledge structured in this fashion will aid in faster development, release, and improved quality, helping the life science industries achieve their Six Sigma quality goals.

Introduction

Manufacturing science is the terminology used at Pfizer to describe the body of scientific knowledge, laws, and principles involved in the transformation of materials and information into products. One of the key components of manufacturing science is the development of scientifically sound recipe parameters based on investigations and studies.

The application of manufacturing science across multiple sites is challenging as a result of the significant variability that exists between sites. For example, unit configuration, material handling, and master recipes are seldom consistent. Thus the information gained at one site cannot easily be compared against information from another. One method for performing multisite and multiple process cell investigations involves defining a consistent way to describe investigation information independently from the underlying equipment. This information needs to address differences that may include different unit layouts, different unit sizes, different material transfer methods, and different levels of automation at sites.

This problem is similar to the problem addressed by the ISA-88.03 standard, in describing the manufacturing process without reference to the specific target equipment. Not surprisingly, general and site recipes are good starting points for documenting the information needed for multisite manufacturing science. In order to use GRs for manufacturing science investigations, several elements need to be extended or customized in the following areas: process parameters, process reports, and observed operational modes.

Process Parameters

General and site recipes contain process parameters, process inputs, and process outputs:

- Process inputs and process outputs are defined as materials.
- Process parameters are defined as "information that is needed to manufacture a material but does not fall into the classification of process input or process output."

While these definitions are very open-ended, more rigor is needed for manufacturing science documentation and investigations.

Quality Attributes

We use the term *quality attributes* to signify a specific type of process parameter. Quality attributes define the target qualities of the produced material. Quality attributes are the physical, chemical, or microbiological properties of a material that directly or indirectly impacts product specifications. For example, a quality attribute for a "Blend" process action may be a measure of the target homogeneity and the appearance of the blend (e.g., that it has no visible discolorations or clumping).

Quality attributes may be further categorized as *critical* or *key*. Critical quality attributes are registered with regulatory authorities and are a characteristic of the product. Key quality attributes are not registered.

Process Parameters

Process parameters define values that must be controlled in the process because they influence quality attributes. Process parameters must be controlled within predefined limits to ensure the product meets its predefined quality attributes. Critical process parameters influence critical quality attributes. Key process parameters influence key quality attributes. Figure 21.1 illustrates a quality attribute (i.e., sterility) and the parameters that must be controlled (i.e., time and temperature) to achieve the desired attribute. Process parameters and quality attributes become some of the primary inputs used in converting a GR to a master recipe.

Process Action Definition

ID:	Sterilize with Heat
Version:	V01
Process Level:	Process Action
Description:	Sterilize material using heat.
Lifecycle State:	Draft

Author:	J. Smith
Effective Date:	25-Nov-05
Withdrawn Date:	
Replaces Version:	

Critical and Key Quality Attributes

Attribute Name	Description	Attribute Value	Critical/Key	Req/Opt
Sterile	Sterility of the material	True	Critical	Required

Critical and Key Process Parameters

Parameter Name	Description	Default Value	Ranges	Critical/Key	Req/Opt
Temperature	Temperature for sterility	90 UOM: Deg C	Click here to insert add Range	Critical	Requir
Time	Time to maintain at temperature	20 UOM: Min	Click here to insert add Range	Critical	Requir

Figure 21.1. Quality attributes and process parameters.

Process Reports

In addition to defining attributes and parameters, investigations also need a place to document information that needs to be reported as a result of the execution of the process action. These are called "process reports." They define the information that must be collected and reported on while the process is operating. Figure 21.2 illustrates some typical process reports.

Process reports should be collected regardless of the equipment layout or level of automation. They provide the raw material used in multisite investigations. The most time-consuming part of an investigation is the collection of the process data. By formalizing the minimum amount of data that should be collected and made available, the time to perform multisite investigations should drop by a substantial amount.

Process Reports		
Report Name	**Description**	**Req/Opt**
Actual % of active to total	Actual percentage of active ingredient to total material	Require
Density	Density of material	Require
Particle Size	Average and medium particle size	Option:

Figure 21.2. A process report definition.

Observed Modes

Other manufacturing science information can also be added to a GR. The GR can be used as a form of process knowledge repository. The GR Other Information (GROI) category in the ISA-88 standard is used to contain information that is not captured in other recipe elements. Additional structure can be added to GROI to record information that includes but is not limited to the following:

- *Observed failure modes.* Any failure modes that have been observed in actual operation
- *Observed alarm modes.* Any modes that have resulted in alarm conditions that have been observed in typical operations
- *Observed exception modes.* Any modes that have resulted in exception conditions that have been observed in typical operations

Figure 21.3 illustrates some of the information associated with different modes, including known examples and discovered causes.

Observed Failure Modes			
Mode Name	**Description**	**Examples**	**Causes**

Figure 21.3. Documenting observed modes.

Parameter and Attribute Relationships

Once causal relationships have been established between parameters that can be controlled and attributes that can be measured, these relationships must be captured and maintained. This information is obtained from validation batches, process capability studies, and investigations. This critical information also provides a good framework for risk management. Again, the GR can be used as a repository for this information. Many of the relationships are product or material specific, such as the relationships between blending speed, blending time, and output material homogeneity. Even though much of this information may be unit or equipment specific, it is a valuable resource for investigations. While this information is often maintained at a site level, keeping this at the corporate GR level supports multisite investigations.

Multiple-site Manufacturing Science

The previous extensions to GROI capture corporate process knowledge in GRs. This knowledge must be combined with site knowledge, production knowledge, and manufacturing science investigations for a more complete manufacturing science environment, capable of the following:

- GR information can be expanded to become a repository of manufacturing science.
- The extensions defined for GRs can also be applied to site recipes for the capture of site knowledge.
- Production history can be recorded as described in ISA-88.04.

The relationships among these list items are shown in Figure 21.4. These relationships allow a cycle of continuous improvement in manufacturing. The elements of the extended GR and site recipes are shown in Figure 21.5. Master recipes can also be extended to include different observed modes as aids to operations and as aids to process engineering.

Production history and manufacturing science investigations are composed of the elements illustrated in Figure 21.6. Production history involves more than just the batch history recorded in a batch production record; it may also include information from other sources, such as Laboratory Information Management Systems (LIMS) and Certificate of Analysis (CoA) systems. Often, the important information needed for manufacturing science is not in the batch records but in

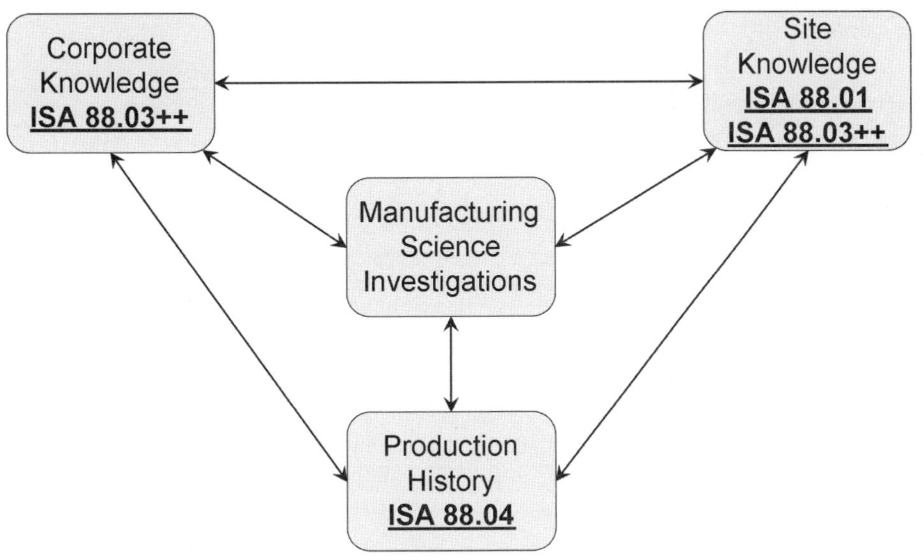

Figure 21.4. Elements of multiple-site manufacturing science.

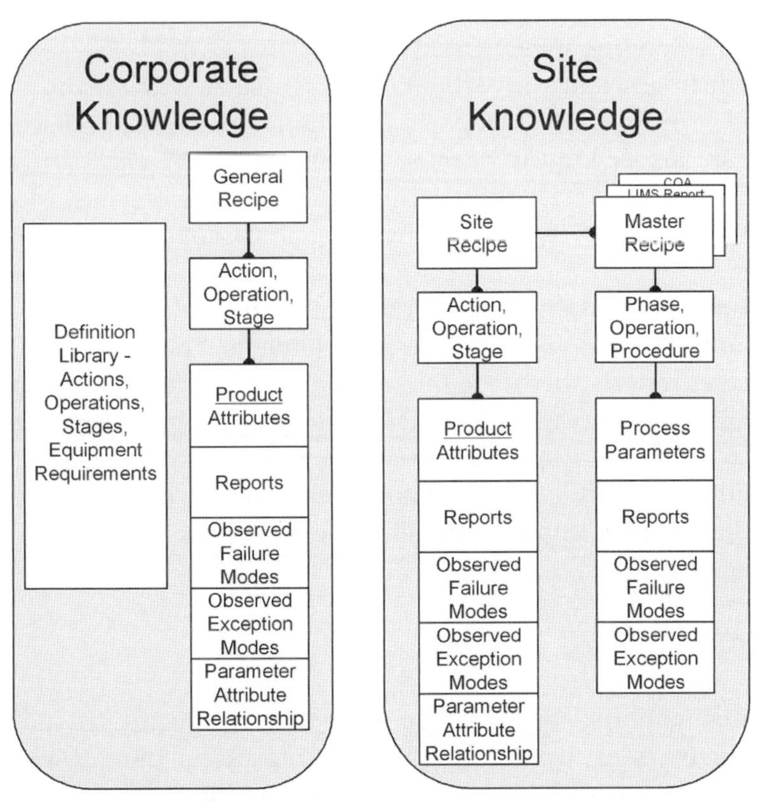

Figure 21.5. Corporate and site knowledge.

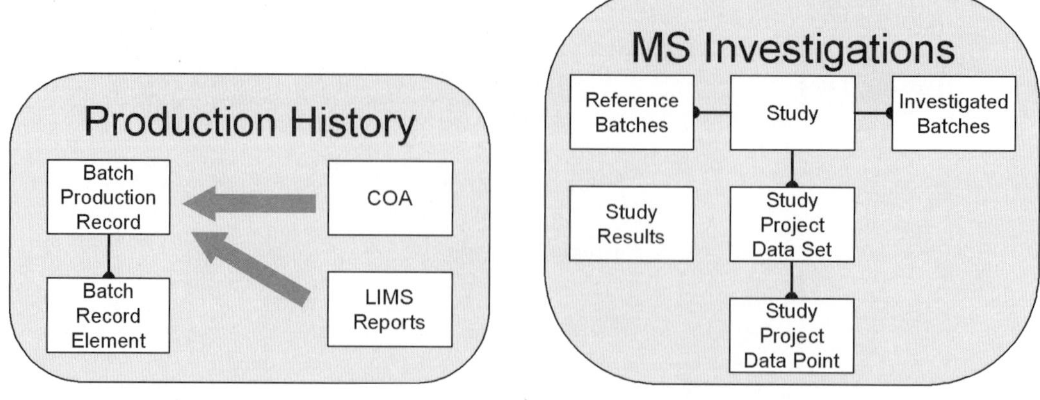

Figure 21.6. Production history and manufacturing science investigations.

the systems that support production. This includes information about the storage history of raw, intermediate, and final products; lab tests on intermediates; and information about the environmental conditions of production (e.g., what material was produced before the investigated batch).

Manufacturing science investigations often start with a study. Studies may be performed to optimize processes, to investigate incidents, to discover improvement opportunities, or for any number of reasons. Often, each study is related to a set of investigated batches and may also use reference batches. Most studies try to encompass at least thirty batches in order to discover statistically significant relationships.

A study project contains a study data set—a multidimensional data set composed of different batches and different data elements. The study data set is derived from the process reports that were initially defined in the GR. The complete set of relationships is shown in Figure 21.7. The structure shown in Figure 21.7 is conceptual and not related to any specific software, product, or database repository, but it does illustrate how extended general and site recipes, used with batch production records, can be used to support multisite investigations.

Conclusion

Multisite investigations need process data that is specific to individual sites' equipment peculiarities. By defining process reports in the GR and letting each site determine how to collect the data, the information can be made available for multisite investigations without knowing the details of the specific master recipes.

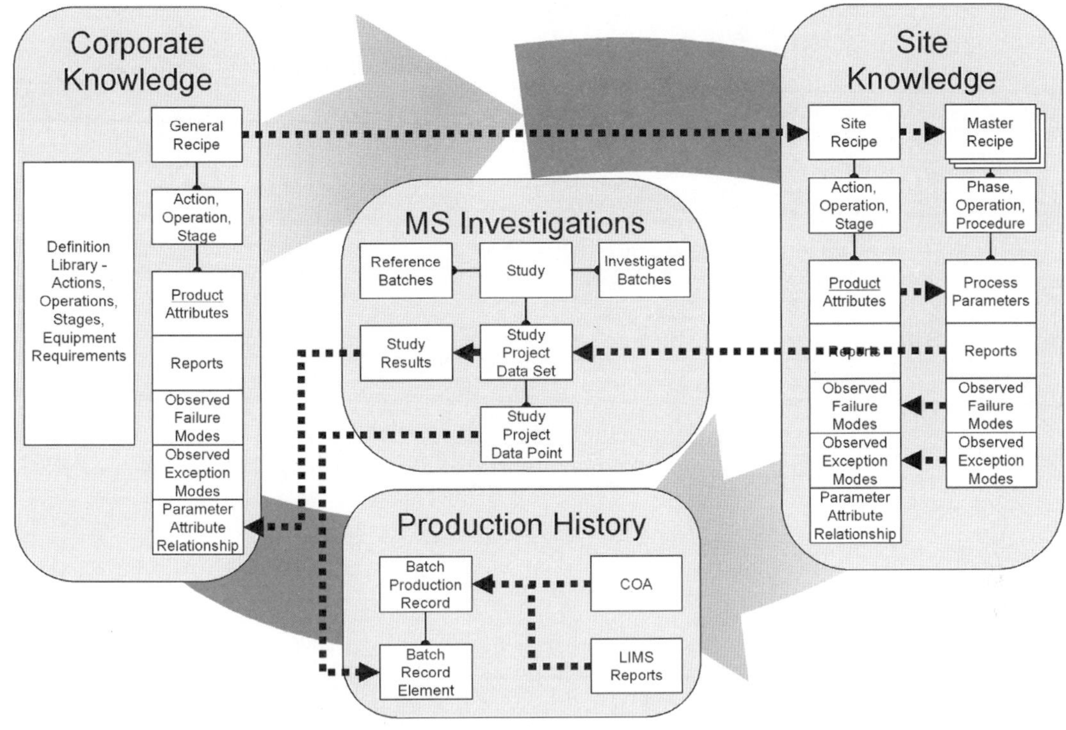

Figure 21.7. Elements of manufacturing science.

Even if the same operation is performed in different process cell layouts in different sites with different numbers of units, the information across sites can be compared. However, applying this to real-world multisite problems requires rigor in defining the process- and product-specific information reports that must be collected. This rigor should pay off in much faster multisite investigations, with reduced effort for site personnel and corporate investigators.

CHAPTER 22

Manufacturing Science Model Extensions to Address Lean Manufacturing and Supply Chain Optimization

Presented at the WBF European Conference, November 10–12, 2008, by

Velumani Pillai
velumani.pillai@pfizer.com
Pfizer, MSC 722, 100 Route 206 North
Peapack, NJ 07977 USA

Rob Burrows
rob.burrows@pfizer.com
Pfizer, MSC 420, 100 Route 206 North
Peapack, NJ 07977 USA

Dennis Brandl
dnbrandl@brlconsulting.com
BR&L Consulting, 208 Townsend Ct.
Cary, NC 27511 USA

Abstract

Innovation in manufacturing, through faster development, faster new product release, robust process understanding, and vastly improved quality by design is critical for advanced manufacturing organizations. Lean manufacturing and supply chain optimization projects are increasingly important elements of advanced manufacturing organizations, and capturing information to support these projects is an important part of product and process definitions.

Manufacturing science is the body of scientific knowledge, regulations, and principles involved in the transformation of materials and information into products. In this chapter, we propose Manufacturing Science Informatics (MSI) as the new information model framework to support the process and product knowledge that is discovered, shared, retained, and applied over a product's entire life cycle. An MSI framework enables established business processes such as production sourcing, material sourcing, new product introduction, continuous process improvement, process optimization, and product investigations, and it can also be extended to address new business processes. The MSI framework uses ISA, IEC, and ISO standards as foundational information entities. In this chapter, we propose extensions to the ISA-88 General Recipe (GR) model and other product and process information entities in the ISA-95 standards to handle Lean manufacturing and supply chain optimization requirements.

Lean manufacturing is the elimination of non-value-added operations and the elimination of waste in production. Lean manufacturing requires an unambiguous definition of production and process options, expected yields, and maximum expected waste, so that site-specific Lean decisions can be made.

Supply chain optimization involves the sourcing of materials, the selection of sites for production, and the elimination of logistics bottlenecks. This requires an unambiguous definition of material substitutions, production and process options, and expected processing times and throughputs, so that site-sourcing decisions (i.e., decisions about where to manufacture a product) can be made.

MSI Extensions for Lean Manufacturing and Supply Chain Optimization

Manufacturing science is the body of scientific knowledge, regulations, and principles involved in the transformation of materials and information into products. An information model framework—MSI—has been proposed to address the evolving information needs of manufacturing.[1] This framework contains conceptual models that support all processes that contribute to manufacturing processes and is based on the recognized ISA, IEC, and ISO standards, such as ISA-88.03 for GRs, ISA-88.04 for batch production records, and ISA-95-IEC/ISO 62264-1 for capability models.

One of the key components of manufacturing science is the development of scientifically sound recipe parameters and control strategies through investigations and studies. The MSI framework can easily be extended to handle investigations and studies in support of Lean manufacturing projects and supply chain optimization projects.

Explaining Lean Manufacturing

The essence of Lean manufacturing (as defined in the Toyota Production System [TPS][5]) focuses on the elimination of all nonmanufacturing wastes[3] (e.g., unneeded inventory; wasted motion; unneeded transportation of materials, equipment, or personnel; quality defects; and changeovers). There is often a main focus on the reduction of three types of waste (using the Japanese terms): *muda* (non-value-adding work), *muri* (overburden), and *mura* (unevenness), to expose both systematic and randomly occurring problems.

Lean manufacturing is a relative term. Some best-in-class manufacturing companies operate with zero waste, no non-value-added activities, and even process flows. Unfortunately, these companies are few and far between. Every other company must compare its operations against its own best practices and against industry benchmarks in order see how Lean they really are and how much they can improve.

Lean and optimization projects often follow the Define, Measure, Analyze, Improve, and Control (DMAIC) process. DMAIC is a Six Sigma methodology for improving an existing or established process. Six Sigma seeks to identify and remove the causes of defects and errors in manufacturing or business processes. The basic DMAIC process consists of the following five steps:

- **D**efine process improvement goals that are consistent with customer demands and the enterprise strategy.
- **M**easure key aspects of the current process and collect relevant data.
- **A**nalyze the data to verify cause and effect relationships. Determine what the relationships are, and attempt to ensure that all factors have been considered.
- **I**mprove or optimize the process based on data analysis using techniques like Design of Experiments (DoE).
- **C**ontrol to ensure that any deviations from targets are corrected before they result in defects.

When the ISA-88 standards are overlaid on the DMAIC process, there is a natural place where the ISA-88 data elements fit, as shown in Figure 22.1:

- The ISA-88.03 standard defines GRs as the repository of product knowledge.
- The ISA-88.01 standard defines the recipe and equipment elements required for control.
- The ISA-88.04 standard defines production history that can be used to record and preserve measured information to be used for analysis.

GRs have been proposed as a method to document cross-site reporting requirements.[2] They can also include benchmark and best practice information. A GR can specify which information should be obtained during execution and associate it with operations and actions. This information has been defined in Chapter 21 as "process reports," which contain specific types of process parameters, as defined in the ISA-88.03 and IEC 61512-3 GR standards. This same method can be used to define information required for Lean benchmarking and comparisons of work centers or sites.

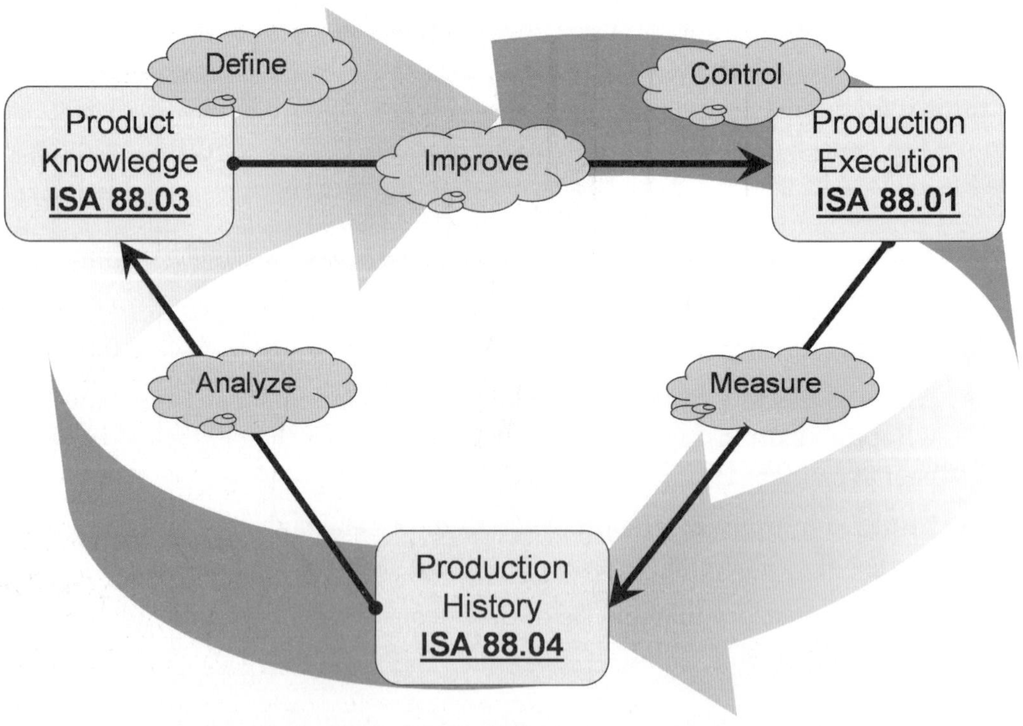

Figure 22.1. DMAIC elements and ISA-88 elements.

Process Reports

Process reports define the information that must be collected and reported during the execution of the process. Process reports for Lean manufacturing projects may be associated with specific process actions, process operations, or entire stages. For example, there may be a best practice yield that can be associated with a process operation but cannot be directly associated with any specific contained process action. Figure 22.2 illustrates some typical process reports.

Process reports are to be collected regardless of the equipment layout or level of automation. They provide the raw material used in Lean investigations. The most time-consuming part of an investigation is the collection of the process data. Formalizing the minimum amount of information that must be collected and made available will significantly reduce the time and effort required to perform Lean improvement investigations. Seven of the common criteria evaluated for Lean improvement include the following:

- Measures of transportation time and cost (for the moving of materials that are not actually required to perform the processing)
- Measure of inventory (including all components, work in progress, and finished product not being processed)
- Measure of motion (for people or equipment that are moving or walking more than is required to perform the processing)
- Waiting time (for the next production step)
- Overproduction (i.e., production performed ahead of demand)
- Overprocessing (i.e., rework due to poor tool or product design)
- Defects (i.e., the effort involved in inspecting and fixing defects)[4]

These criteria help specify specific process reports for Lean manufacturing:

Process Reports — Add | Delete | Hide | Show

Report Name	Critical/Key	Req/Opt		Description
Defects	na	Optional		Number of sample defects. For example, are ideally 0.
Yield	na	Required		Yield. For example, are ideally 100% but most often 98-100%

Figure 22.2. Process reports for Lean optimization.

- *Industry benchmarks of best practices.* These are the best practice values within the industry or similar industries. They provide targets for each site and allow sites to identify areas for improvements, such as the following:
    - Production time, waiting time, and setup and cleanup time
    - Yield
    - Defects
    - Waste
    - Energy usage
    - Best batch size
- *Company best practices.* These are the company's best practices. They identify the site with the best practice and values. This allows comparisons of sites and aids in the allocation of capital projects and improvement projects, focusing on the same areas for improvement as listed previously.

Supply Chain Optimization

Lean improvements can also be applied to supply chains and supply systems. A "Lean supply system" will optimize plant and supply chain performance, with fewer buffers and a segmented approach to managing the supply network and internal manufacturing operations based on product characteristics (e.g., volume, demand volatility, product value).

The efficiency of a manufacturing process is usually determined by choices made early in the development of a product or a process. The efficiencies may be based on sourcing decisions (i.e., where to manufacture a product), as well as the basic production process used to make the product. Typically, the choices made early in development, such as alternate production pathways or estimated production Key Performance Indicators (KPI), are not formally captured and therefore information is not available for later use in supply chain optimization or Lean manufacturing initiatives. General and site recipes can be a repository for the information discovered during development, and this information can be very valuable for later decisions about sourcing production.

Alternate Production Processes

During the development of a product or process, there may be several alternate production processes defined. These may be alternate reaction synthesis pathways for chemical processes (e.g., Active Pharmaceutical Ingredient [API] production) or alternate processes (e.g., alternate drying methods for pharmaceutical unit dosage production). In some industries (e.g., biotechnology, pharmaceutical), a single production process is selected because of regulatory or validation concerns, but in other industries (e.g., food, beverage, consumer packaged goods, chemical) there may be multiple acceptable production processes for a single product.

The GR structure provides a method to describe both the primary and the alternate production processes. For example, there may be a pathway that uses minimal energy and produces the least waste but is a slow process, while an alternate process may use more energy and product waste but production time is significantly shorter, as shown in Figure 22.3. In this example, one plant may have implemented the "least energy" pathway, another may have implemented the "least time" pathway, and a third may have implemented both pathways.

The alternate pathways provide flexibility in supply chain optimization, allowing sourcing decisions to be made based on delivery dates and energy and waste costs. If delivery time is a critical factor, then production may be sourced in the second or third plant in the example shown in Figure 22.3. If minimal cost is the critical factor, then production may be sourced in the first or third plant.

Alternate pathways may also be defined in a site recipe to indicate a modified production method that is suitable within a site. For example, there may be differences in local regulations for emissions or waste production or alternate methods for disposal or use of by-products. Different process may be defined with additional operations, such as cleaning of process streams and recording of waste generated.

Alternate pathways can be defined for different batch sizes. This is especially useful when the chemistry or physics is based on the surface area of a vessel or container, due to the reaction of degradation issues.

Using Process Stages for Sourcing Decisions

Process stages within a GR can be used to improve sourcing decisions. For example, there may be three stages for a product: API production, tableting, and packaging. If there are sites that can implement different stages, then there may be up to four sourcing alternatives:

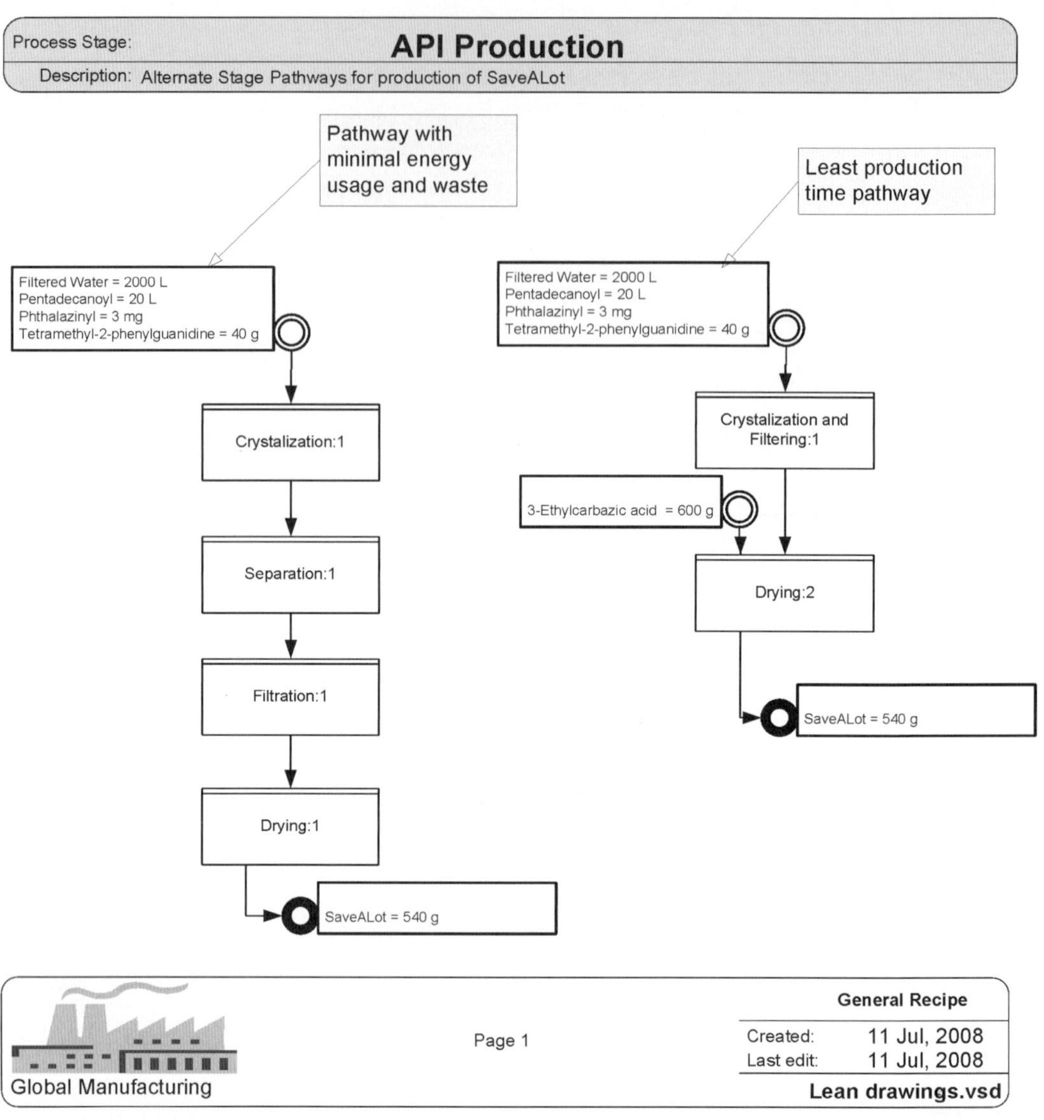

Figure 22.3. Alternate process operations in a GR.

- All stages performed in a single facility
- Each stage performed in a separate facility
- API and tableting in one site and the packaging in a another site
- API in one site and tableting and packaging stages in a single site, as shown in Figure 22.4

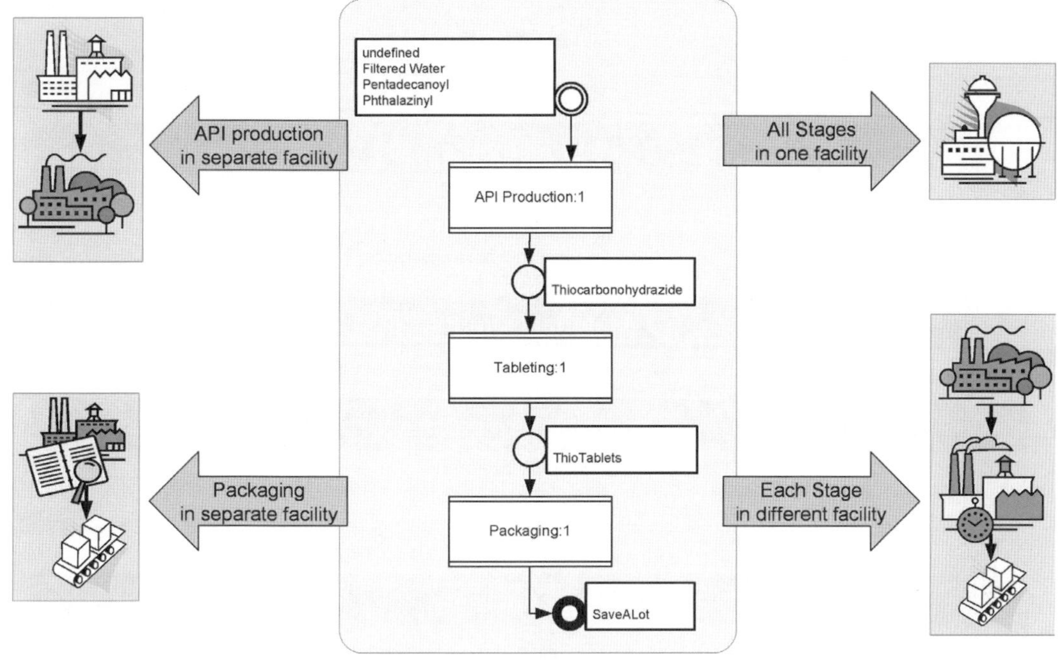

Figure 22.4. Alternate sourcing decisions from GR information.

Alternate Materials

Lists of alternate materials and amounts may also be included in general or site recipes. While alternate materials are often maintained in corporate material master information, not all materials may be equivalent in all processes, so the alternates may be product (or operation) specific. Figure 22.5 illustrates alternate materials that are included in the material information of a GR. Additionally, alternate materials may have different activities, and therefore, different amounts would be required. In this case, the alternate materials may require alternate production process pathways, as shown in Figure 22.3.

Summary

Lean manufacturing and supply chain optimization are increasingly important elements of advanced manufacturing organizations, and capturing information to support these efforts is an important part of product and process definitions. MSI is an information model framework to support process and product knowledge that is discovered, shared, retained, and applied over a product's entire life cycle.

Materials

Material Inputs		
23379	2000 L	Filtered Water
8892	20 L	Pentadecanoyl
99827	3 mg	Phthalazinyl
99888	40 g	Tetramethyl-2-phenylguanidine
19645	600 g	3-Ethylcarbazic acid

Material Outputs		
88397	540 g	SaveALot

Material Intermediates		
5566	600 g	Thiocarbonohydrazide
88393	10000 tablets	ThioTablets
88726	2000 8-Blister Packs	Thio

Material Alternates					
8892	20 L	Pentadecanoyl	77627	21.5 L	Pentadecanoyl-2
8892	20 L	Pentadecanoyl	77628	19.85 L	Pentadecanoyl-3

Figure 22.5. Alternate materials listed in a GR.

This framework contains conceptual models that support all processes involved with manufacturing and is based on recognized ISA, IEC, and ISO standards, such as ISA-88.03 GRs, ISA-88.04 batch production records, and the ISA-95-IEC/ISO 62264-1 capability models.

Information for Lean manufacturing and supply chain optimization projects can be included in general and site recipes and preserved in batch production records. This information can define target values for yields, production times, setup times, and maximum defects and can define specific information to be collected at each production facility, to be used for global optimization projects. This information can also be used for sourcing decisions, alternate production pathways, and material selection optimization.

The MSI framework for product and production information extends the ISA-88.03 models to include quality attributes, process reports, sustainable manufacturing information, Lean manufacturing information, and supply chain optimization information. These extensions provide a valuable manufacturing information repository for 21st-century manufacturing.

References

1. Burrows, R., D. Brandl, and V. Pillai. 2005. Information model framework for manufacturing science. Paper presented at the WBF North American Conference, March 2005.
2. ———. 2006. Using general recipes for standardized multiple plant manufacturing science. Paper presented at the WBF North American Conference, March 2006,
3. Ohno, Taichi. 1988. *Toyota production system*. New York: Productivity Press.
4. Womack, James P., and Daniel T. Jones. 2003. *Lean thinking*. London: Simon & Schuster.
5. Womack, James P., Daniel T. Jones, and Daniel Roos. 1991. *The machine that changed the world*. New York: Harper Perennial.

CHAPTER 23

Manufacturing Science Model Extensions to Address Product and Process Sustainability

Presented at the WBF North American Conference, March 24–26, 2008, by

Velumani Pillai
velumani.pillai@pfizer.com
Pfizer, MSC 722, 100 Route 206 North
Peapack, NJ 07977 USA

Rob Burrows
rob.burrows@pfizer.com
Pfizer, MSC 420, 100 Route 206 North
Peapack, NJ 07977 USA

Dennis Brandl
dnbrandl@brlconsulting.com
BR&L Consulting, 208 Townsend Ct.
Cary, NC 27511 USA

Abstract

Innovation in manufacturing with faster development, faster new product release, robust process understanding, and vastly improved quality by design is critical for advanced manufacturing organizations. Sustainability awareness (i.e., the environment and societal impact of the product and production processes) is also permeating into consumers' minds when they evaluate products.

Manufacturing science is the body of scientific knowledge, regulations, and principles involved in the transformation of materials and information into products. In this chapter, we propose Manufacturing Science Informatics (MSI) as the new information model framework to support process and product knowledge discovered, shared, retained, and applied over a product's entire life cycle. The MSI framework enables established business processes (e.g., production sourcing, material sourcing, new product introduction, continuous process improvement, process optimization, and product investigations) but can also be extended to address new business processes. The MSI framework uses ISA-88 General Recipes (GR) as one of the foundational information entities.

Emerging sustainability and green directives (e.g., EU Directive 2005/32/EC) are placing new reporting and assessment demands, such as total and actual energy use and greenhouse gas emissions, to be tracked and reported not only during design but also during manufacturing of a product. Sustainability criteria is increasingly finding its way into development and sourcing decisions. Here we propose extensions to the ISA-88.03 GR model, and other product and process information entities in the ISA-95 standards, to handle these evolving needs.

Introduction

Sustainability is defined as a characteristic of a process or state that can be maintained at a certain level indefinitely. The United Nations (UN) World Commission on Environment and Development defines it as "meeting the needs of the present without compromising the ability of future generations to meet their own needs."[5]

For example, sustainability in agriculture is defined as being able to have agricultural systems (e.g., land, water, planting, harvesting) that can be expected to last indefinitely. Indefinitely is a long time, but most sustainability is often defined as having systems in place that can last for hundreds of years, barring catastrophic conditions.

According to Jim MacNeil, secretary general of the UN World Commission on Environment and Development, sustainability is "growth based on forms and processes of development that do not undermine the integrity of the environment on which they depend."

Being *sustainable* also means creating and maintaining conditions under which humans and nature can coexist in productive harmony that also fulfill the social, economic, and other requirements of present and future generations of humanity.[2]

Sustainability in manufacturing is the development and deployment of processes that conserve energy, reduce waste materials, reduce the use of nonsustainable materials in the production of processes, and increase the recyclability of products and waste materials. Sustainability is creeping into products and production through legislation and consumer pressure. Emerging sustainability and green directives (e.g., EU Directive 2005/32/EC) are placing new reporting and assessment demands in areas such as total and actual energy use and greenhouse gas emissions to be tracked and reported not only during design but also during manufacturing of a product. There are also increasing discussions about "Carbon Taxes"—a tax on a company's energy use.

Sustainable manufacturing is also an increasingly important element of a company's societal commitment. Sustainability requires a balancing act of economic needs, protection for environment, and our social responsibility to leave the world a better place for future generations.

Sustainability Impacts on Pharmaceutical Process Development and Manufacturing

In many cases, pharmaceutical product manufacturing involves hazardous substances. Traditional approaches to this issue involved dealing with pollution itself, but in the last decade, pollution prevention approaches started gaining traction. In the life science industries (e.g., pharmaceutical and biotechnology), the concept of sustainability is closely related to the following:

- Use of safer solvents
- Design of manufacturing processes based on energy efficiency
- Use of renewable feed stocks
- Identification of recyclable materials
- Identification of recycling and disposal costs

Some of these elements are addressed as *green chemistry,* as defined by the American Chemical Society's Green Chemistry Institute (CGI) and the U.S. Environmental Protection Agency (EPA). Other elements are addressed in manufacturing through the selection of sustainable production methods.

Green chemistry is the design of chemical product and processes to reduce or eliminate the use of hazardous substances. The twelve principles of green chemistry, identified in the seminal work, *Green Chemistry: Theory and Practice*[1] by Paul Anastas and John Warner, are the underpinning of the EPA's green chemistry program and provide process development scientists and chemists a robust road map for pollution prevention. The green chemistry principles are as follows:

1. *Prevent waste*. This involves the design of chemical synthesis routes to prevent waste, leaving no waste to treat or clean up.
2. *Design safer chemicals and products*. This involves the design of chemical products to be fully effective yet have little or no toxicity.
3. *Design less hazardous chemical syntheses*. This involves design syntheses to use and generate substances with little or no toxicity to humans and environment.
4. *Use renewable feedstock*. This is the use of raw materials and feed stocks that are renewable rather than depleting. Renewable feed stocks are often made from agricultural products or are the wastes of other processes; depleting feed stocks are made from fossil fuels (e.g., petroleum, natural gas, coal) or are mined.
5. *Use catalysts, not stoichiometric reagents*. This minimizes waste by using catalytic reactions. Catalysts are used in small amounts and can carry out a single reaction many times. They are preferable to stoichiometric reagents, which are used in excess and work only once.
6. *Avoid chemical derivatives*. This is the avoidance of blocking or protecting groups or any temporary modifications, if possible. Derivatives use additional reagents and generate waste.
7. *Maximize atom economy*. This is using design syntheses so that the final product contains the maximum proportion of the starting materials. There should be few, if any, wasted atoms.
8. *Use safer solvents and reaction conditions*. This avoids using solvents, separation agents, or other auxiliary chemicals. If these chemicals are necessary, then use innocuous chemicals.
9. *Increase energy efficiency*. This involves running chemical reactions at ambient temperature and pressure whenever possible.
10. *Design chemicals and products to degrade after use*. This is the design of chemical products that break down to innocuous substances after use, so that they do not accumulate in the environment.

11. *Analyze in real time to prevent pollution*. This is the in-process, real-time monitoring and control during syntheses to minimize or eliminate the formation of undesired byproducts.
12. *Minimize potential for accidents*. This is the design of chemicals and their forms (e.g., solid, liquid, gas) to minimize the potential for chemical accidents, including explosions, fires, and releases to the environment.

Compliance or Good Business

The U.S. Toxic Substances Control Act (TSCA-1976) authorizes the EPA to secure information on new and existing chemical substances and control substances determined to cause unreasonable risk to human health and the environment. In the United States, certain states have also started enacting recycling and substance restriction legislations.

The environmental regulations related to sustainability have emerged from European Union. Among them are the following:

1. *EU REACH (EU Regulation 1907/2006)*. Registration, Evaluation, Authorization, and Restriction of Chemicals (REACH), effective 2007, regulates that chemical substances made in the European Union or imported to the European Union be tested for health and safety by a new agency, the European Chemical Agency (ECA).
2. *EU RoHS (2002/95/EC)*. The Restriction of Hazardous Substances (RoHS) directive, which became law in July 2006, restricts the use of six hazardous substances, namely, lead, mercury, cadmium, hexavalent chromium, polybrominated biphenyls (PBB), and polybrominated diphenyl ether (PBDE), for several categories of products. Unique to this directive is that the responsibility for compliance is placed on the company that puts the product on the market and not placed on the components or subassemblies.
3. *EU WEEE (2002/96/EC)*. The Waste Electrical and Electronic Equipment (WEEE) directive, enacted in 2003, sets specific targets for collection, recycling, and recovery for all electrical goods.
4. *EU EuP (2005/32/EC)*. The Energy-using Products (EuP) directive provides a framework for setting eco-design parameters for energy-using products.

All regulations include reporting and tracking in some form or fashion. Staying in concert and meeting compliance requirements can be a daunting task when the sourcing supply chains for products are increasingly becoming more diverse and global. Most companies still focus on compliance first, rather than green design.

There is a ground swell of interest among consumers who are willing to pay more for products that minimize emissions or use recycled materials. E-factor is an industry norm that measures the ratio of organic waste to final product by weight. The pharmaceutical industry values typically range from 25 to 100. This seems large, but the annual tonnage of pharmaceutical active ingredients is low compared to oil refining or bulk chemicals.

Pfizer has an ongoing green chemistry initiative that focuses on sustainable development for active pharmaceutical ingredients. Here we offer four specific examples related to sustainable development:

1. Pfizer redesigned the sertraline (Zoloft) commercial process to eliminate the number and volume of solvents and reduce titanium dioxide waste and 35% of hydrochloric acid waste. The new process uses 26 L/kg of product versus the original 98 L/kg of the first commercial route. Pfizer received the 2002 Presidential Green Chemistry Award for Greener Synthetic Pathways for this redesign.
2. Pfizer improved the sildenafil (Viagra, Revatio) manufacturing process by reducing solvent usage of 22 L/kg to 7 L/kg, improving the synthesis route and efficient solvent recovery. Pfizer received the UK Green Technology Award in 2003 for their efforts that improved the E-factor to 6.
3. Pfizer's atoravastin calcium (Lipitor) product uses a key building block, hydroxynitrile (HN). Codexis devised three bioengineered enzymes to improve volumetric productivity by 100-fold of reduction reaction and cyanation reaction by 4000-fold of the HN process. This has resulted in a newer process that has fewer unit operations, improved yield, reduced formation of by-products and waste, better equipment, and improved worker safety. Codexis received the 2006 Presidential Green Chemistry Award for Greener Reaction Conditions.
4. Pfizer developed a new enzymatic route in the manufacturing of pregabalin (Lyrica) that eliminated organic solvents from three reaction steps, while increasing the overall yield. This also minimized waste significantly.

Common among these examples is the redesign of the Active Pharmaceutical Ingredient (API) process, leveraging these green chemistry principles with a primary focus on the use of less hazardous solvents, better synthesis routes, and biocatalysts to improve yield and reduce waste.

The active pursuit of green chemistry can be quite rewarding and provide payback as well. According to Dr. Beverly Cue, Pfizer's former vice president of research and development, "If you have a drug on the market for eight, ten years, and you're saving ten to fifteen million dollars a year, that's as much as a hundred fifty million dollars of manufacturing cost reduction over the lifetime of the patent." These efforts and other similar pharmaceutical industry work have shown that there is also a large economic benefit from applying sustainability and reducing waste generation and energy use.

Manufacturing Science

Manufacturing science is the term used at Pfizer to describe the body of scientific knowledge, laws, and principles involved in the transformation of materials and information into products. In addition to chemistry, physics, and biology, manufacturing science also includes the bodies of knowledge covered by chemical engineering, environmental engineering, electrical engineering, mechanical engineering, industrial engineering, and software engineering. Each of these disciplines has its own terminology, knowledge base, and standard ways to represent information.

Manufacturing science includes all of these disciplines and must share information and knowledge across disciplines; a framework or model for the cross-discipline information is a vital part of implementing advanced manufacturing strategies. The management of information is commonly called *informatics*. Informatics includes the application of information technology and statistical techniques to the management of information; it is defined as the science of information and the practice of information processing. It studies the structure, behavior, and interactions of natural and artificial systems that store, process, and communicate information. Since computers, individuals, and organizations all process information, informatics has computational, cognitive, and social aspects.

MSI is defined as the application of information technology to manufacturing science.[3] MSI must handle a wide range of information types to support manufacturing science activities. This information should also support sustainable process development and manufacturing. We propose specific extensions to the ISA-88 GR information models to handle these evolving sustainability needs.

The proliferation of regulations, different environmental reporting requirements, and the need to enable sustainable design processes places the following additional requirements on existing MSI models:

1. Track and report energy efficiency (carbon footprint) to make sourcing decisions.
2. Track and calculate waste (hazardous, both toxic and nontoxic, and nonhazardous), by-products, and coproducts.
3. Track waste-handling costs, E-factor, and energy costs.
4. Access authoritative material property information for raw materials that includes all toxicology information and limits on volatile organic compounds.
5. Import or export information from public databases, expert systems for reactions, solvent databases, model-based simulation applications (e.g., Catalyst, BatchPlus, ProcessONE), and biocatalysis libraries during process development.
6. Report by batch or by production run on the energy use; E-factor; and use of solvents, reagents, and catalysts.
7. Provide a single view that provides consistency of information and compliance with all regulations, including environmental regulations.
8. Offer specific guidance on approved or desirable materials according to green thinking.
9. Define real-time monitoring methods using Process Analytical Technology (PAT) to monitor the formation of hazardous substances during manufacturing.

Based on these requirements, extensions are proposed to the process elements (e.g., process action, operation, stage), material definition, and GR Other Information (GROI).

Focus on Products and Processes

Many sustainability decisions are made when process methods are chosen and tested in the labs and pilot plants. The synthesis process is then usually replicated in production-scale processes. This means that sustainability efforts must focus on both the product and the processes. This is illustrated in Figure 23.1, which relates different sustainability efforts to the two main tasks required to develop process descriptions.

Waste can be generated at any process stage, operation, or action, and E-factors can be calculated for each process stage, operation, or action in the production process. A process stage is defined in ISA-88.03 as follows: "A process stage defines a part of a process that usually operates independently and usually results in a planned sequence of chemical or physical changes in the material being processed. Process inputs and process outputs depicted in a recipe should always be identified as resources or materials. In the case where a process output from one stage is a process input to another stage; this material is called a process intermediate and does not need to be otherwise identified. Process stages can generate multiple process outputs or process intermediates."

Within each process stage is a set of process operations. These are defined as follows: "A process operation is a major processing activity that is defined without specification of the actual target equipment configuration and usually results in a chemical or physical change in the material being processed. Process operations are generally defined based on chemical or physical considerations." The following is a list of common process stages in pharmaceutical processes:

- Intermediate API production
- Final API production

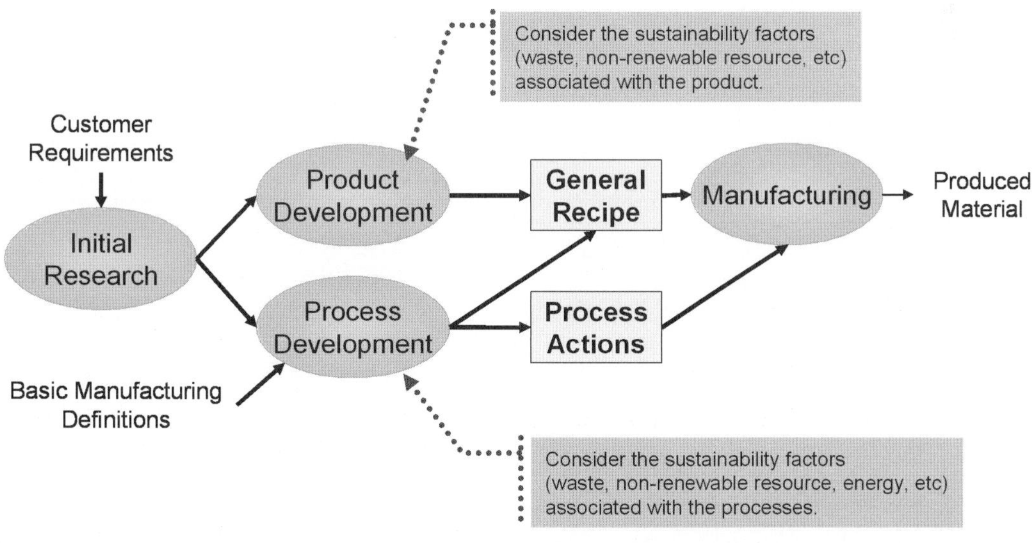

Figure 23.1. Where sustainability decisions are made.

- Unit dosage production
- Packaging

Intermediate API and final API production typically use chemical processes for pharmaceutical production and production efforts; therefore they are covered under green chemistry initiatives (Fig. 23.2).

The following is a list of common process stages in biotechnology processes:

- Seeding and growth
- Harvesting, separation, and recovery
- Unit dosage production
- Packaging

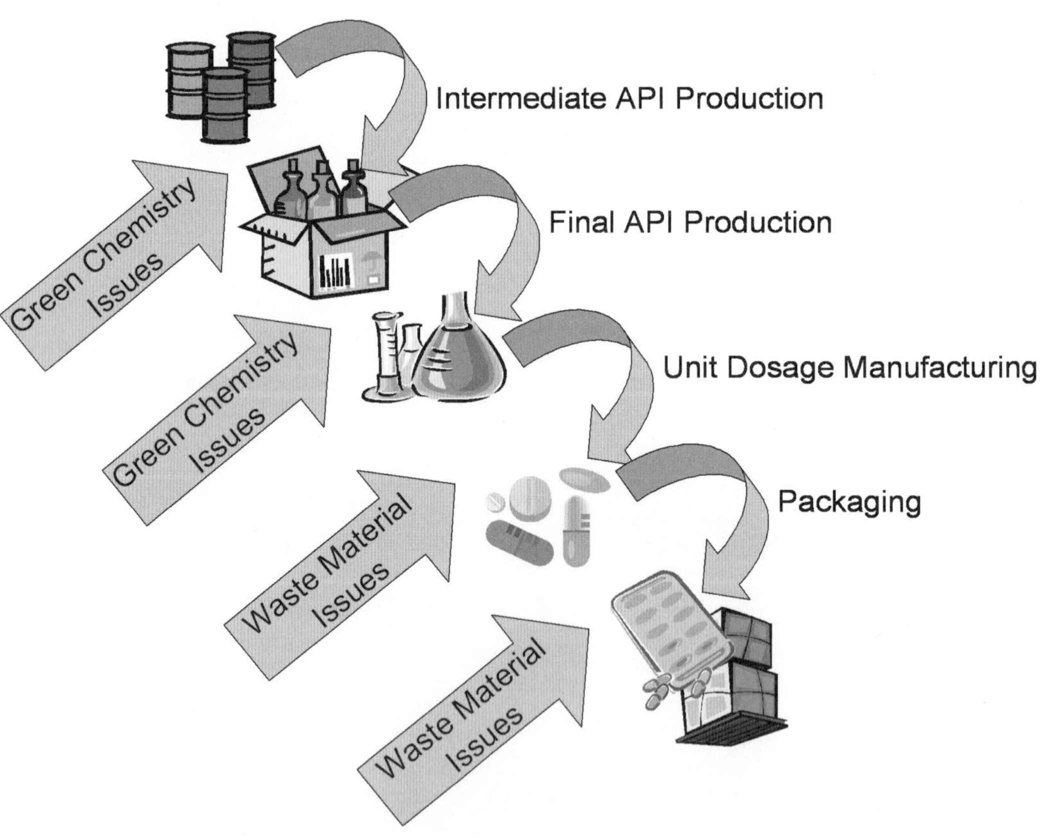

Figure 23.2. Common process stages in pharmaceutical manufacturing.

Seeding and growth are fermentation processes. Sustainability efforts in these stages typically deal with issues associated with waste water and greenhouse gas emissions. Harvesting and recovery stages typically deal with green chemistry and the reduction of toxins.

Waste in unit dosage production is often associated with nonrecoverable materials used in the production of powders, tablets, solutions, and filled vials or ampoules. The waste may be unused and nonreusable material dispensed for production (e.g., extra material provided to handle spoilage situations or to correct physical properties), such as acids and bases in the batch kit used to adjust the pH of a solution.

Waste in packaging is often associated with cutoff or discarded material generated by the primary packaging, such as scrap foil in the production of blister packs. Waste can be generated in any process operation, and E-factors can be calculated for any process operation. Often waste, energy usage, and E-factors can be associated with specific process operations.

Manufacturing Science GRs

Tying sustainability to manufacturing science GRs can be accomplished by two principle means—material properties and GR elements.

Process Stages, Process Operations, Process Actions, and Parameters

Each process stage, process operation, or process action may have an estimated E-factor, cost, and waste. This means that this extra information should be added to the baseline ISA-88 data models, preferably as attributes or properties to the process operation.

A library of process operations and process actions can be built for each primary process. The primary and alternate process definitions can include an estimated energy requirement and E-factor. The process definitions should also identify the waste materials associated with the process, but the specific disposal cost of the material should be maintained in a material library.

Often, costs and waste cannot be directly assignable to any specific process action. The energy use can occur as a result of the chemistry (exothermic or endothermic). Waste may occur due to material separation or physical properties of the equipment (e.g., wasted blister foil when the roll is cut into blister shapes). In these cases, the waste and E-factor can be rolled up to the process operation or process stage.

Process Inputs, Process Outputs, and Process Intermediates

Materials are defined as process inputs and process outputs. Materials can also be identified as *process intermediates*. These are materials produced within one stage, operation, or action that are later consumed by another stage, operation, or action. Each process input and output corresponds to a material definition (ISA-88 and ISA-95). The material definition should include sustainability attributes, such as the following:

- *Material role.* The material can be a feedstock, product, by-product, coproduct, solvent, catalyst, or waste. It is conceivable that the material role could be different for the same material when used as a process input, process output, or process intermediate. If the material is a process intermediate or process output, then the yield should be included; this attribute defines the expected or calculated yield for a specific process operation or process action.
- *Amount recovered.* This attribute defines the material recovered as a result of the execution of a process operation or process action. This can be shown as a process output from the operation or action (Fig. 23.3).
- *Percent waste.* This defines the expected waste percentage of the material. This can be shown as a process output from the operation or action (Fig. 23.3).

Process parameters are defined in ISA-88.01 as "information that is needed to manufacture a material but does not fall into the classification of process input or process output."

Sustainability Attributes

We propose the use of the term *sustainability attributes* to refer to a specific type of process parameter for a process element. This is similar to a quality attribute, which directly or indirectly impacts product specifications.

If these attributes do not apply at a process action level, they can be rolled up to a process operation or process stage level. Sustainability attributes can define the target values, minimum values, or maximum values, such as desired maximum carbon footprint, target waste, and minimum transportation costs and energy. The list is not comprehensive but can also include specific green attributes such as:

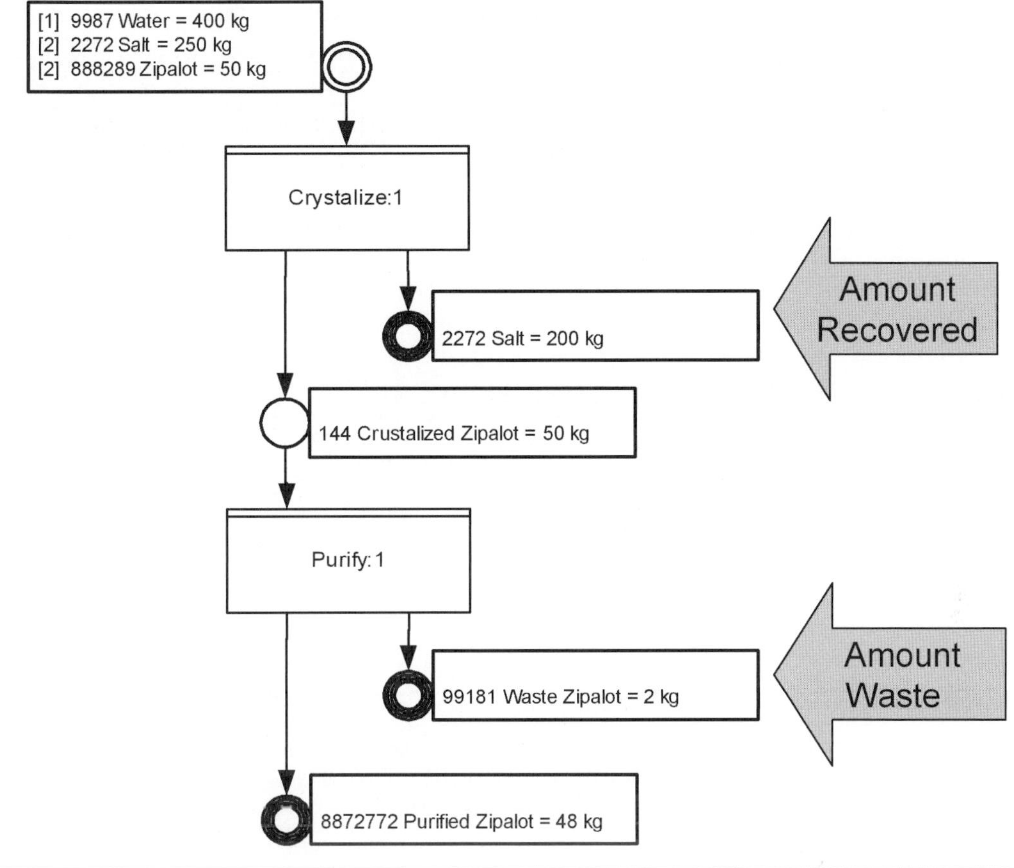

Figure 23.3. Recovered and waste material in a GR.

- Amount of organic waste produced
- Atom economy
- Reaction mass efficiency
- Chemical yield
- Aqueous waste
- Vapor emissions
- Energy usage (renewable and nonrenewable)
- E-factor

Sustainability attributes can also be further categorized as *critical* or *key*. Critical attributes can be shared with regulatory authorities as a characteristic of the product. Process reports can include actual data related to sustainability attributes.

Process parameters define values that must be controlled in the process because they influence quality attributes. Process parameters must be controlled within predefined limits to ensure that the product meets its predefined quality or sustainability attributes. Critical process parameters influence critical quality or sustainability attributes. Key process parameters influence key quality or sustainability attributes.

Figure 23.4 illustrates a sustainability attribute (e.g., vapor emissions) and the parameters that must be controlled (e.g., time and temperature) to achieve the desired attribute.

For each process parameter, we also propose adding an additional attribute for the method used to measure the associated quality attribute. When using PAT to measure or verify a quality or sustainability attribute, a reference to a measurement method should be included. For this, each process parameter can reference measurement method definitions that are not explicitly defined in the GR model but can easily be added to the information models. Additional categories of specifications may include supply chain specification patterns, regulation patterns, and testing method definitions.

Extensions to Process Reports

In addition to defining sustainability attributes, process development teams or environmental stewards also need a place to define or specify information that needs to be reported as a result of the execution of each ISA-88.01 process element. These are called *process reports*.[4] They define the information that must be collected and reported on during the execution of the process. Figure 23.5 illustrates some typical process reports that are specific to sustainability attributes. Process reports related to sustainability can be collected regardless of the sourcing method used.

Material Definitions

Material definitions should also include sustainability attributes. These can be represented in the information exchange models that are defined in ISA-95 and B2MML as properties. The information may include unique reference locators to look up specific information in expert systems, knowledge bases, or Master Data Management (MDM) systems. Some specific proposed attributes are the following:

- Chemical Abstract Service (CAS) reference number
- Molecular weight

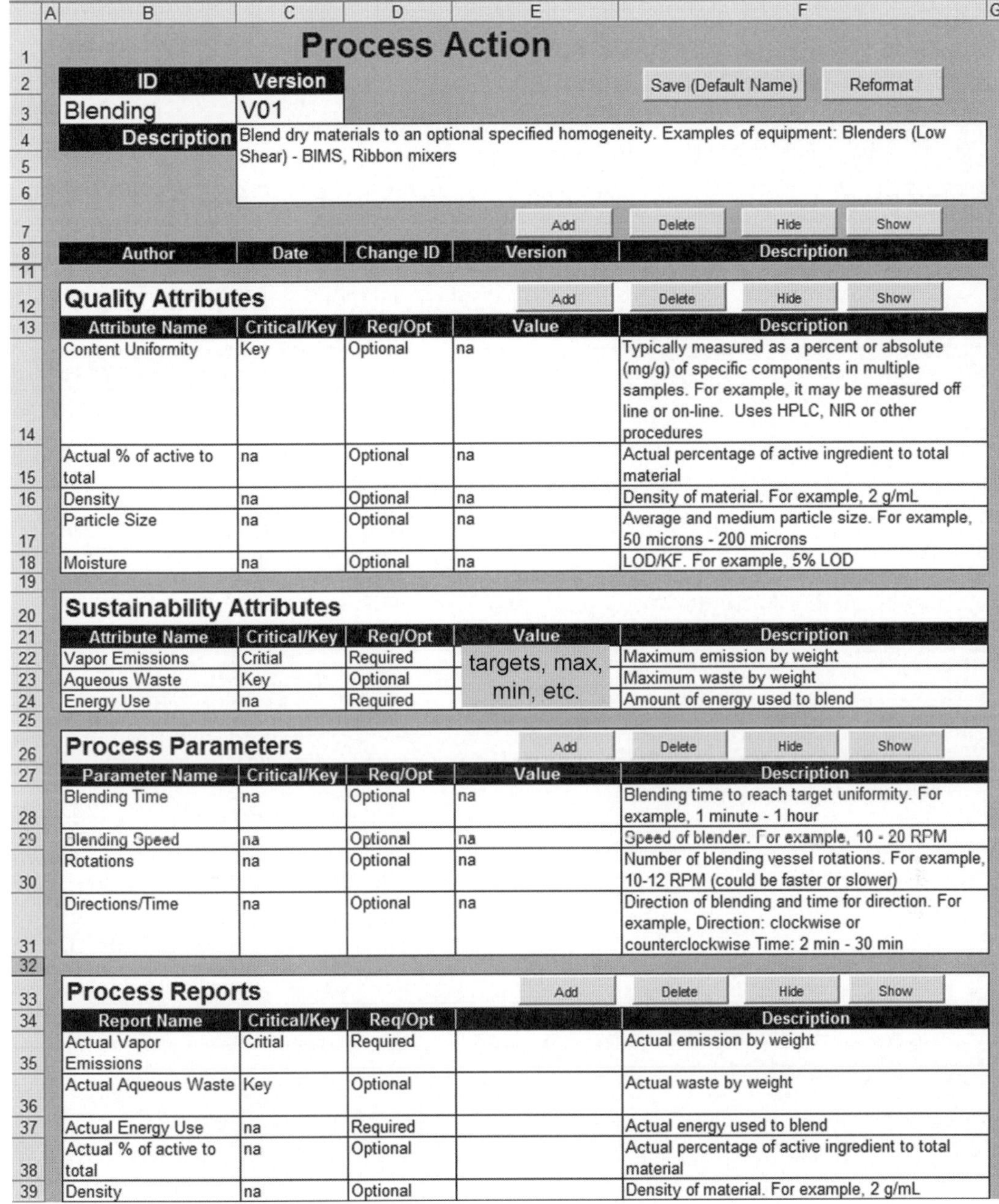

Process Action

Save (Default Name) | Reformat

ID	Version
Blending	V01

Description: Blend dry materials to an optional specified homogeneity. Examples of equipment: Blenders (Low Shear) - BIMS, Ribbon mixers

Add | Delete | Hide | Show

Author	Date	Change ID	Version	Description

Quality Attributes

Add | Delete | Hide | Show

Attribute Name	Critical/Key	Req/Opt	Value	Description
Content Uniformity	Key	Optional	na	Typically measured as a percent or absolute (mg/g) of specific components in multiple samples. For example, it may be measured off line or on-line. Uses HPLC, NIR or other procedures
Actual % of active to total	na	Optional	na	Actual percentage of active ingredient to total material
Density	na	Optional	na	Density of material. For example, 2 g/mL
Particle Size	na	Optional	na	Average and medium particle size. For example, 50 microns - 200 microns
Moisture	na	Optional	na	LOD/KF. For example, 5% LOD

Sustainability Attributes

Attribute Name	Critical/Key	Req/Opt	Value	Description
Vapor Emissions	Critial	Required	targets, max, min, etc.	Maximum emission by weight
Aqueous Waste	Key	Optional		Maximum waste by weight
Energy Use	na	Required		Amount of energy used to blend

Process Parameters

Add | Delete | Hide | Show

Parameter Name	Critical/Key	Req/Opt	Value	Description
Blending Time	na	Optional	na	Blending time to reach target uniformity. For example, 1 minute - 1 hour
Blending Speed	na	Optional	na	Speed of blender. For example, 10 - 20 RPM
Rotations	na	Optional	na	Number of blending vessel rotations. For example, 10-12 RPM (could be faster or slower)
Directions/Time	na	Optional	na	Direction of blending and time for direction. For example, Direction: clockwise or counterclockwise Time: 2 min - 30 min

Process Reports

Add | Delete | Hide | Show

Report Name	Critical/Key	Req/Opt		Description
Actual Vapor Emissions	Critial	Required		Actual emission by weight
Actual Aqueous Waste	Key	Optional		Actual waste by weight
Actual Energy Use	na	Required		Actual energy used to blend
Actual % of active to total	na	Optional		Actual percentage of active ingredient to total material
Density	na	Optional		Density of material. For example, 2 g/mL

Figure 23.4. Additional process parameter types for sustainability attributes.

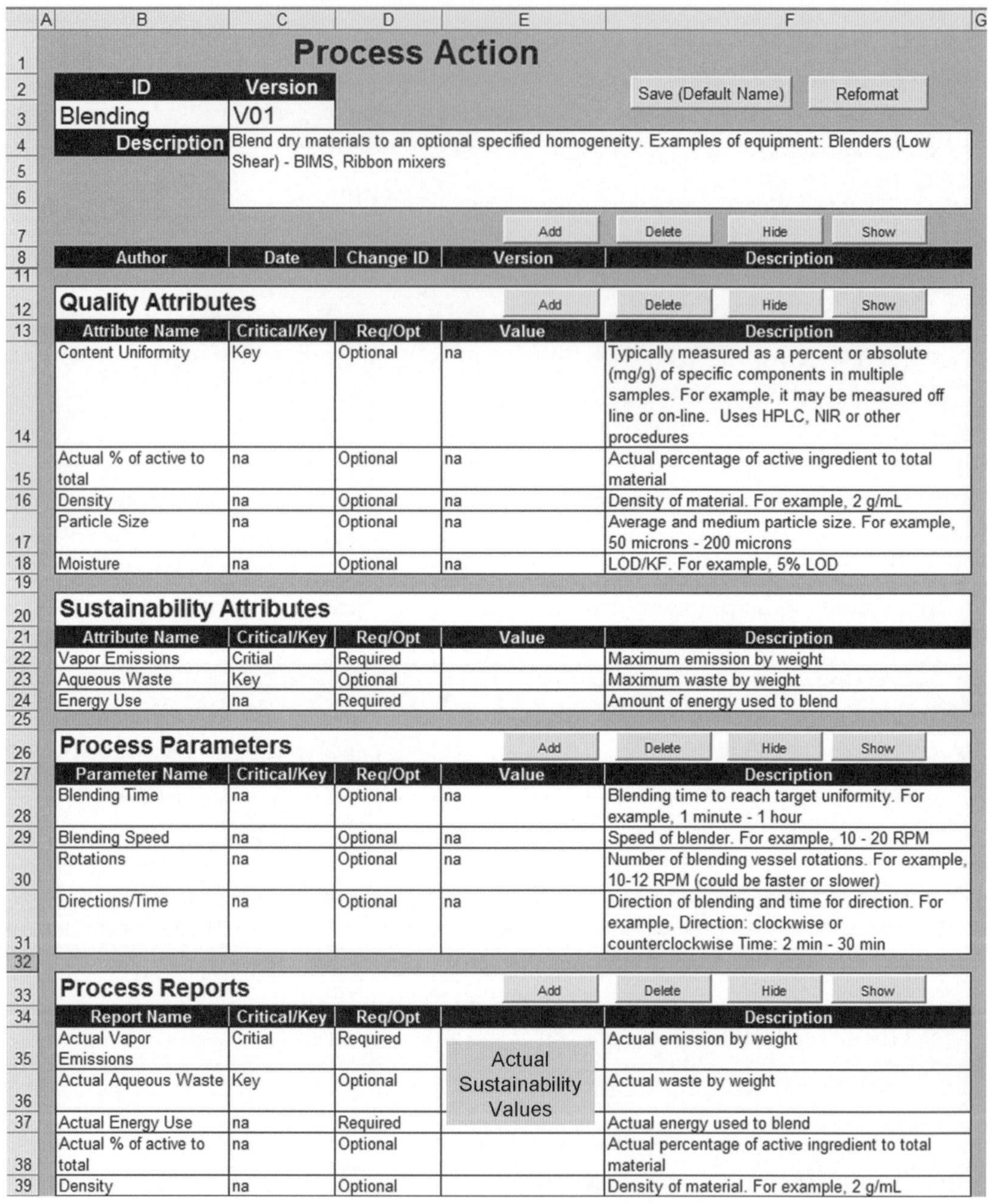

Process Action

ID	Version
Blending	V01

Description: Blend dry materials to an optional specified homogeneity. Examples of equipment: Blenders (Low Shear) - BIMS, Ribbon mixers

Save (Default Name) | Reformat

Add | Delete | Hide | Show

Author	Date	Change ID	Version	Description

Quality Attributes — Add | Delete | Hide | Show

Attribute Name	Critical/Key	Req/Opt	Value	Description
Content Uniformity	Key	Optional	na	Typically measured as a percent or absolute (mg/g) of specific components in multiple samples. For example, it may be measured off line or on-line. Uses HPLC, NIR or other procedures
Actual % of active to total	na	Optional	na	Actual percentage of active ingredient to total material
Density	na	Optional	na	Density of material. For example, 2 g/mL
Particle Size	na	Optional	na	Average and medium particle size. For example, 50 microns - 200 microns
Moisture	na	Optional	na	LOD/KF. For example, 5% LOD

Sustainability Attributes

Attribute Name	Critical/Key	Req/Opt	Value	Description
Vapor Emissions	Critial	Required		Maximum emission by weight
Aqueous Waste	Key	Optional		Maximum waste by weight
Energy Use	na	Required		Amount of energy used to blend

Process Parameters — Add | Delete | Hide | Show

Parameter Name	Critical/Key	Req/Opt	Value	Description
Blending Time	na	Optional	na	Blending time to reach target uniformity. For example, 1 minute - 1 hour
Blending Speed	na	Optional	na	Speed of blender. For example, 10 - 20 RPM
Rotations	na	Optional	na	Number of blending vessel rotations. For example, 10-12 RPM (could be faster or slower)
Directions/Time	na	Optional	na	Direction of blending and time for direction. For example, Direction: clockwise or counterclockwise Time: 2 min - 30 min

Process Reports — Add | Delete | Hide | Show

Report Name	Critical/Key	Req/Opt		Description
Actual Vapor Emissions	Critial	Required	Actual Sustainability Values	Actual emission by weight
Actual Aqueous Waste	Key	Optional		Actual waste by weight
Actual Energy Use	na	Required		Actual energy used to blend
Actual % of active to total	na	Optional		Actual percentage of active ingredient to total material
Density	na	Optional		Density of material. For example, 2 g/mL

Figure 23.5. Actual sustainable attributes values in process reports.

- Ozone depletion potential
- Global warming potential
- Environmental regulation impact
- Toxicity

One issue to consider with material information is the different rules associated with materials for different countries or even different states. For example, a chemical often used to extract proteins in biotechnology processes is considered toxic in California, while other jurisdictions do not consider it to be hazardous.

Even disposal costs may vary based on location. In some sites, material can be recovered and recycled; in others, it may need to be destroyed or buried or require long-term disposal. Other sites may allow the resale of the waste for use in other processes.

This information should be part of the corporate material library and be made available to chemists and engineers when they are developing the products and processes. Use of the B2MML standard for data exchange allows this information to be sourced from a single location but made available to other tools and services.

Information Model

Figure 23.6 (from ISA-88.03) illustrates where GR attributes should be added to ISA-88 data models. Figure 23.7 illustrates where sustainability factors can be applied to GR development in a process defined in ISA-88.03.

If sustainability information is available from process stages, operations, and materials, then reports on the relative actual total costs, including produced material, consumed material, produced waste, waste disposal, and energy use can be generated. This allows for comparisons of different production methods (e.g., GRs) to be made, perhaps at the site level, to determine the optimal production method.

Summary

As pharmaceutical and biotechnology yields increase, the cost of waste materials and unneeded energy use becomes significant. Avoiding costs and meeting a company's environmental and societal commitment are also important driving factors for implementing new products and processes. There is also significant economic benefit obtained from sustainable manufacturing. Additional information must

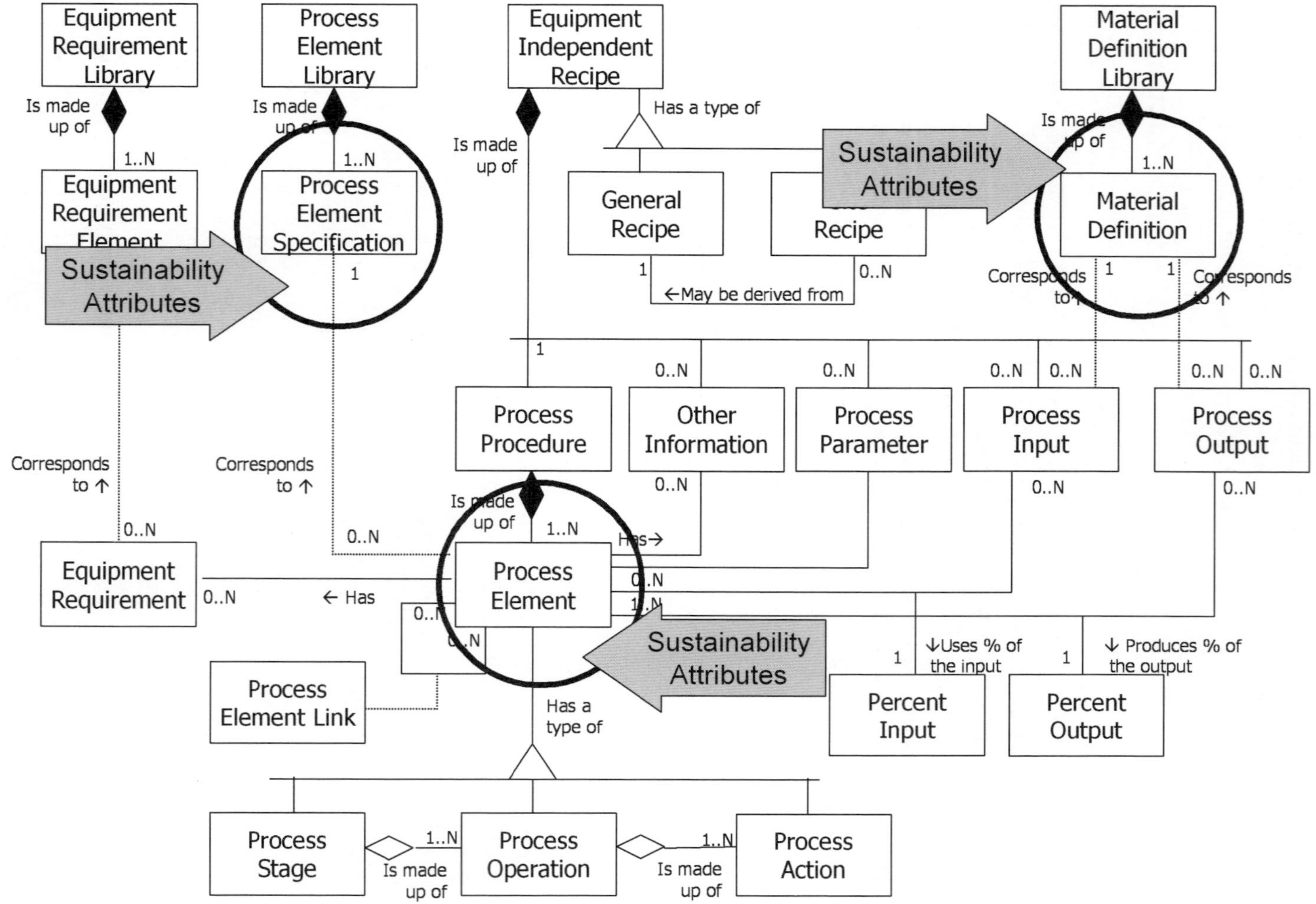

Figure 23.6. Sustainability attributes added to GR information.

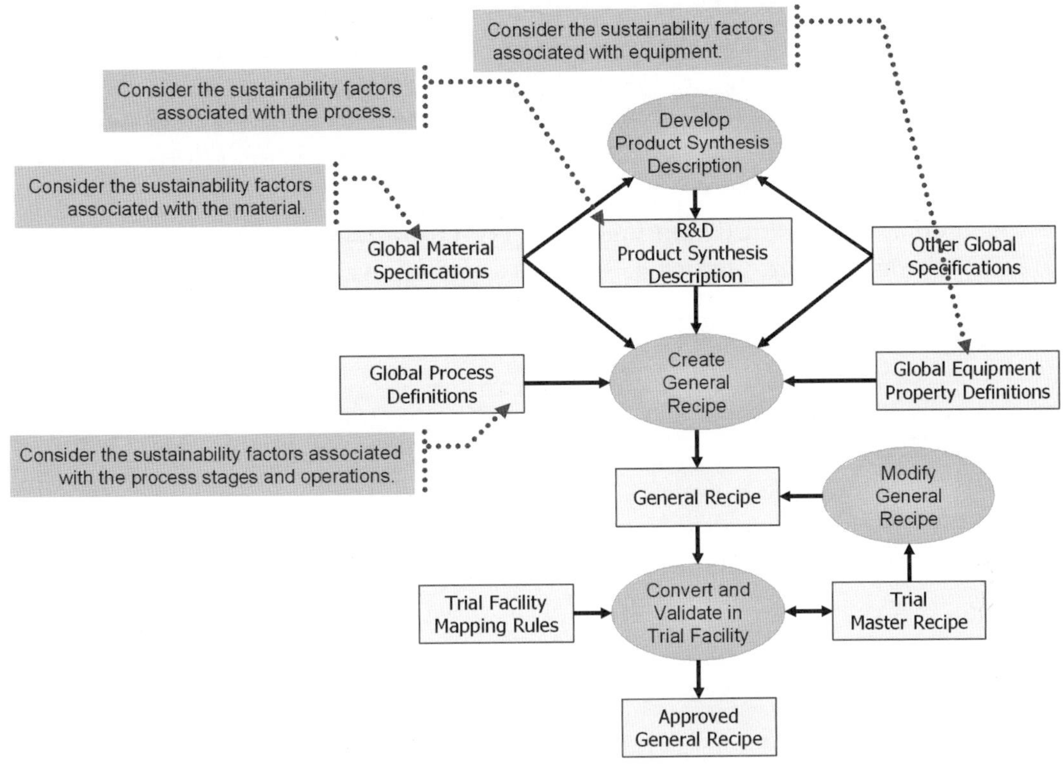

Figure 23.7 Sustainability factors in recipe development.

often be kept on material disposal costs, waste materials, and E-factors. This information is related to materials with libraries of process definitions, and alternative process definitions should be maintained over the life cycle of the product or process. GRs provide a good basis for documenting sustainability information and determining manufacturing and sourcing alternatives. The considerations defined here can also apply to other types of production, such as chemical or consumer packaged goods.

References

1. Anastas, Paul T., and John C. Warner. 1998. *Green chemistry: Theory and practice*. Oxford, UK: Oxford University Press.
2. Bush, George W. 2007. Executive order 13423: Strengthening federal environmental, energy, and transportation management. January 24.

3. Pillai, Velumani, and Dennis Brandl. 2006. An information model framework for manufacturing science. Paper presented at the WBF European Conference, November 2006.
4. Pillai, Velumani, Rob Burrows, and Dennis Brandl. 2006. Standardized multiple plant investigations. Paper presented at the WBF North American Conference, March 2006.
5. United Nations World Commission on Environment and Development. 1987. *Our common future*. Oxford, UK: Oxford University Press.

Manufacturing Science Model Extensions to Address Quality by Design and Risk Assessments

Presented at the WBF Make2Profit Conference, May 24–26, 2010, by

Velumani Pillai
velumani.pillai@pfizer.com
Pfizer, MSC 722, 100 Route 206 North
Peapack, NJ 07977 USA

Dennis Brandl
dnbrandl@brlconsulting.com
BR&L Consulting, 208 Townsend Ct.
Cary, NC 27511 USA

Abstract

Pharmaceutical manufacturing is globalizing as never seen before. Innovation in manufacturing with faster development, faster new product release, robust process understanding, and vastly improved Quality by Design (QbD) is critical for advanced manufacturing organizations. The goal of QbD is to design a quality product and manufacturing process to consistently deliver the intended performance of the product. The U.S. Food and Drug Administration (FDA) issued its current thinking on QbD, "Guidance for Industry: Q8 Pharmaceutical Development," in June 2009, which defines it as process understanding, risk assessment, and design space.

Manufacturing science is the body of scientific knowledge, regulations, and principles involved in the transformation of materials and information into products. In this chapter, we propose Manufacturing Science Informatics (MSI) as the new information model framework to support the process and product knowledge discovered, created, shared, retained, and applied over a product's entire life cycle.

Design space is the multidimensional combination and interaction of input variables and process parameters that have been demonstrated to provide assurance of quality. Commonly used risk management methods and tools such as Failure Mode and Effect Analysis (FMEA), Fault Tree Analysis (FTA), and Hazard and Operability (HAZOP) analysis also help in documenting product definitions.

The MSI framework enables established business processes (e.g., process development, production sourcing, material sourcing, product introduction, continuous process improvement, process optimization, product investigations) but can also be extended to address new business processes. The MSI framework uses General Recipes (GR) that are described in ISA-88.03 as one of the foundational information entities. In this chapter, we propose extensions to the GR model and other product and process information entities in the ISA-95 standards to handle design space and risk assessments.

Introduction

QbD and risk assessments are a critical part of modern pharmaceutical and biotechnology manufacturing; however, they are also important in all forms of manufacturing in the continual drive for higher-quality products. The concept of QbD is not new; its goal is to design higher final product quality through a strong understanding of the physical production process. This is the alternative process that tries to achieve quality through testing. All engineering disciplines have recognized that high productivity is achieved by understanding and controlling the production process. This minimizes rework and continual testing to discover errors that were introduced in production.

Within pharmaceutical and biotechnology manufacturing, this concept has been given the name QbD by the FDA. QbD is defined in the FDA International Council on Harmonization (ICH) guideline "Q8 Pharmaceutical Development."[9] A QbD approach to product development may involve the following elements:

- A systematic evaluation, understanding, and refining of the formulation and manufacturing process that includes two parts:

- Identifying (through prior knowledge, experimentation, and risk assessment) the material attributes and process parameters that can have an effect on product Critical Quality Attributes (CQAs)
- Determining the functional relationships that link material attributes and process parameters to product CQAs

- Using the enhanced product and process understanding in combination with quality risk management to establish an appropriate control strategy that can, for example, include a proposal for a design space, real-time release testing, or both.

In the life science industries, QbD means the following:

- The product is designed to meet patient requirements.
- The process is designed to consistently meet product CQAs.
- The impact of formulation components and process parameters on product quality is understood.
- Critical sources of process variability are identified and controlled.
- The process is continually monitored and updated to assure consistent quality over time.

One of the key aspects of a QbD process is to assess the risks to product or process quality. Risk assessment should help determine the potential sources of variability; how they can affect critical quality attributes; and how they can be prevented, detected, or mitigated. Risk assessments are used to discover and document what can go wrong in the production process, what will be the impact, how it relates to the design and control space, how it can be avoided, and how it can be detected. Commonly used risk management methods and tools can be used to help build robust production processes.

Quality by Design

QbD requires process knowledge. This knowledge is initially gathered during the Research and Development (R&D) phase of a product's development. Usually, knowledge gained on a bench-scale process must be augmented with scale-up, pilot, and full-scale production knowledge. QbD knowledge is used in the development of control strategy, in the creation of master recipes, and to determine the control parameters in the recipe. This knowledge is represented as a design space in the QbD model. There is a sequence of activities involved in the creation of a design space, and knowledge of the design space may grow through process development and process performance validation, as shown in Figure 24.1. Risk assessments may also occur at multiple places in the development life cycle and may impact the design space decisions, adding additional risk-based constraints to the design space.

A design space, as defined in ICH Q8, is "the multidimensional combination and interaction of input variables (e.g., material attributes) and process parameters that have been demonstrated to provide assurance of quality."

The key point here is "multidimensional." The design space concept recognizes that there are interdependences across process parameters and depicts how they relate to quality attributes. Figure 24.2 illustrates a design space that relates process parameters to quality attributes.

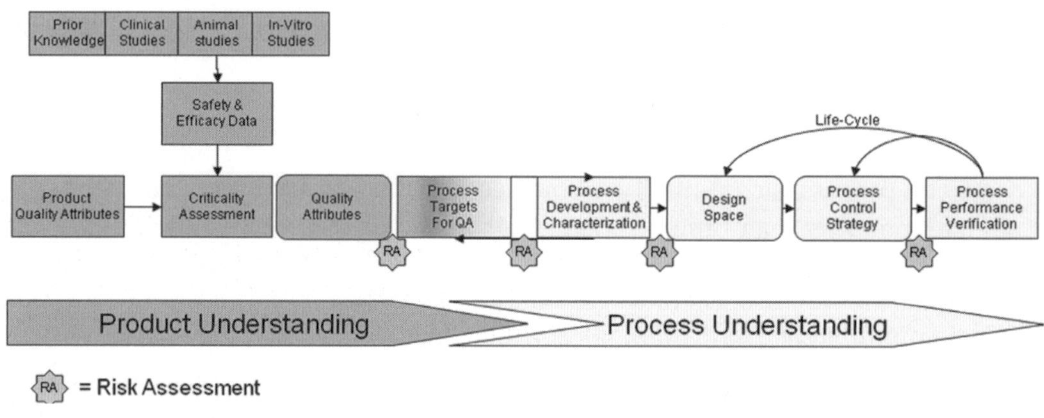

Figure 24.1. Design space development and risk assessment in the development cycle.

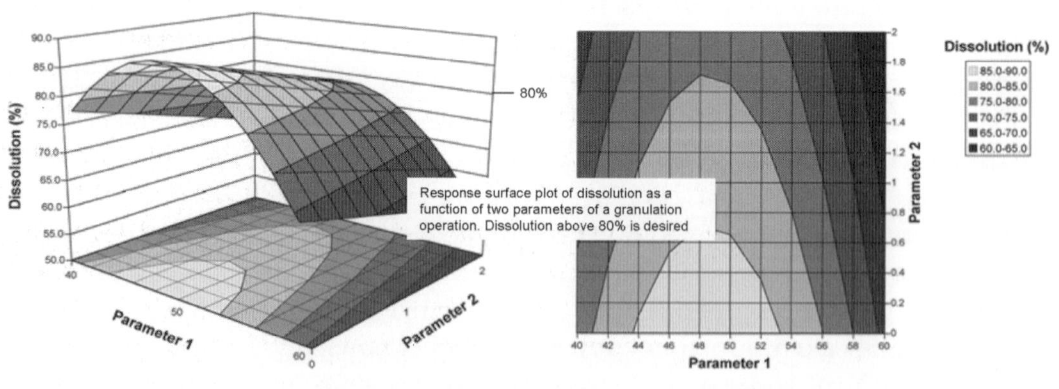

Figure 24.2. Example of a multidimensional design space from ICH Q8.

Quality Attributes

Quality attributes define the target qualities of the produced material. Quality attributes are the physical, chemical, or microbiological properties of a material that directly or indirectly impact product specifications. For example, quality attributes for a "Blend" process action may be a measure of the target homogeneity and the appearance of the blend (e.g., no visible discolorations or no clumping).

Quality attributes may be further categorized as *critical* or *key*. Critical quality attributes, which are registered with regulatory authorities, are a characteristic of the product. Key quality attributes are not registered.

Process Parameters

Process parameters have values that must be controlled within limits because they influence quality attributes. Critical process parameters influence critical quality attributes, and, as you might guess, key process parameters influence key quality attributes.

Design Space

A design space represents the degrees of flexibility available to manufacturing. As long as the manufacturing system can maintain control of the system within the design space, then the net result should be a quality product. The FDA does not require any changes to the control process that remain within the design space to be registered. This is a major change for the pharmaceutical and biotech industries, since previously it needed to be proven that any change to the production process did not change the critical quality attributes. This barrier to change significantly constrained innovation and continuous improvements in life science manufacturing. The new regulations allow life science manufacturers to make continual improvements to their processes without fear of regulatory issues.

A single dimension of variability (e.g., control of a temperature) is not a design space. Few, if any, processes have final quality variability in a single dimension. For example, in a drying process, there are variables such as air flow, air temperature, air humidity, time, incoming product water content, and method or amount of agitation on the material. These variables relate to the final quality attribute of product humidity. The design space is the relationship between these variables for acceptable final product humidity.

Figure 24.3 illustrates an example of a complex design space in which there are two independent process parameters (i.e., compression pressure and compression time) and two independent quality attributes (i.e., dissolution and

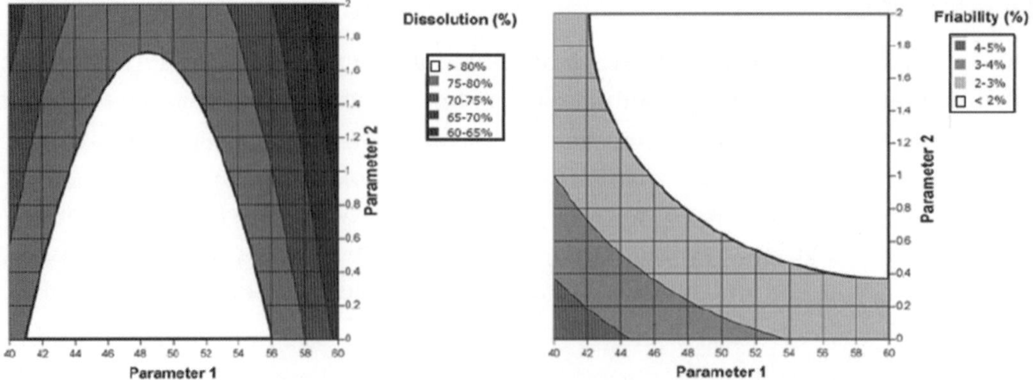

Contour plot of dissolution as a function of parameters 1 and 2.

Contour plot of friability as a function of parameters 1 and 2.

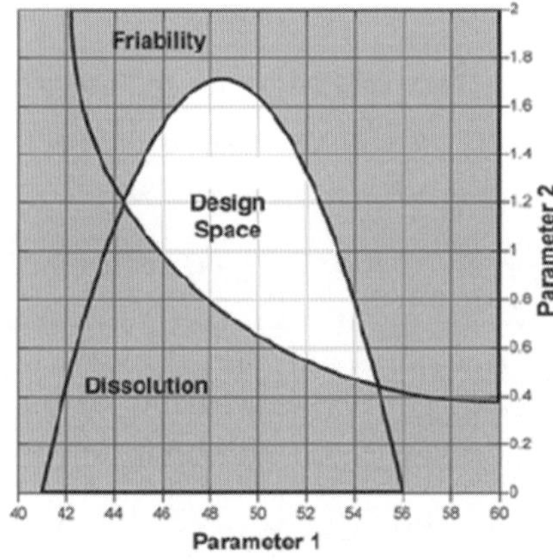

Proposed design space, comprised of the overlap region of ranges for friability and dissolution

Figure 24.3. Examples of a complex from ICH Q8.

friability). As long as the manufacturing process remains within the design space shown in Figure 24.3, then there will be acceptable values for dissolution and friability. A Design Space Formula (DSF) will often be one or more mathematical expressions that relate a quality attribute's value to parameter values. If there are many quality attributes, then the formula may be defined using a linear or nonlinear matrix. No matter what representation is used, the formula can be used to define the alert and control limits for a process automation system. FDA submissions with design spaces can vary from two to hundreds of dimensions, but a reasonable design space is usually limited to a handful of independent dimensions that have the most impact.

Design of Experiments

Often, a design space is defined through a Design of Experiments (DoE) process. In a DoE process, a set of test batches are run on the production process, each one varying for several of the design space variables. The quality attributes are measured at the end of each run. Information from the batch runs allows the discovery of the formal mathematical relations between the parameters and attributes.

Figure 24.4 shows a simple example of two control parameters (i.e., time and temperature) and one attribute (i.e., yield). Ten experiments were run to characterize the design space and to determine the conditions for the optimal yield. DoE

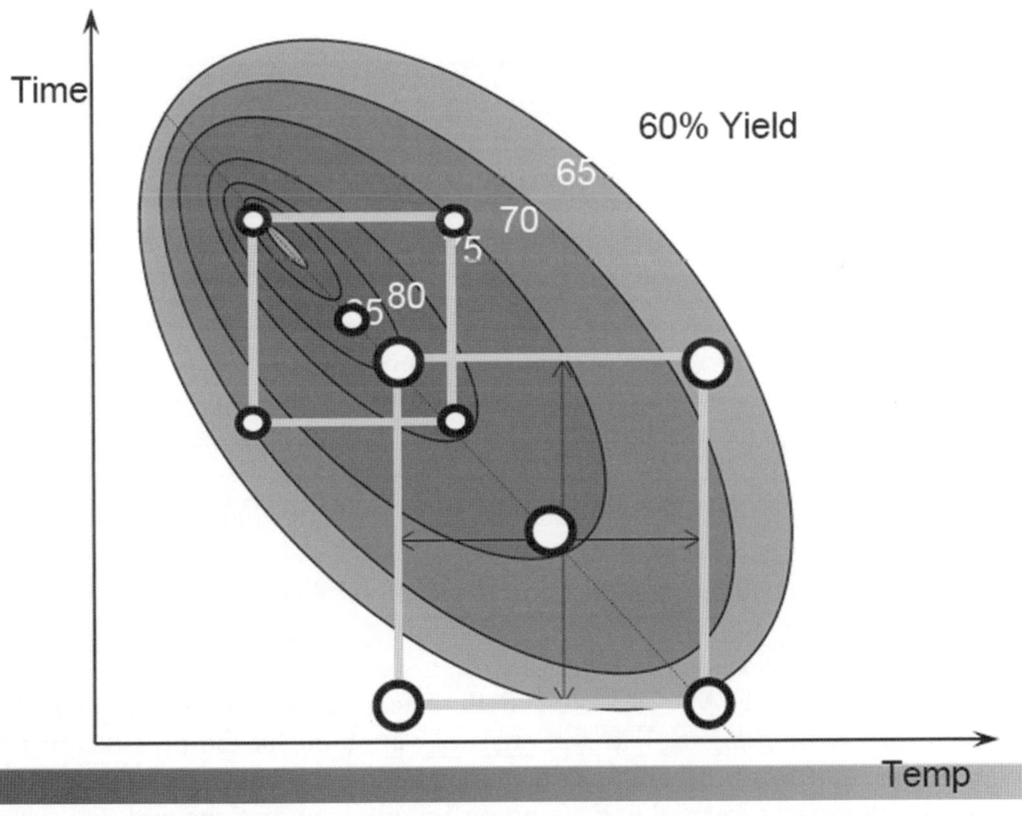

Figure 24.4. Sample DoE with cooperating factors, from Umetrics.

includes methods for determining the minimal number of tests required to characterize the design space.

Risk Assessment

The design space also requires a thorough understanding of all factors that can affect the process during its execution. Risk assessments can be performed at various stages of process development.

The risk assessments contain knowledge that can be critical during process investigations. This information should be preserved and maintained. It can be maintained as additional General Recipe Other Information (GROI), as mentioned in Chapter 21, or the recipe may simply contain a reference to the risk assessment results.

The risk assessments may be done at the process stage, process operation, or process action levels. The example shown in Figure 24.5, from ICH Q8, illustrates the use of a risk assessment tool and Ishikawa diagram that would be defined for a process stage (e.g., tableting). There may be another risk assessment made for the coating process and the packaging process.

During actual production, this provides a roadmap for what to check during an investigation of a process upset. The diagram can contain links to sensitivity formulas, discovered using DoE for each arm of the diagram. ICH Q8 states the following about the use of risk assessment tools:

> For example, a cross-functional team of experts could work together to develop an Ishikawa (fishbone) diagram that identifies potential variables that can have an impact on the desired quality attribute. The team could then rank the variables based on probability, severity, and detectability using failure mode effects analysis (FMEA) or similar tools based on prior knowledge and initial experimental data. Design of experiments or other experimental approaches could then be used to evaluate the impact of the higher ranked variables, to gain greater understanding of the process, and to develop a proper control strategy.

Fault Tree Analysis

Fault Tree Analysis (FTA) is a visual method for performing risk assessments. While an Ishikawa diagram lists the areas of variability of the process, an FTA identifies all relevant events and conditions leading to an "undesired" event. It can be used to determine parallel and sequential event combinations and their interrelationships (Fig. 24.6).

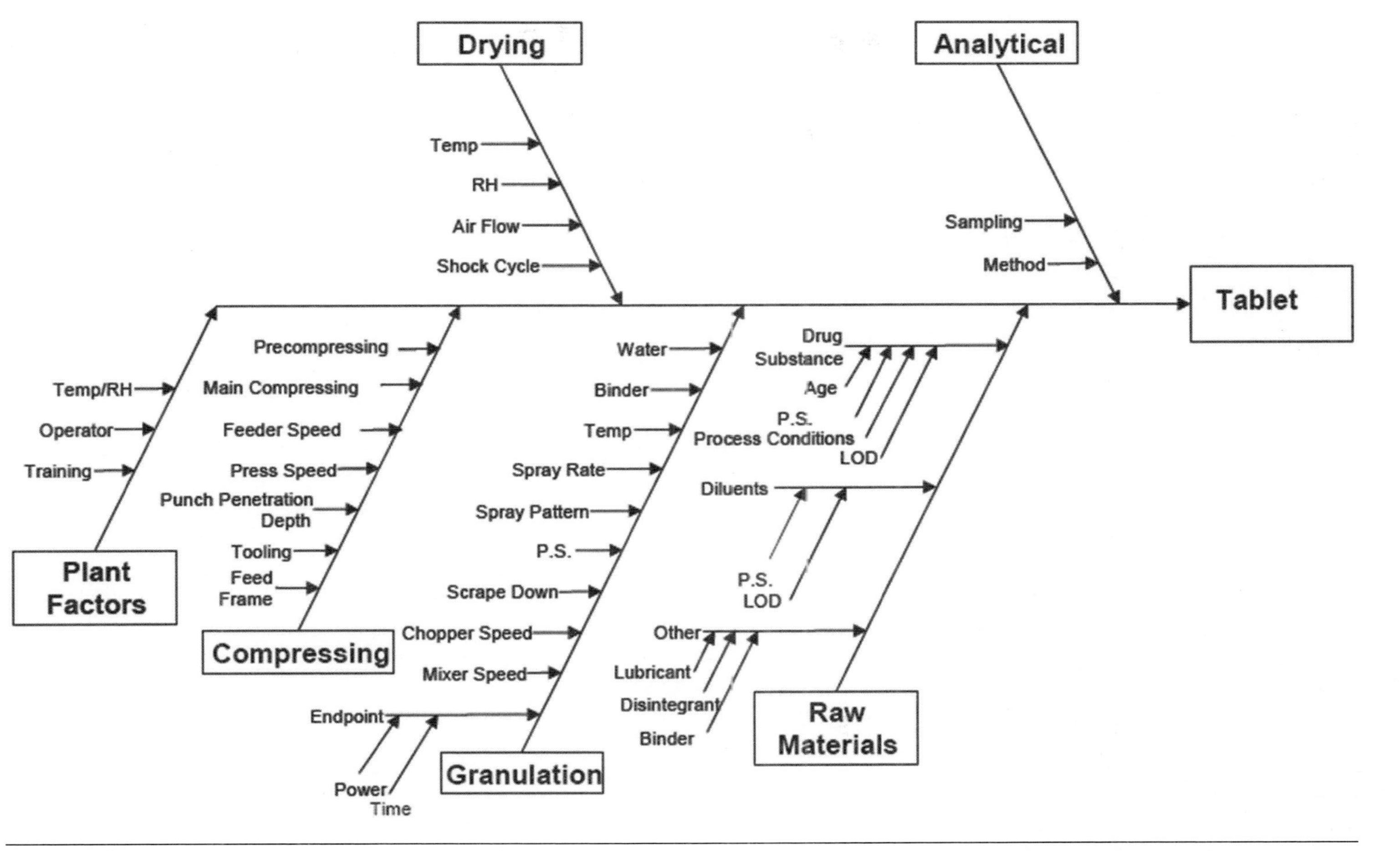

Figure 24.5. Example of a risk assessment Ishikawa diagram from ICH Q8.

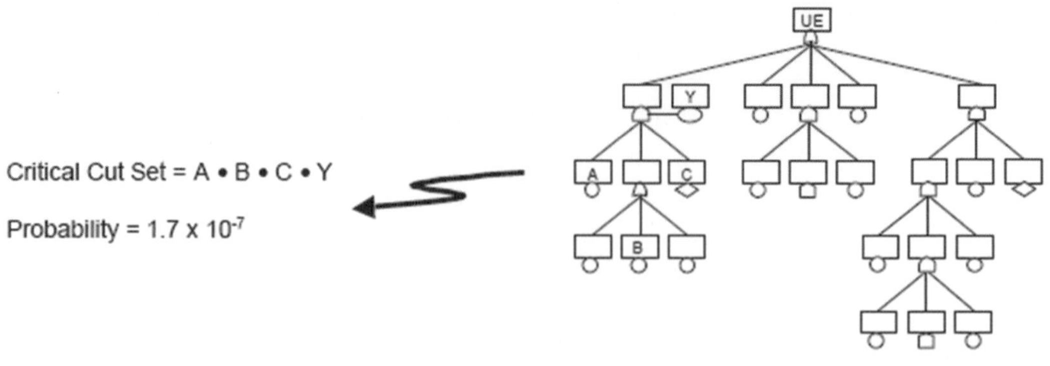

Figure 24.6. Example of a fault tree, identification of critical sets, and probability calculation.

An FTA allows for a calculation of the probability of an event occurring, allowing the effective assignment of priorities and resources to avoid "undesirable" events. The end result of an FTA is often a DoE used to determine the likelihood of an occurrence. An FTA can be helpful during process upsets if it contains sufficient information to determine which branch of the tree was associated with the upset; it contains the type of information that is obtained during process development and could be referenced by a GR.

HAZOP

HAZOP analysis is a widely recognized qualitative method for systematically studying a process using industry-specific guide words. It is used to uncover how deviations from the desired result can occur in equipment, actions, or materials and if the consequences of these deviations can result in a hazard to the product, equipment, or personnel. The study is usually carried out by a team who is primarily focused on the goal of a safe and operable manufacturing process. Traditionally, HAZOP studies have focused on process units.

Guide words are simple words (e.g., high, low, reverse, leak, no) that are used to modify the design intent and to guide and stimulate the brainstorming process for identifying process hazards. The guide word and measurable process condition provide the elements for study. A combination of guide word and process condition defines the area for study (e.g., high flow, no flow, reverse flow, misdirected flow, high pressure, low pressure, high temperature, low temperature, fluctuating humidity).

The final result for a HAZOP analysis is a set of recommendations for changes or improvements to a process. This information can be helpful during updates or changes to the master recipes—typically to exception handling. However, if product exceptions are added as changes to master recipes or if products are manufactured in a networked process cell, then exception handling approaches need to be well defined, managed, and implemented. One approach is to perform a "master recipe procedure HAZOP" after the initial HAZOP.[8]

At this point in the process, HAZOP analysis has already looked at possible deviations and can provide guidance on what to avoid in changes to recipes. For example, if there is a condition of low flow in a material that could result in reduced process safety, then that information can be used during normal, continuous improvement processes to stay away from changes that reduce the flow rate.

To avoid the scattering of process knowledge, with some product exceptions in a recipe procedure and the rest in equipment entities, HAZOP analysis results need to be individually captured and can be effectively managed or referenced in a GR as exception cases. Possible failure reasons are listed as GROI.

Risk assessments may result in additional DoE to determine the sensitivity of quality attributes to uncontrollable product or process qualities. For example, humidity may not be controllable if the process must occur in the open (e.g., food or bulk chemical operations).

The QbD should be maintained throughout the life cycle of the product and should not end with the development of the initial control strategy. The manufacturing science gained during process development continues to be built with more manufacturing experience through additional sourcing studies, process characterization efforts, scale-up, manufacturing (e.g., reactive and prospective continuous improvement studies), and product performance as it relates to the Quality Target Product Profile (QTPP).

We have already proposed that the GR is the key information entity to manage manufacturing science information over its entire life cycle.[1,2,3] The following sections propose specific extensions to handle design space and risk assessment information.

Storing and Using Design Space Information

Design space knowledge is only useful for a QbD process if it is readily available when the actual production process is defined. In batch and continuous manufacturing, this means that the information must be available when creating master recipes. This manufacturing information is often related to specific steps in the process, based on the scale, and is often independent of a production unit. These are the characteristics of GRs, and the QbD and design space information should be elements of a GR definition.

A GR is a container for manufacturing information about a product. It defines the critical and key process parameters. A GR may contain specific target values for the parameters or acceptable ranges for the parameters; however, these definitions are not a design space.

Single target values do not define the sensitivity of parameter values. Ranges for parameters define a multidimensional rectangular region for operation, but this either (1) does not cover the entire acceptable design space or (2) is larger than the design space.

In the first situation, there may be more efficient operation or higher throughput in the design space that is outside the multidimensional rectangular region. In the second situation, you may be operating within the defined multidimensional rectangular but still be producing unacceptable quality.

Figure 24.7 shows the impact of using a range of parameters instead of a DSF. The graph on the right defines a more constrained operational space than the actual acceptable operations space shown on the left.

In order to use GRs to store QbD and design process parameters, several rules and extensions should be used:

1. Process parameters should be flagged as either critical or key process parameters. This allows for the identification of which process parameters must be documented in production records that should be made available to regulatory authorities.
2. Process parameters that are associated with a design space should be flagged, and the associated DSF should be identified.
3. Process parameters should be specified as ranges of values and not as specific target values. The ranges should define the maximum range of a process parameter, bounding the design space in a multidimensional rectangular space. Applying the bounding space to the

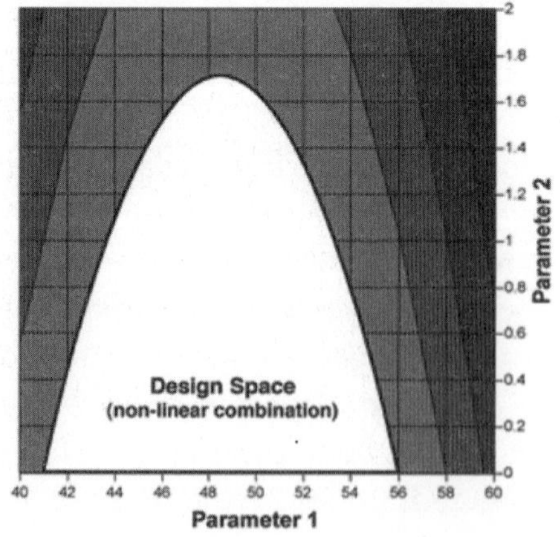

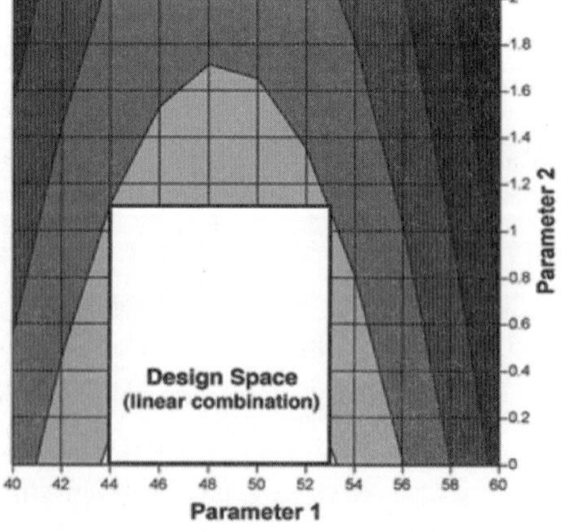

Figure 24.7. Design space examples from ICH Q8.

design space can often simplify the DSF, eliminating unrealistic or unreachable operating conditions.

4. If the DSF is known and mathematically reproducible, then it should be stored as a special type of GROI that can be associated with process stages, operations, and actions. Fortunately, this concept is supported by the object model defined in the ISA-88.03 standard (Fig. 24.8), where any process element may have GROI.
5. There may be a DSF associated with a GR's process stage definition. Often, this DSF will deal with critical process parameters associated with input materials. This DSF can be used to determine if the production of the product is even possible with the available materials. In some industries, this may allow other products to be produced using the available materials because their specifications fall into the material parameter's range. Examples of alternate products may be different delivery mechanisms for a drug (e.g., gelatin, liquid, powder), different grades for a chemical product, or different food products (e.g., raw, prepared, animal food). According to ISA-88.03, "A process stage defines a part of a process that usually operates independently and usually results in a planned sequence of chemical or physical changes in the material being processed."
6. There may be a DSF associated with a process operation definition. Typically, the design of elements is performed around process operations. The goal of a DoE is often to determine the design space for a critical quality parameter of an intermediate material, such as the friability of a tablet or the average density of an active ingredient. Often, the DoE will coordinate the timing of actions (e.g., drying time, cooking time, mixing time) and number of steps (e.g., drying steps, mixing vessel reversals). According to ISA-88.03, "A process operation is a major processing activity that is defined without specification of the actual target equipment configuration and usually results in a chemical or physical change in the material being processed."
7. There may be a DSF associated with a specific process action, such as a Mix, Blend, or Dry. Some process actions can be very complex, such as lyophilization (i.e., freeze-drying), where there can be a complex set of temperature and humidity steps. Often, DoE is used for the complex actions that are represented in a GR as a single process action. According to ISA-88.03, "A process action is a minor processing activity, like grind, cool, heat, delay, test or mix. Many

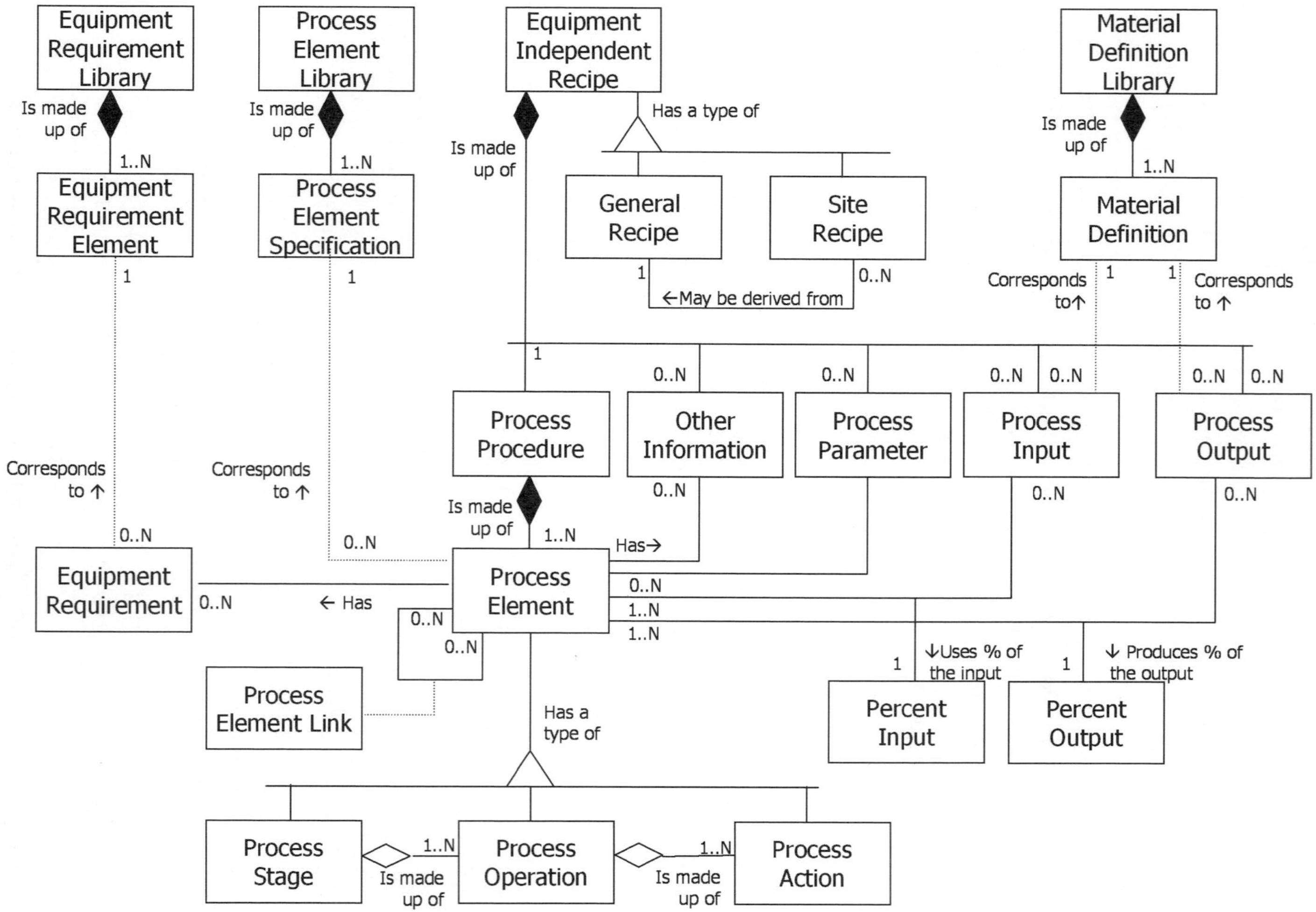

Figure 24.8. Figure 8 from ANSI/ISA-88.00.03 2003.

actions are simply the addition of material, removal of material, addition of energy, or removal of energy."

8. DoE data may also be stored with the GR. Raw DoE data are often important to have when looking at process upsets and exceptions. They may be used to indicate the expected change in quality-attribute values. This can aid in determining if the batch can be salvaged through rework, if it can be used for a different product, or if it must be scrapped. DoE data is often large and sometimes unstructured. If the GR management system does not support this type of information, then the GR should at least contain a reference to the DoE data.
9. The GR should include references to exception handling cases, although it is typically managed at a process unit level.

Extending GROI

In order to make the GR a true container for product knowledge, the GROI category must be augmented to include DoE, design space, and risk analysis information.

The GROI models in the first three ISA-88 standards[4,5,6] do not define a structure or standard attributes for GROI; however, the ISA-88.04 standard for batch production records[7] does include a structure for GROI, as shown in Figure 24.9.

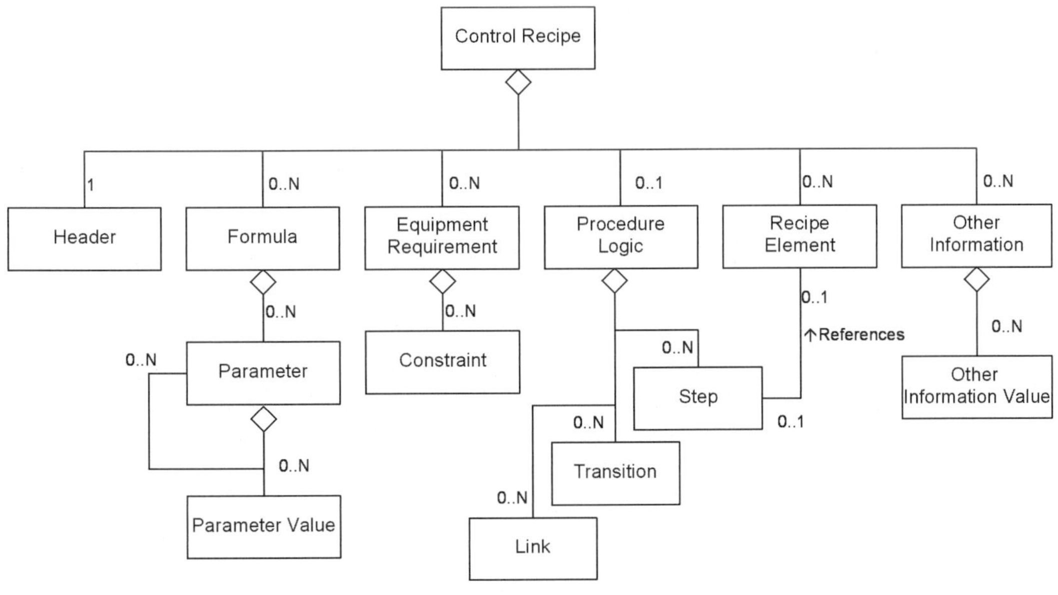

Figure 24.9. Figure 20 from ISA-88.00.04.

ISA-88.04 also defines some attributes for the GROI value, as shown in Figure 24.10. While these are a start, a more complete list of required attributes for the GROI value can be found in Table 24.1.

Figure 24.11 shows a process flow diagram that represents a granulation process operation. Because this involves multiple actions and two independent materials (until mixed during granulation), there can be dozens of parameters and

Name	Description	Example
Value String	The value of the parameter	127 Red A*(B+C+D)
Data Interpretation	Identification of how to interpret the value string. Choices are "Constant", "Reference", "Equation", "External"	Constant Reference Equation
Data Type	Identification of the data type contained in the value string. Each implementation of this model will define the allowed data types. If one supported data type is an enumeration, then the data type shall be called "Enumeration".	Float Date String
Unit of Measure	Unit of measure associated with the value string	KG Tons KL

Figure 24.10. Section 5.16.14 from ISA-88.00.04.

Table 24.1. Required attributes for the GROI value	
Name	*Description*
Value	This identifies the value of the parameter. This may be a direct value or may be a reference to the location of the actual data, depending on the Data Interpretation.
Data Interpretation	This identifies how to interpret the value string. The four choices follow: *Constant.* The string in the "Value" attribute is to be interpreted as a constant value based on the Data Type attribute. *Reference.* The string in the "Value" attribute is to be interpreted as a Universal Resource Identifier (URI) to an external data source whose meaning is defined in the Data Category attribute. *Equation.* The string in the "Value" attribute is to be interpreted as an equation to be parsed and solved, using references for the variables. *External.* The string in the "Value" attribute is to be interpreted as a reference in a format that is determined by the engineering tool.

(Continued on page 309)

Table 24.1. Required attributes for the GROI value (*continued*)	
Name	*Description*
Data Type	This identifies the data type contained in the value string, using the UN/CEFACT data types defined in their Core Components Technical Specification (CCTS). The following are allowed: *Amount.* A number of monetary units specified in a currency where the unit of currency is explicit or implied *BinaryObject.* A data type representing graphics, pictures, sound, video, or other forms of data that can be represented as a finite length sequence of binary octets *Code.* A character string that is used to represent an entry from a fixed set of enumerations *DateTime.* A particular point in time together with the relevant supplementary information to identify the time zone information *Identifier.* A character string to uniquely distinguish one instance of an object in a name space from all other objects in the same name space *Indicator.* A list of two mutually exclusive Boolean values that express the only possible states of a property (e.g., true or false) *Measure.* A numeric value determined by measuring an object along with the specified unit of measure *Numeric.* A numeric value *Quantity.* A counted number of nonmonetary units, possibly including fractions *Text.* A character string (i.e., a finite set of characters) The following are not UN/CEFACT data types: *uriReference.* A URI that may be a Universal Resource Locator (URL) and/or or a Universal Resource Name (URN) *SVG.* A Scalable Vector Graphics object; a language for describing drawings in terms of the shapes that compose them, so that these can be rendered as well as possible *MathML.* Mathematical Markup Language (MathML), an Extensible Markup Language (XML) document type
Data Category	This can be identified as one of the following values: *Product.* The value contains product information. *Equipment.* The value contains equipment-specific information not related to equipment constraints. *Design Space.* The value defines design space information. *Risk Assessment.* The value contains risk assessment information. *DoE.* The value contains Design of Experiment information. *Exception Cases.* The value that contains HAZOP results. *Other.* A vendor or end-user specified category.
Data Subcategory	The Data Subcategory provides a further breakdown for each specific data category.

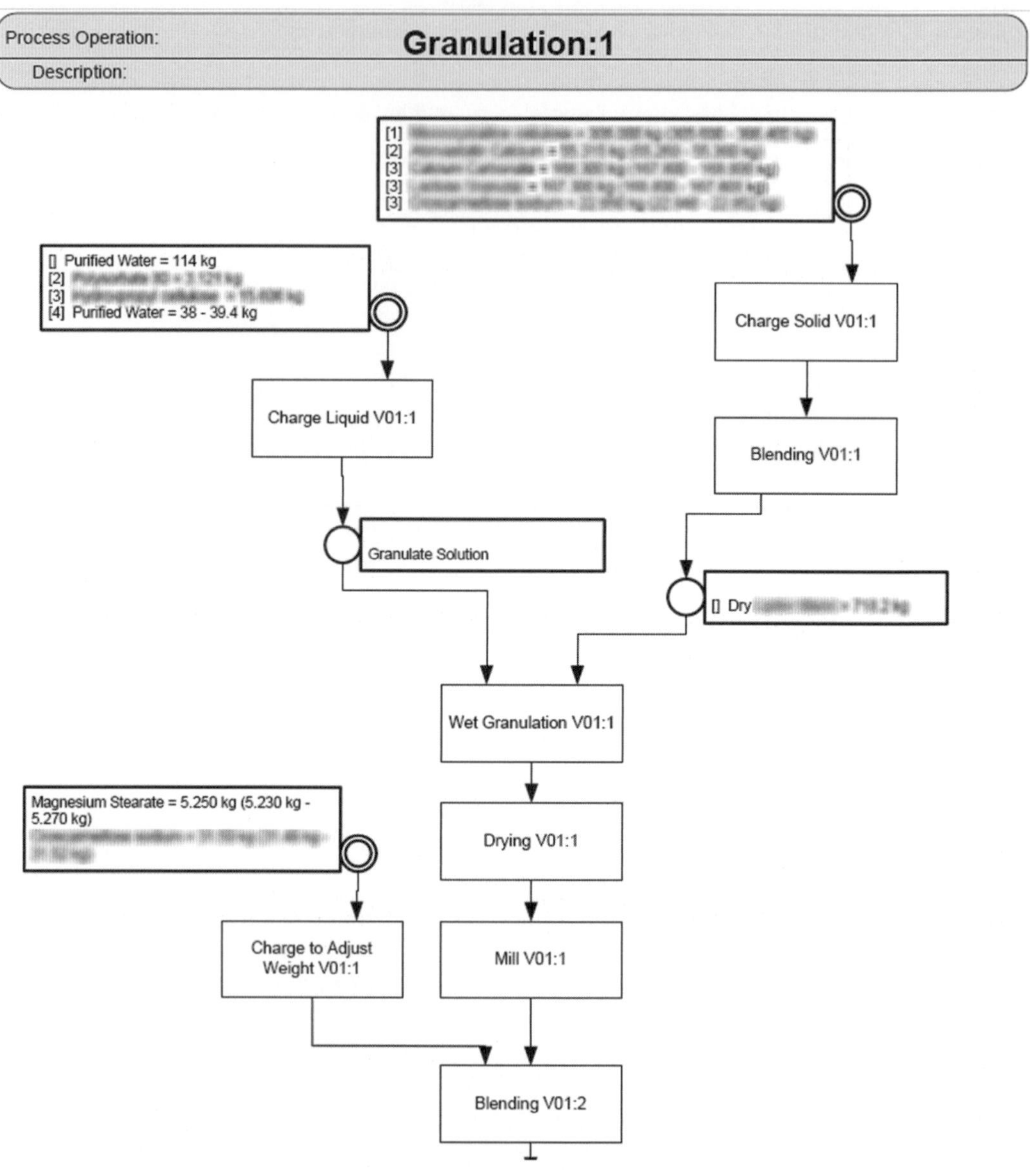

Figure 24.11. Process operation for granulation.

incoming material attributes in the design space. The final quality attributes may just be a density measurement and potency, as shown in Figure 24.12.

Figure 24.12 illustrates the process operation details with associated design space and risk analysis information. The last two rows in the right spreadsheet reference design space and risk assessment information. The detailed format for the external information files will usually be based on the tools used to generate and store the data.

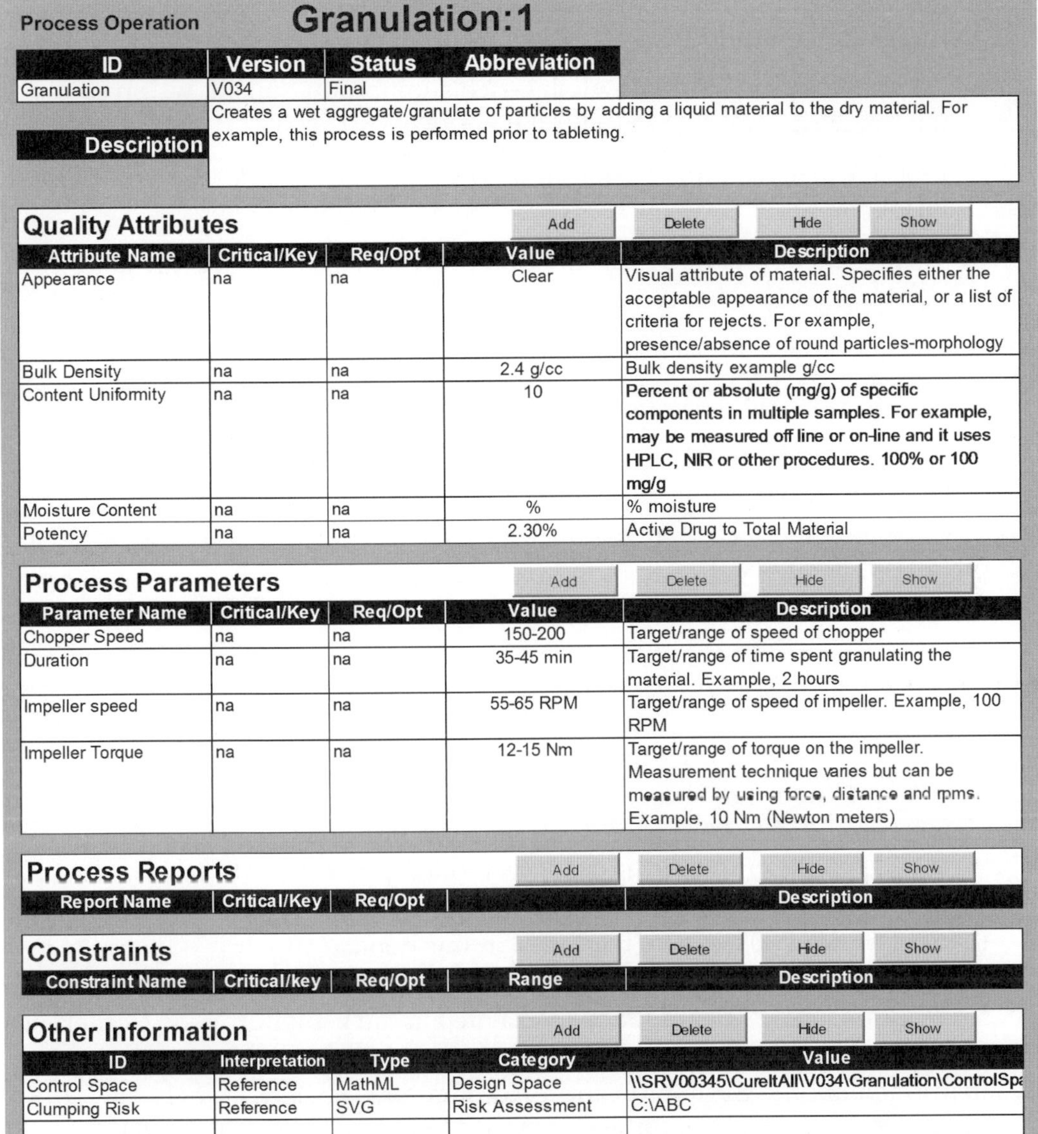

Process Operation **Granulation:1**

ID	Version	Status	Abbreviation
Granulation	V034	Final	

Description: Creates a wet aggregate/granulate of particles by adding a liquid material to the dry material. For example, this process is performed prior to tableting.

Quality Attributes (Add / Delete / Hide / Show)

Attribute Name	Critical/Key	Req/Opt	Value	Description
Appearance	na	na	Clear	Visual attribute of material. Specifies either the acceptable appearance of the material, or a list of criteria for rejects. For example, presence/absence of round particles-morphology
Bulk Density	na	na	2.4 g/cc	Bulk density example g/cc
Content Uniformity	na	na	10	**Percent or absolute (mg/g) of specific components in multiple samples. For example, may be measured off line or on-line and it uses HPLC, NIR or other procedures. 100% or 100 mg/g**
Moisture Content	na	na	%	% moisture
Potency	na	na	2.30%	Active Drug to Total Material

Process Parameters (Add / Delete / Hide / Show)

Parameter Name	Critical/Key	Req/Opt	Value	Description
Chopper Speed	na	na	150-200	Target/range of speed of chopper
Duration	na	na	35-45 min	Target/range of time spent granulating the material. Example, 2 hours
Impeller speed	na	na	55-65 RPM	Target/range of speed of impeller. Example, 100 RPM
Impeller Torque	na	na	12-15 Nm	Target/range of torque on the impeller. Measurement technique varies but can be measured by using force, distance and rpms. Example, 10 Nm (Newton meters)

Process Reports (Add / Delete / Hide / Show)

Report Name	Critical/Key	Req/Opt		Description

Constraints (Add / Delete / Hide / Show)

Constraint Name	Critical/key	Req/Opt	Range	Description

Other Information (Add / Delete / Hide / Show)

ID	Interpretation	Type	Category	Value
Control Space	Reference	MathML	Design Space	\\SRV00345\CureItAll\V034\Granulation\ControlSpa
Clumping Risk	Reference	SVG	Risk Assessment	C:\ABC

Figure 24.12. An example process operation definition with design space information.

Summary

QbD is a scientific approach to obtaining high-quality products in a manufacturing environment. It relies on a deep understanding of the process and the impact of process parameters and material characteristics on the final product quality and on process efficiency. In QbD, the following is true:

- Quality is built into product and process by design, based on scientific understanding.
- Companies provide a knowledge-rich submission that shows product knowledge and process understanding.
- Product specifications are based on product performance requirements.
- There is flexibility of the production process within the design space, allowing continuous improvement efforts.
- There is a focus on robustness, which involves understanding and controlling variation in the production processes.

DoE is used in a QbD process to build the formal mathematical models of a design space. The DoE information, risk assessment, and the design space can be captured in a GR and made available to process engineers during the development of equipment-specific master recipes.

References

1. Brandl, Dennis, Rob Burrows, and Velumani Pillai. 2006. Standardized multiple plant investigations. Paper presented at the WBF North American Conference, March 2006. Also, Chapter 21 in this book.
2. ———. 2008. Manufacturing science model extensions to address product and process sustainability. Paper presented at the WBF North American Conference, March 2008. Also Chapter 23 in this book.
3. Brandl, Dennis, and Velumani Pillai. 2006. An information model framework for manufacturing science. Paper presented at the WBF European Conference, November 2006.
4. Instrumentation, Systems, and Automation Society. 1995. *ANSI/ISA-ISA-88.01-1995: Batch control part 1: Models and terminology*. Research Triangle Park, NC: ISA.
5. ———. 2001. *ANSI/ISA-88.02-2001: Batch control part 2: Data structures and guidelines for languages*. Research Triangle Park, NC: ISA.
6. ———. 2003. *ANSI/ISA-88.03-2003: Batch control part 3: General and site recipe models and representation*. Research Triangle Park, NC: ISA.
7. ———. 2006. *ANSI/ISA-88.04-2006: Batch control part 4: Batch production records*. Research Triangle Park, NC: ISA.

8. Korkmaz, Baha, Velumani Pillai, and Rajagopalan Srinivasan. 1999. Guidelines for exception handling design for batch processes. Paper presented at the WBF North American Conference, April 1999.
9. U.S. Food and Drug Administration. 2009. *Q8 (R2) pharmaceutical development*. Washington, DC: Government Printing Office.

CHAPTER 25

Batch Release and Material Use Reporting: A Case Study

Presented at the WBF North American Conference, May 15–18, 2005, by

Scott C. Clark
Sr. Project Engineer
scott_clark@merck.com
Merck & Co., Inc, PO Box 7,
Elkton, VA 22801 USA

Abstract

This chapter will discuss and explore the enhanced benefits gained from fully utilizing the ISA-88 standard when applying it to batch release and raw material usage reporting. During the development of a complex bulk pharmaceutical facility's automation system, batch reporting was integrated into the procedural models to provide Critical Process Parameter (CPP) and critical alarm reporting that used native control system data historians, kept in lockstep with the master recipes. This fully leveraged the structure of ISA-88 to provide simple yet effective batch quality reports. After the batch release reporting system was validated, components of this system were reused to develop recipe-based raw material usage reports on a batch-by-batch basis, complete with economic metrics. Significant cost savings were realized by not porting data or the procedural models outside the automation system. Reduced manpower was required to release batches and calculate raw material cost per batch, and reusability was demonstrated during the development of the material tracking system, which further enhanced the bottom-line benefit of the initial investment.

Introduction

In the pharmaceutical industry, batch reporting for quality release has been a major focus. This is becoming increasingly important with the advent of the Process Analytical Technology (PAT) initiative, release by exception, critical alarm compliance, and quality by design. It is a difficult nut to crack in that one must deal with the inherent pitfalls in robustly managing start and stop triggers, adapting legacy systems, and the unclear need to prevent the comingling of batch quality data with troubleshooting or optimization data. One is faced with ever-stronger data security requirements, the challenge of managing and synchronizing report changes with the ISA-88 procedures, and the challenge of dealing with the diverse teams that are involved in developing reports.

During a large batch automation project, batch release reporting was successfully integrated into the procedural models to provide focused Critical Process Parameter (CPP) and Critical Alarm (CA) reporting. The minimum and maximum CPP values were reported, along with CAs that were triggered during the critical batch processing stages. This reporting system was well received by its users and became the foundation for the second phase of batch reports, which included raw material tracking. In this chapter, highlights from and the design of the CPP/CA and material tracking reporting will be reviewed, as will the benefits of leveraging the CPP/CA reporting model to develop a recipe-based material tracking report capable of calculating material usage and cost.

Batch Release Report

ISA-88 provides an excellent framework to insert batch reporting "hooks" in parallel with batch processing automation. By keeping the requirements of a batch release report as fundamental as possible and leveraging the ISA-88 procedural models, it is possible, with thoughtful design, to develop a robust batch release reporting system for a process automation system. Some of the key factors in making this successful include the following:

- A master clock to ensure synchronized data stamps
- Embedding the batch report triggers into the ISA-88 procedural model, thereby making the reports recipe based and keeping the report in lockstep with the procedural model
- Using ISA-88 Control Modules (CMs) and phases to perform batch reporting monitoring and calculations

- Keeping data in their native format (i.e., embedded historians)
- Open programming standards (e.g., Extensible Markup Language [XML], Visual Basic [VB]) and database interfaces (e.g., Open Database Connectivity [ODBC])
- Providing a simple, Graphical User Interface (GUI) to generate reports

Design Approach

A successful approach is demonstrated in Figure 25.1, which depicts a stage in a unit procedure where vessel temperature is considered critical. Temperature will be monitored as a CPP (i.e., a max and min will be calculated), and the high alarms will be considered critical as well. CPP/CA monitoring is turned on in the CM via recipe by the Batch Monitor phase, and the temperature extremes are calculated in the vessel temperature CM. The batch historian is told, using report parameters, that the critical stage has started making it easier for the reporting system to look for CA events at a later date. Temperature monitoring will continue in the CM until told to be turned off at the completion of critical phase 2 or when the unit procedure is stopped or aborted. CPP/CA monitoring continues in the CM during a batch hold or restart. When critical phase 2 (in this case, the Age phase) completes, certain end-of-monitoring events, such as the calculated extremes, are written to the batch historian, and the CPP/CA monitoring is turned off in the CM by the Batch Monitor phase. The Batch Monitor phase will change the priority of the critical alarms, directly within the CM at the start and end of the phase, to identify them as critical.

Figure 25.2 depicts an approach to managing CPP monitoring functionality within the CM. The Batch Monitor phase directly writes to the function block On_Off input to start or stop monitoring during the critical stage and reads the results at the end of the critical stage from the CPP Monitoring function block outputs. The Hold input to the function block allows unique logic to handle nonsteady state processing situations and does not need to be linked to the phase logic. The hold command essentially "pauses" CPP monitoring to handle nonsteady state (but normal) situations, such as a batch processing restart.

The GUI employs the embedded security of the database where the native data resides, thereby reducing external security configuration; it was designed to run on a standard corporate desktop computer, thus making the data and reports very accessible.

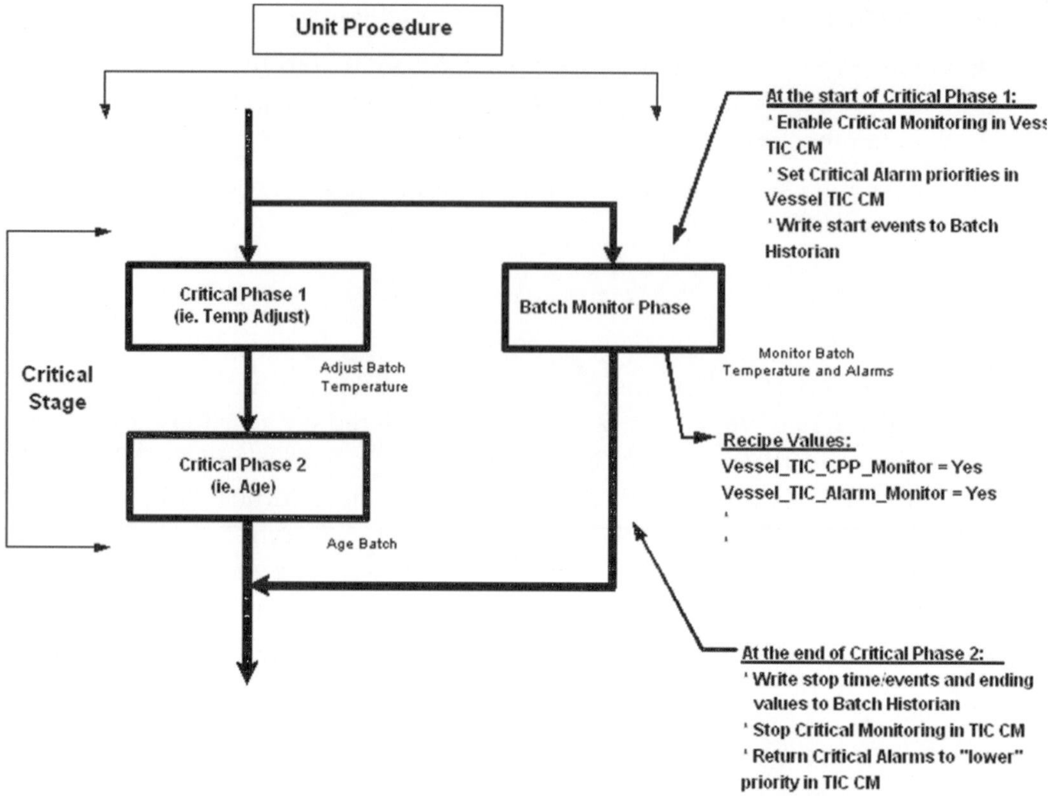

Unit Procedure
At the start of Critical Phase 1:
' Enable Critical Monitoring in Vess
TIC CM
' Set Critical Alarm priorities in Vessel TIC CM
' Write start events to Batch Historian
Critical Phase 1 (ie. Temp Adjust)
Batch Monitor Phase
Critical Stage
Adjust Batch Temperature
Monitor Batch Temperature and Alarms
Recipe Values:
Vessel_TIC_CPP_Monitor = Yes
Vessel_TIC_Alarm_Monitor = Yes
Critical Phase 2 (ie. Age)
Age Batch
At the end of Critical Phase 2:
' Write stop time/events and ending values to Batch Historian
' Stop Critical Monitoring in TIC CM
' Return Critical Alarms to "lower" priority in TIC CM

Figure 25.1. Example procedural model with integrated CPP/CA reporting.

TIT-1234-01
AI
AI1
30
SIMULATE_IN
OUT
39
#1
Engineering Units for CPP Monitoring.
UNITS
Deg. C
MODULE_TAG
TIC-1234-01
CPP Monitoring Block.
CPP_MONITOR
CPP_MONITORING
ON
ON_OFF
STATUS
ON
PV
HOLD_STAT
OFF
39
UNITS
MAX
50.5
Deg. C
OFF
HOLD
MIN
27.5
AVG
38.324
TIME
0.041944
#11
TIME Output is how long you've been monitoring in hours.

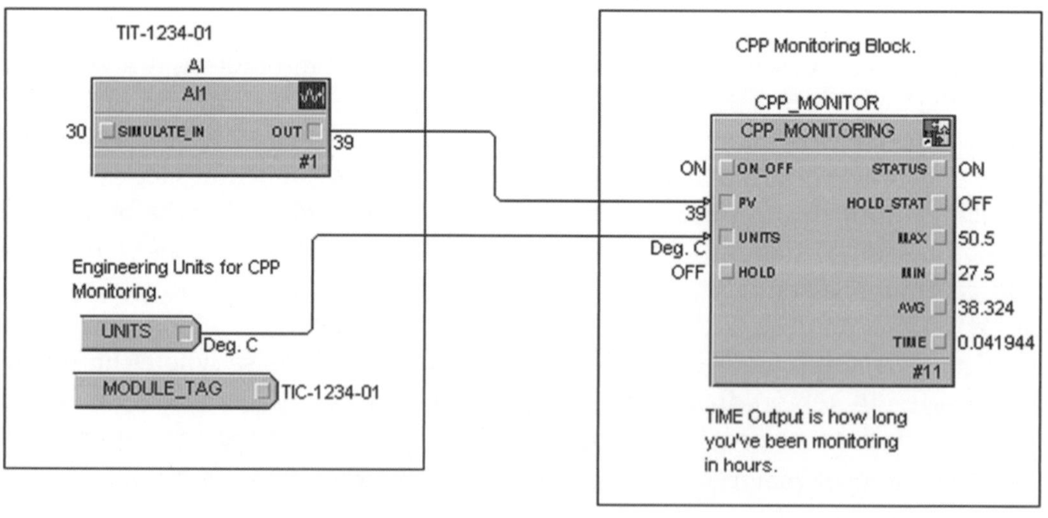

Figure 25.2. Example CPP Monitoring function block.

System Benefits

In this model, the CMs and phases are performing calculations within the control system controllers, providing a redundant environment to handle the calculations and exceptions that might occur during batch processing. Since the batch monitor operation lives in the procedural model, all batch reporting events are logged in direct context with the procedural model and are very visible to the recipe developer. With the batch reporting operation as an integral part of the procedure, there is a reduced chance that the "batch reporting" procedural model could be placed out of sync with "batch processing" procedural model. The generation of reports is a matter of picking the appropriate batch ID through the GUI, and generation is performed quickly. The data is reported in an exception format for the CA, while the extremes are always reported for CPPs. If a critical alarm or an out-of-range CPP is reported, then the investigation is left to the production and quality groups. Critical batch data does reside within the historian, alongside noncritical batch data, but it is uniquely identified as a CPP or CA event.

Material Tracking Report

After the successful validation of the CPP/CA report, the next logical step was to automate the tracking of material into and out of the batch process. Material accounting is a laborious task, based on the complexity of the process and the sheer number of batch sheet pages from which raw material quantities need to be extracted. A recipe-based system is required to allow the end users to pick the materials they want to track within the operations that transferred the respective material and to allow the tabulation of the totals and costs for those materials on a batch-by-batch basis. The infrastructure to generate such a report was well established, as material transfers were already coded into the automation system, and many components of the GUI, developed to generate the batch release reports, could be reused.

Design Approach

The batch automation system contains operations to transfer recipe-based quantities of material, including a methodology to log and report this information to the batch historian. These preexisting operations are robust and include the charging or discharging of material, automatically or manually. The only reporting components that needed to be developed were a way to identify the materials at the appropriate stages in the process and associate a cost with them. The material names were specified by recipe and made available in all operations that transfer material. An external database table was used to match these material names with an associated cost. This approach allows a recipe developer to "turn on" material tracking for any transfer done by the control system and update the cost data as necessary without being "hard coded."

Figure 25.3 provides a graphical depiction of how material tracking is configured via recipe and how the data makes its way to the historian in context with the procedural model. In Receive Operation 1, the value of recipe parameter "Track_Material" is configured as the name of the material that is to be charged—in this example, "Material 1." At the end of the receive operation, the actual quantity charged is written to the batch historian. The material tracking report interface will later search the batch historian for the material name specified (e.g., "Material 1"), grab the actual quantity charged from the appropriate batch report parameter, match it to a cost table, and tabulate these material quantities and costs.

The report output includes a minimum of two data sheets—the transfer report summary sheet and the material transfer sheet. The transfer report summary sheet displays the report name, totals for all materials reported, and a list of batches reported. The material transfer sheet displays the quantities and costs for all transfers of the material for all the selected batches. Figure 25.4 provides an example of the transfer report summary sheet output, listing a summary and total of all

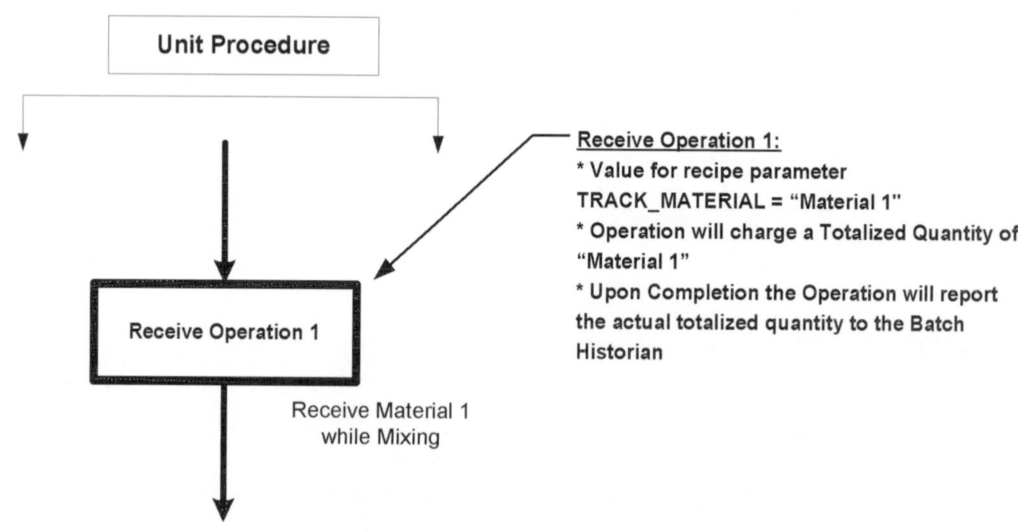

Figure 25.3. Example procedural model with integrated material tracking.

Transfer Report Summary

Report Date:	**08-Feb-2005**
Process Step:	**1**

Material Totals

Material Part Number	Total Transfer Quantity	Total Material Cost
Material 1	6341.41	$31,888.14
Material 2	6436.55	$40,035.32
Material 3	110.87	$416.89
Material 4	433.18	$14,043.44

Included Batch IDs

Batch ID	Start Time	End Time
Batch 01	11-Apr-2004 07:59:58 AM	11-Apr-2004 08:40:13 AM
Batch 02	14-Apr-2004 11:44:22 AM	14-Apr-2004 09:12:51 PM
Batch 03	04-May-2004 04:03:55 AM	07-May-2004 08:46:03 PM

Figure 25.4. Transfer report summary sheet output.

materials being tracked for the chosen batch IDs. Figure 25.5 provides an example of the material transfer sheet output, listing a summary and total of one material (i.e., Material 1) being tracked for the chosen batch IDs.

As with the batch release report, user security is handled within the embedded batch historian under the same security strategy as the control system, minimizing user account management. Exception and error handling are accomplished by the use of an additional "untracked" transfers table, to list transfers that are not specifically tracked via recipe, and by the ability to remove any suspicious material quantity from the material transfer sheet.

Benefits

The most prevalent economic benefit of this reporting system is the reduction in the resource time required to generate material tracking reports; the overall effort has been reduced by approximately 4 hours per batch. This system eliminates human errors in the reports that were previously generated manually. Since the report is configured directly into procedural recipes, maintenance is low and is very visible

Transfer Report for Material 1

Part Number: **Material 1**
Date: **08-Feb-2005**

Batch ID	Material	Transfer Quantity	Conversion Factor	Actual Quantity	Units	Unit Cost	Total Cost
Batch 01	Material 1	8.30	1.0000	8.30	kg	$1.25	$10.38
Batch 01	Material 1	155.48	1.0000	155.48	kg	$1.25	$194.34
Batch 02	90% Material 1, 0.15% Material 2	453.24	0.9000	407.91	gal	$15.25	$6,220.68
Batch 02	Material 1	941.00	1.0000	941.00	kg	$1.25	$1,176.25
Batch 02	90% Material 1, 0.15% Material 2	12.16	0.9000	10.95	gal	$15.25	$166.95
Batch 02	90% Material 1, 0.15% Material 2	2.27	0.9000	2.04	gal	$15.25	$31.18
Batch 02	90% Material 1, 0.15% Material 2	2.26	0.9000	2.04	gal	$15.25	$31.06
Batch 02	90% Material 1, 0.15% Material 2	2.78	0.9000	2.50	gal	$15.25	$38.14
Batch 03	0.15% Material 1	3893.22	0.2642	1028.59	kg	$1.25	$1,285.74
Batch 03	0.15% Material 1, 45% Material 3	3782.60	1.0000	3782.60	gal	$6.01	$22,733.42
			Totals	**6341.41**	----	----	**$31,888.14**

Figure 25.5. Material transfer sheet output.

to those reviewing or modifying recipes. A 50% development cost savings has been realized for the material tracking report by leveraging many components of the batch release reporting system, and since the reporting system has a similar GUI, user training has been minimized. Report data is permanently attached to batch IDs, thus enhancing the ability to retrieve reports in the future. Although the original design of the system was to calculate the cost of raw materials for batches, any "material" can be reported; for example, the system could easily be configured to track waste streams.

Conclusion

This chapter discussed the enhanced benefits gained from fully utilizing the ISA-88 standard for batch release and raw material usage reporting, as batch reporting was tightly integrated into the procedural models to provide CPP and CA reporting. Embedded historians were leveraged that kept data in lockstep with the master recipes; these provided simple but effective batch quality reports. Components of the batch release reporting system were reused to develop recipe-based raw material usage reports on a batch-by-batch basis, which provided itemized and total raw material costs. Significant cost savings were realized by not porting data or the procedural models outside the automation system, thus keeping the reporting configuration within the procedural models. Finally, less manpower was required to release batches and calculate raw material cost per batch, which also resulted in significant cost savings.

Index

alarm modes, 255–59
alignment strategies, 145–47
all-in-one architecture, 237–39
area layout, 97–98
As-Low-As-Reasonably-Possible (ALARP) distribution, 215–19
auto-generation, 104
automation
  engineering tools, 190
  global configuration and integration of, 183–97
  multiple batch systems integration, 150–51
  Product Life-cycle Management (PLM), 121–25
  risk-based engineering assessment guidelines, 202–6
Automation Execution Framework (AEF), 188–90
Automation-Project Activity Model (A-PAM), 184, 188–90
auxiliary equipment, 37

Basic Design (BD)
  batch processing strategies, 131–34
  global automation integration and configuration, 185–88
batch execution, Manufacturing Execution System (MES), 165–66
batch independence and genealogy, 33
Batch Markup Language (BatchML) standard
  global business systems recipe development management, 231–40
  process definition enablement, 92
  transfer of, 82
batch processing
  bulk pharmaceutical batch process case study, 13–27
  complex equipment management, 30–37
  design and implementation case study, 95–111
  engineering and planning strategies, 129–42
  on-line analytics, 57–69
  release operations case study, 315–22
  software upgrades, validated applications, 39–46
  start-to-finish solution, 92–93
batch server, batch processing design, 103
bioreactor temperature perturbation, 66–68
Building Automation System (BAS), 202–6
bulk pharmaceutical batch process case study, 13–27

Business To Manufacturing Markup Language (B2MML)
  global business systems recipe development management, 234
  Manufacturing Execution Systems (MES), 174–82
  material definitions, sustainability issues and, 286–89

cell culture technology, 229–30
Certificate of Analysis (CoA) systems, 256–59
change control, 106, 108, 228
class-based equipment, 37
Clean In Place (CIP) operations
  Manufacturing Execution System (MES), 160–63, 166–69
  modularization, pharmaceutical plant construction, 51–56
clinical trial phase, 119
Code of federal regulations (CFR), 21 CFR Part 11, 50–56, 99–100, 157
commercial manufacturing
  life science development operations versus, 74–76
  WBF technology transfer to, 82
Commercial Off The Shelf (COTS) applications
  complex equipment management, 31
  multiple batch systems integration, 146–47
communication, batch processing engineering and planning strategies, 129–42
complex equipment management
  global automation integration and configuration, 186–88
  GMP environment, 30–37
computer-controlled production processes, 137–39
Conceptual Design (CD)
  batch processing strategies, 131–34
  global automation integration and configuration, 185–88
configuration parameters, 36
construction period, pharmaceutical plant, 48–56
content reuse, 88–89
Contract Manufacturing Organizations (CMOs), 89–90
Control Modules (CM)
  batch processing design and development, 102, 141–42
  batch release and material use reporting, 316–22
  bulk pharmaceutical batch process case study, 14–15
  core operations, 4
  identification and development, 7–8
  multiple batch systems, 224–29
  risk-based engineering assessment guidelines, 206
Cooperative Research And Development Agreement (CRADA)
  process definition management, 88–89
  Quality by Design initiative, 77
core process operations, 3
Corrective And Preventative Action (CAPA) plan
  Manufacturing Execution System (MES), 177–82
  risk-based engineering assessment guidelines, 201–6
Critical Alarm (CA) reporting, 315–22
criticality assessment, 203–6
Critical Process Parameter (CPP), 315–22
Critical to Quality (CTQ) attributes, 87
Customer Relationship Management (CRM), 114–27

database development
  batch processing design, 102–3
  bulk pharmaceutical batch process case study, 19–20
data management and analysis
  batch processing design and control, 104–5
  complex equipment status, 32–33
  Manufacturing Execution System (MES), 159–60
  Process Analytical Technology configuration and data preprocessing, 59–62
Define, Measure, Analyze, Improve, and Control (DMAIC), 263–64
Design of Experiments (DoE)
  design space information storage and application, 307
  manufacturing science and, 299–300
design space
  information storage and applications, 303–7
  manufacturing science guidelines on, 295–300
  Quality by Design guidelines and, 297–98
Design Space Formula (DSF)
  information storage and application, 305–7
  principles of, 298–99
discrepancy issues during multiple batch systems integration, 145–47
distributed architecture, 237–39
Distributed Control System (DCS)
  bulk pharmaceutical batch process case study, 14
  Manufacturing Execution System (MES), 169–70
  PAT configuration and data preprocessing, 59–62
  process definition management, 89
documentation guidelines
  bulk pharmaceutical batch process case study, 25–27
  lean computer validation, 218–19
  Manufacturing Execution System (MES), 163–64
  software upgrades for batch applications, 42
downtime analysis
  multiple batch systems, 227–28
  multiple batch systems integration, 150–51
Dynamic Time Warping (DTW)
  on-line analytics, 58–59
  PAT configuration and data preprocessing, 60–62

E-factor
  process operations, sustainability in, 283–84
  sustainability compliance and, 278–79
Electronic Batch Record (EBR), batch processing design and implementation, 99, 108–11
electronic Common Technical Document (eCTD) regulatory submissions, 91
electronic record, 37
electronic signature, 37
Enterprise Resource Planning (ERP)
  global business systems recipe development management, 234
  Manufacturing Execution System (MES), 169–70, 174–82
  product life-cycle stages, 114–27
Environmental Protection Agency (EPA), 277–79

equipment constraints, 241–42, 248–50
equipment definitions, 31–32
equipment genealogy
complex equipment status management, 33
defined, 37
equipment modeling, 164
Equipment Modules (EMs)
bulk pharmaceutical batch process case study, 14–15, 17–19
defined, 31
global automation integration and configuration, 186–88
pharmaceutical process module library, 7–8
pharmaceutical process module library core operations, 4
equipment status management
complex equipment, batch processing, 30–37
definitions, 37
solution components, 35–36
value rules, 34–35, 38
European Union (EU) sustainability regulations, 277–79
exception modes, 255–59
Extensible Markup Language (XML)
global business systems recipe development management, 231–40
process definition enablement, 91–92
regulatory submissions, 91

Factory Acceptance Test (FAT), 105–7
Failure Mode Effects Analysis (FMEA), 200–206
failure modes, 255–59
fast-track engineering, 170–71
Fault Tree Analysis (FTA)
risk-based engineering assessment, 300–302
risk-based engineering assessment guidelines, 200–206
feedback loop, 133–34
Food and Drug Administration (FDA)
drug development initiatives, 76–77
lean computer validation, 212–18
Process Analytical Technology program, 58–59
product life-cycle management, 117
Quality by Design initiative, 73, 76–77
risk-based engineering assessment guidelines, 201–6
front-end definition, 131–34
functional requirements
batch processing design and development, 102
software systems, 40–41
Functional Specification (FS)
batch processing planning strategies, 134–39
content and communication, 137–39
pharmaceutical process module library core operations, 4–6
FVIIa hemophilia medicine case study, 193–95

GAMP 4 V-Model frames, 135–39
general execution principle, 159–60
General Recipe (GR)
alternate materials selection, 269–70
alternate production processes, 267–70
design space information storage and application, 303–7
global business systems recipe development management, 233–34
lean manufacturing science model, 262–70
manufacturing contracts, 241–50

manufacturing science, sustainability and, 283–84
pharmaceutical process module library, 2–3
product life-cycle management, 118–19
standardized multiple plant manufacturing, 251–59
sustainability production and processing, 274–91
transfer units, 8–9
General Recipe Other Information (GROI) category
design space information storage and application, 305–7
extensions of, 307–11
HAZOP analysis, 302–3
risk assessment, 300–12
standardized multiple plant manufacturing, 255–59
global authoring architecture, 237–39
global business systems
automation integration and configuration, 183–97
multiple batch systems integration, 143–54
recipe development management, 231–40
global supply network, 190–93
Good Automation Manufacturing Practices (GAMP)
Automation-Project Activity Model, 188–90
batch processing design and development, 100–11
lean computer validation, 212–18
manufacturing automation project planning and execution, 123–25
Manufacturing Execution System (MES), 175–82
product introduction, 48–49
good business practices, sustainability compliance and, 277–79
Good Laboratory Practice Regulations (3rd ed.), 181–82
Good Manufacturing Practices (GMPs)
complex equipment management and, 30–37
lean computer validation, 212–18
granularity, 10, 21
green chemistry, 275–77

harvesting, sustainability and, 282–84
HAZOP analysis, 302–3
Human Machine Interface (HMI), 165–66
hybrid unfolding, 62–65

IEC 61512-01 standard, 242–50
IEC/ISO 62264-1 standard, 262–70
implementation
batch processing design and control, 104–5
global automation integration and configuration, 193–95
industry benchmarks, lean manufacturing, 266
information model, sustainability management, manufacturing science and, 289–91
information storage and applications, 303–7
Installation Qualification (IQ)
batch processing design and control, 106
global automation integration and configuration, 186–88
lean computer validation, 217–19
manufacturing automation project planning and execution, 125
software systems, 41–42

Insulin Bulk Plant (IBP) project case study, 156–71
integration systems
batch release and material use reporting, 320–22
global automation integration and configuration, 183–97
Manufacturing Execution System (MES), 166–68
Intermediate Bulk Carriers (IBCs), 148–51
International Council on Harmonization (ICH), Q8 Pharmaceutical Development guideline, 294–308
International Society for Pharmaceutical Engineers (ISPE), product introduction guidelines, 48–49
ISA-88.03 standard
design space information storage and application, 305–7
lean manufacturing and supply chain optimization, 262–70
life science development operations, 74–76
multiple batch systems, monoclonal antibody case study, 225–26
standardized multiple plant manufacturing, 251–59
ISA-88.04 standard
General Recipe Other Information (GROI) extensions and, 307–11
lean manufacturing and supply chain optimization, 262–70
ISA-88 and ISA-88.01 standards
automation project planning and execution, 122–25
batch processing design and implementation case study, 95–111
benefits, in pharmaceutical process module library, 11–12
bulk pharmaceutical batch process case study, 13–27
construction guidelines, 2–10
development concepts, 78–83
General Recipes, manufacturing contracts, 241–50
General Recipe Other Information (GROI) extensions and, 307–11
global automation integration and configuration, 186–88
life science development operations, 71–83
Manufacturing Execution System (MES), 174–82
multiple batch systems, monoclonal antibody case study, 222–29
pharmaceutical plant construction time line, 47–56
pharmaceutical process module library framework, 1–12
Process Analytical Technology guidelines and, 126–27
product life-cycle stages, 113–27
recipe representation standard, 86
start-to-finish solution, 92–93
ISA-95 standard
global automation integration and configuration, 186–88
global business systems recipe development management, 234
lean manufacturing and supply chain optimization, 262–70
Manufacturing Execution System (MES), 174–82
material definitions, sustainability issues and, 286–89
multiple batch systems, 225–26

multiple batch systems integration, 143–54
product life-cycle stages, 113–27

knowledge management, 88–89

Laboratory Information Management System (LIMS) functions
multiple-site manufacturing science, 256–59
product life-cycle stages, 115–27
risk classification, 181–82
lean computer validation, 209–19
lean manufacturing
basic principles, 263–64
manufacturing science model, 261–70
LEGO brick model
global automation integration and configuration, 185–88
modularization, 49–56
library control, 248–50
life-cycle management, ISA-88 and ISA-95 standards, 113–27
life science development operations
ISA-88 implementation, 71–83
process definition management, 86–94
locking operations, 163

Manufacturing Control Systems (MCS), standards development, 174–82
Manufacturing Execution System (MES)
data model and general execution principle, 159–60
global business systems recipe development management, 234
highlights and benefits, 168–69
integration experience, 155–71
lean computer validation, 210–11
process definition management, 89
product life-cycle management, 120–21
risk assessment-based validation costs, 173–82
system architecture, 157–68
Manufacturing Operations Management (MOM), 149–54
manufacturing schema, 147–51
manufacturing science
lean manufacturing and supply chain optimization, 261–70
multiple-site manufacturing science, 256–59
Pfizer green chemistry initiative and, 279–80
quality by design and risk assessments, 293–312
sustainability production and processing, 273–91
Manufacturing Science Informatics (MSI)
development of, 262–70
lean manufacturing and supply chain optimization, 262–70
Quality by Design and risk assessment, 293–312
sustainability and, 279–80
manufacturing systems
contracts, General Recipes, 241–50
Product Life-cycle Management, 121–25
standardized multiple plants, 251–59
market forces
global automation integration and configuration, 184
product introduction, 48–49

master recipes
  Manufacturing Execution System (MES), 160–63
  multiple batch systems, 224
  product life-cycle management, 119–21
material definition and control
  Manufacturing Execution System (MES), 164–65
  sustainability issues and, 284, 286–87
material tracking report, 319–22
material transfer sheet output, 320–22
material use reporting case study, 315–22
mechanical data sheet templates, 6–7
modularization
  global automation integration and configuration, 185–88
  multiple batch systems integration, 147–51
  product introduction, 48–55
monitoring operations, 316–22
monoclonal antibody case study, 221–29
multidimensional design space, manufacturing science guidelines on, 295–300
multiple batch systems
  global business system integration, 143–54
  monoclonal antibody case study, 221–29
  standardized manufacturing, 251–59
multiple-site manufacturing science, 256–59
multisite authoring model, 238–39
multi-tier validation plan, risk classification, 180–82

National Electric Code (NEC), 99
New Molecular Entities (NMEs), 143–45
non-class-based equipment, 34
Nonlinear Iterative Partial Least Squares (NIPALS) algorithm, 63–65
non-unit-based equipment, 38
"Not Invented Here" (NIH) syndrome, 147

objectives definition, 132–34
observed modes, 255–59
Off The Shelf (OTS) units
  batch processing design and implementation case study, 107–8
  modularization, 50–56
on-line analytics, 57–69
on-line process monitoring, 65–66
OPC
  Manufacturing Execution System (MES), 166–68
  PAT configuration and data preprocessing, 60–62
OPC-XML standards, 174–82
Operation Qualification (OQ)
  batch processing design and control, 106
  global automation integration and configuration, 186–88
  lean computer validation, 217–19
  manufacturing automation project planning and execution, 125
  software systems, 41–42
operator access control, 166
optimization
  modularization, 52–56
  process development, 81
  supply chains, 261–70
overhead requirements, 224
overwriting principle, 160–63

packaging technology, sustainability and, 282–84
parallel activity testing, 54–56
parallel engineering, 48–49

parameter-attribute relationships, 256–59
Personal Computer (PC) clones, 214–18
Pfizer green chemistry initiative, 278–79
pharmaceutical plant construction, ISA-88 and ISA-88.01 guidelines and, 47–56
pharmaceutical process development and manufacturing, 275–77, 281–83
pharmaceutical process module library
  applications, 10–11
  component development, 10
  core process operations, 3–4
  Equipment Module identification and development, 7–8
  Functional Specification, 4–6
  GR library construction, 2–3
  ISA-88.01 benefits, 11–12
  ISA-88 framework, 1–12
  mechanical features, 6
  physical model development, 4
  Piping and Instrumentation Diagram (P&ID), 4
  procedural model development, 9–10
  transfer units, 8–9
  workflow, 2–3
phase development, 21
phase requirements, 20–21
physical model development
  bulk pharmaceutical batch process case study, 15–17
  pharmaceutical process module library, 4
pilot recipes, 119
Piping and Instrumentation Diagram (P&ID)
  batch processing design and development, 101–11
  bulk pharmaceutical batch process case study, 16
  pharmaceutical process module library core operations, 4
plant utility Refrigerated Water (RW) system, 204–6
PLS model
  Dynamic Time Warping algorithm, 61–62
  on-line process monitoring, 65–66
preassembled facilities, 50–56
Preliminary Hazard Analysis (PHA), 215–19
Principal Component Analysis (PCA)
  batch process modeling, 62–65
  Dynamic Time Warping algorithm, 61–62
  functionality testing, 66–68
  on-line analytics, 57–69
probability estimation, 215–19
probability of detection, 178–82
procedural models
  batch processing design and development, 102
  batch release and material use reporting, 320–22
  bulk pharmaceutical batch process case study, 19–20
  pharmaceutical process library development, 9–10
process actions and methods
  batch processing design and implementation case study, 98
  General Recipes, manufacturing contracts, 241–50
  life science development operations, 75–83
  multiple batch systems, 222–23
  product life-cycle management, 117–18
  Quality by Design guidelines and, 297
  standardized multiple plant manufacturing, 253–59
  sustainability and, 280–84

Process Analytical Technology (PAT) guidelines
  batch release and material use reporting, 316–22
  configuration and data processing, 59–62
  ISA-88 standard and, 126–27
  on-line analytics, 57–69
  on-line process monitoring, 65–66
  product life-cycle stages, 115–27
Process Control System (PCS)
  Manufacturing Execution System (MES), 155–56, 159–60, 166–69
  product life-cycle management, 120–21
  risk-based engineering assessment guidelines, 204–6
process definition management
  barriers to, 90–91
  General Recipes, manufacturing contracts, 241–50
  ISA-88 and BatchML applications, 85–94
  XML enablement, 91–92
Process Flow Diagrams (PFDs)
  bulk pharmaceutical batch process case study, 16
  General Recipe Other Information (GROI) extensions, 308–11
  global automation integration and configuration, 194–95
process intermediates, sustainability issues and, 284
process models
  bulk pharmaceutical batch process case study, 14–15
  product life-cycle management, 118–19
Process Module (PM)
  construction period, pharmaceutical plant, 49–56
  global automation integration and configuration, 186–88
process performance qualification, 42
Process Qualification (PQ)
  global automation integration and configuration, 186–88, 194–95
  manufacturing automation project planning and execution, 125
process reports
  lean manufacturing, 265–66
  standardized multiple plant manufacturing, 254–59
  sustainability issues and, 286
process transfer, 225–26
process variables contribution plot, 66–68
Product Alpha, 3–4
Product and Process Life-cycle Management (PPLM), 74–76
product development
  life science development operations, 74–83
  sustainability and, 280–83
production history, multiple-site manufacturing science, 257–59
product launching, 184
Product Life-cycle Management (PLM)
  ISA-88 and ISA-95 standards, 113–27
  manufacturing system automation, 121–25
product performance qualification, 42
product recovery operations, 223–24
product transfer, 226–27
Programmable Logic Controllers (PLCs)
  batch processing design and control, 103–4
  bulk pharmaceutical batch process case study, 14
  lean computer validation, 210–11
Project Activity Model (PAM)
  batch processing strategies, 131–34
  modularization, 54–56

project execution, 170–71
Projection to Latent Structures (PLS), 57–69
Project Life-Cycle Document (PLCD), 134
project planning
batch processing strategies, 131–34
lean computer validation, 213–18

qualification concept
batch processing design and control, 106
global automation integration and configuration, 186–88
modularization, 54–56
quality assurance and control
Manufacturing Execution System (MES), 163
Process Analytical Technology guidelines, 126–27
software systems, 41
quality attributes
Quality by Design guidelines and, 297
standardized multiple plant manufacturing, 253–59
Quality by Design (QbD)
basic principles, 295–300
defined, 294–95
design space information storage and application, 303–7
HAZOP analysis, 302–3
life science development operations, 73, 76–77
manufacturing science and, 293–312
process definition enablement, 91–92
process definition management, 87
regulatory submissions, 90
Quality System Inspection Technique (QSIT), 201–6
Quality Target Product Profile (QTPP), 303
real time parameters
batch processing design and control, 103
equipment status management, 36
recipe management
global business systems recipe development management, 233–37
multiple batch systems, 224
product life-cycle management and, 117–21
recovery amounts, sustainability issues and, 284
regression testing, 41
regulatory requirements
batch processing design and implementation case study, 98–99
multiple batch systems integration, 144–45
process definition management, 90
sustainability compliance, 277–79
XML-based procedures, 91
report definition and generation
batch release and material use reporting, 316–22
Manufacturing Execution System (MES), 166
report server and database, 102–3
retired recipes, 121
Return On Investment (ROI)
global automation integration and configuration, 195–97
start-to-finish solution, 93
reusability
bulk pharmaceutical batch process case study, 20–21
global business systems recipe development management, 235–37
process definition management, 88–89
revalidation process, software systems, 41, 44–46

risk-based engineering assessment
case study, 199–206
fault tree analysis, 300–302
HAZOP analysis, 302–3
lean computer validation, 209–19
Manufacturing Execution System (MES) validation costs, 173–82
manufacturing science and, 293, 300–12

scalability
batch processing design and control, structural foundation, 104
equipment status management, 36
security issues, equipment status management, 36
seeding and growth processes, 282–84
severity of risk metrics
lean computer validation, 215–19
Manufacturing Execution Systems (MES) validation, 177–82
Singular Value Decomposition (SVD), 63–65
Site Acceptance Testing (SAT), 105–7
Six Sigma
lean manufacturing and, 263–64
on-line analytics, 58–59
skid control
bulk pharmaceutical batch process case study, 24
modularization, 50–56
Skip command, 80–83
"SMART" objectives, 132–34, 138–39
Software Development Specification (SDS), 139
Software Module Design Specification (SMDS), 136–39
software systems
requirements, 40
upgrades, 39–46
validation, 40
sourcing decisions, 267–69
specifications, 40
Square Prediction Error (SPE), 64–65
standardized multiple plant manufacturing, General Recipes applications in, 251–59
standard modules, 186–88
standards development
Manufacturing Execution System (MES), 174–82
multiple batch systems integration for global business systems, 151–53
status repository, 38
Structured Product Labeling (SPL) standards, 91
structured recipe repository, 81–83
Subject Matter Experts, manufacturing automation project planning and execution, 125
Supervisory Control And Data Acquisition (SCADA)
batch processing design and implementation case study, 99–100, 104, 107–8
lean computer validation, 209–19
Supply Chain Management (SCM), 114–27
supply chain optimization, 261–70
sustainability attributes, manufacturing science and, 284–86
sustainability production and processing
basic principles and definitions, 274–75
compliance and good business practices, 277–79
manufacturing science and, 273–91
pharmaceutical process development and manufacturing, 275–77
system architecture
batch processing design and development, 100–102

Manufacturing Execution System (MES), 157–68
system flexibility
batch processing design and control, 106–7
batch release and material use reporting, 319–22

technology transfer
batch processing design and implementation case study, 99–100
General Recipes, manufacturing contracts, 242–50
multiple batch systems, 225–26
process definition management, 89–90
WBF transfer to commercial manufacturing, 82
temperature control tools, 140–42
testing results
batch processing design and control, 106
batch process modeling, 66–68
manufacturing automation project planning and execution, 124–25
three-dimensional array, 62–65
time to market driver, 48–49
tool utilization
batch processing planning strategies, 140–42
global automation integration and configuration, 190
Toyota Production System (TPS), 263–64
traceability, 41
transfer units, pharmaceutical process module library, 8–9
turnkey biotechnology plants, 193–95

unit, 31
unit class status, 33–34
unit dosage production, sustainability and, 282–84
unit operations, ISA-88 mapping guidelines for, 80–81
units, batch processing design and development, 102
unit-to-unit coordination, 21–23
Upper Control Limit (UCL), 64–69
User Requirement Specification (URS)
batch processing planning strategies, 136–39
lean computer validation, 215–19
multiple batch systems, 226–27, 229
U.S. Toxic Substances Control Act, 277–79

validation
complex equipment management, 30
lean computer approach, 209–19
Manufacturing Execution System (MES) validation costs, 173–82
risk-based engineering assessment case study, 199–206
software upgrades, batch applications, 39–46
Vendor-Packaged Equipment (VPE), 148–51
version control
global business systems recipe development management, 235–37
process definition management, 88
V-Model XT model, 175–82

waste management
process operations, sustainability in, 283–84
sustainability and, 281–83
weighing system interface, 107–8
Work-In-Progress (WIP) material, 80–83

XML Schema Definition (XSD), 92